Building Food Security and Managing Risk in Southeast Asia

OECD
BETTER POLICIES FOR BETTER LIVES

This work is published under the responsibility of the Secretary-General of the OECD. The opinions expressed and arguments employed herein do not necessarily reflect the official views of OECD member countries.

This document and any map included herein are without prejudice to the status of or sovereignty over any territory, to the delimitation of international frontiers and boundaries and to the name of any territory, city or area.

Please cite this publication as:
OECD (2017), *Building Food Security and Managing Risk in Southeast Asia*, OECD Publishing, Paris.
http://dx.doi.org/10.1787/9789264272392-en

ISBN 978-92-64-27238-5 (print)
ISBN 978-92-64-27239-2 (PDF)

The statistical data for Israel are supplied by and under the responsibility of the relevant Israeli authorities. The use of such data by the OECD is without prejudice to the status of the Golan Heights, East Jerusalem and Israeli settlements in the West Bank under the terms of international law.

Photo credits: Cover © Jeffrey Fisher

Foreword

The Association of Southeast Asian Nations (ASEAN) region is both dynamic and diverse. Strong economic growth, continued development and the transformation of the agricultural and fisheries sectors have left their mark on regional food security: whereas in the early 1990s, undernourishment rates for the region were the world's highest, by 2014, these rates had fallen below those seen in the majority of other regions worldwide. However, despite this impressive performance, the region contains more than 60 million undernourished people, indicating that more needs to be done.

In further addressing food security in the region, agriculture and fisheries will continue to play a pivotal role, not only providing the food that people eat, but also forming the main income source for many households. Yet in ASEAN, as in other regions, agriculture and fisheries are being asked to do more with less, and to do it more sustainably. Foremost amongst the challenges that lie ahead will be the creation of an environment for sustainable productivity growth – one where producers have the necessary skills and ability to access markets, and where the signals that they receive from prices enable them to make appropriate decisions about what to produce and where.

To this end, policy makers will need to develop a comprehensive policy framework, directed towards increasing long-term productivity growth and improving the sustainability of the sectors, while at the same time coherently addressing short-term food security concerns, such as those generated by price volatility and natural disasters. Governments will also need to resist the temptation to introduce policies that – even if appealing in the short run – effectively "apply the brakes" to productivity growth and distort production decisions. Domestically, governments will also need to ensure that markets create the right incentives for natural resources to be used sustainably, so that they are available for future generations.

International markets will also play an important role in helping the ASEAN region to realise its full agricultural potential. ASEAN has become a significant player in world agro-food trade, accounting for around 9% and 6% of world agro-food exports and imports respectively in 2014. Together with the opportunities that this engagement presents come the challenges of participating in global value chains and dealing with the dynamic landscape of different market requirements across the world.

This report explores a range of agriculture and fisheries-related issues and policies that are important for the region in its efforts to achieve food security and respond to the risk of food insecurity in the face of multiple challenges. Based on in-depth analysis of the characteristics of food security in the region, the future outlook for agricultural markets and food security; and of the effectiveness and efficiency of policies currently in place, key findings and policy recommendations are presented to both help guide the development of appropriate food security policies and enhance the evidence base on which critical future decisions will be made.

Acknowledgements

This report and project was overseen by Jared Greenville from the Trade and Agriculture Directorate of the OECD. The individual chapters represent contributions from a number of different authors and include: Jared Greenville and Clara Thompson-Lipponen for the Overview; Jared Greenville for Chapter 1; Jared Greenville, Ada Ignaciuk and Grégoire Tallard for Chapter 2; Jared Greenville for Chapter 3; Shingo Kimura, Gen Furuhashi, Graham Pilgrim and Hubertus Gay, in addition to Hoa Hoang of the Food and Agricultural Policy Research Institute (FAPRI), University of Missouri-Columbia, for Chapter 4; Emily Gray, Andrzej Kwieciński and Florence Bossard for Chapter 5; Emily Gray for Chapter 6; and Claire Delpeuch, Ingrid Kelling and Franck Jésus for Chapter 7. The report benefited from a number of comments from staff within the OECD Trade and Agriculture Directorate, valuable input from Senior Officials within agriculture ministries across ASEAN countries, and input from delegations to the OECD Working Party on Agricultural Policies and Markets and the OECD Committee for Fisheries. The publication was edited by Clara Thompson-Lipponen, with the support of Janine Treves. Publication support was provided by Anita Lari and Michèle Patterson. Additional statistical support was provided by Clarisse Legendre. Additional editorial services for Chapter 7 were provided by Robert Akam.

The OECD would like to express particular gratitude to the ASEAN Secretariat, namely Dr. Minh Pham Quang, for his support for the project, and to the Australian Government Department of Foreign Affairs and Trade, which provided financial assistance.

The OECD would also like to thank ASEAN Senior Officials, the Food and Agriculture Organization of the United Nations (FAO), the Asian Development Bank (ADB), World Bank, and others who participated in the regional outreach events that took place to support the research presented in this report. The first of these was held in Korea on 12-13 October 2015, titled *OECD-ASEAN Regional Conference in Agricultural Policies to Promote Food Security and Agro-Forestry Productivity*. The conference was supported by the Ministry of Agriculture, Food and Rural Affairs, the Republic of Korea; the Asian Development Bank; and the Australian Government. The second major regional outreach event, the *OECD-FAO-ASEAN Regional Conference on Policies to Enable Food Security, Agricultural Productivity and Improved Nutrition,* which was supported by the FAO and the Governments of Myanmar and Australia, took place on 14-15 June 2016 in Nay Pyi Taw, Myanmar.

The report was declassified by the OECD Working Party on Agricultural Policies and Markets and the OECD Committee for Fisheries.

Table of contents

Tables

Figures

Abbreviations and acronyms

ACFAF	Strategic Plan for ASEAN Cooperation in Food, Agriculture and Forestry
ACIAR	Australian Centre for International Agricultural Research
ADB	Asian Development Bank
AEC	ASEAN Economic Community
AFSIS	ASEAN Food Security Information System
AGEI	Agricultural Growth Enabling Index
AIAT	Assessment Institutes for Agricultural Technology (Indonesia)
AIDS	Almost Ideal Demand System
AIFS	ASEAN Integrated Food Security Framework
AIS	Agricultural innovation system
APAARI	Asia-Pacific Association of Agricultural Research Institutions
APTERR	ASEAN Plus Three Emergency Rice Reserve
ARDA	Agricultural Research Development Agency (Thailand)
ASEAN	Association of Southeast Asian Nations
ASEAN-CRN	ASEAN Climate Resilience Network
ASFN	ASEAN Social Forestry Network
ATI	Agricultural Training Institute (Philippines)
ATWGARD	ASEAN Technical Working Group on Agricultural Research and Development
BAR	Bureau of Agricultural Research (Philippines)
BERNAS	PadiBeras Nasional Berhad (Malaysia)
BULOG	Bureau of Logistics (Indonesia)
CANSEA	Conservation Agriculture in Southeast Asia
CARDI	Cambodian Agricultural Research and Development Institute
CIAT	International Center for Tropical Agriculture
CIFOR	Center for International Forestry Research
CIMMYT	International Maize and Wheat Improvement Center
CIRAD	Agricultural Research Centre for International Development (France)
CRRI	Cambodian Rubber Research Institute
DA	Department of Agriculture (Philippines)
DAE	Department of Agricultural Extension (Cambodia)
DAO	District Agricultural Offices (Cambodia)
DAR	Department of Agricultural Research (Myanmar)
DOA	Department of Agriculture (Malaysia)
DOA	Department of Agriculture (Thailand)
DOAE	Department of Agricultural Extension (Thailand)
DOST	Department of Science and Technology (Philippines)
EAF	Ecosystem Approach to Fisheries
EEZ	Exclusive economic zone
FAO	Food and Agriculture Organization of the United Nations
GAP-CC	ASEAN German-Programme on Response to Climate Change in Agriculture and Forestry
GCI	Global Competitiveness Index
GDP	Gross domestic product
GFSI	Global Food Security Index
GIZ	Deutsche Gesellschaft für Internationale Zusammenarbeit GmbH
GRA	Global Research Alliance on Agricultural Greenhouse Gases
GSSE	General Services Support Estimate
GT	Gross tonnage
GVA	Gross value added
HEI	Higher education institutes
HSES	Household Socio-Economic Survey (Thailand)
IAARD	Indonesian Agency for Agricultural Research and Development
ICRAF	World Agroforestry Centre
ICRISAT	International Crops Research Institute for the Semi-Arid Tropics
ICT	Information and communications technology

IFAD	International Fund for Agricultural Development
IFPRI	International Food Policy Research Institute
IIAT	International Institute of Tropical Agriculture
ILRI	International Livestock Research Institute
IPR	Intellectual property rights
IRIEC	Indonesian Research Institute for Estate Crops
IRRI	International Rice Research Institute
IUU	Illegal, unregulated and unreported
JICA	Japan International Cooperation Agency
JIRCAS	Japan International Research Center for Agricultural Sciences
KCNSA	Kampong Cham National School of Agriculture (Cambodia)
LBVD	Livestock Breeding and Veterinary Department (Myanmar)
LGU	Local Government Units (Philippines)
MAF	Ministry of Agriculture and Forestry (Lao PDR)
MAFF	Ministry of Agriculture, Forestry and Fisheries (Cambodia)
MARD	Ministry of Agriculture and Rural Development (Viet Nam)
MARDI	Malaysian Agricultural Research and Development Institute
MCB	Malaysian Cocoa Board
MCC	Millennium Corporation Challenge
MMAF	Ministry of Marine Affairs and Fisheries (Indonesia)
MoA	Ministry of Agriculture (Indonesia)
MOA	Ministry of Agriculture and Agro-based Industries (Malaysia)
MOAC	Ministry of Agriculture and Cooperatives (Thailand)
MOAI	Ministry of Agriculture and Irrigation (Myanmar)
MOECAF	Ministry of Environmental Conservation and Forestry (Myanmar)
MLFRD	Ministry of Livestock, Fisheries and Rural Development (Myanmar)
MP3EI	Master Plan for Acceleration and Expansion of Indonesian Economic Development 2011-2025
MPIC	Ministry of Plantation Industries and Commodities (Malaysia)
MPOB	Malaysian Palm Oil Board
MRB	Malaysian Rubber Board
NAEC	National Agriculture Extension Center (Viet Nam)
NAFES	National Agriculture and Forestry Extension Service (Lao PDR)
NAFRI	National Agriculture and Forestry Research Institute (Lao PDR)
NESDP	11th National Economic and Social Development Plan (2012-2016) (Thailand)
NFA	National Food Authority (Philippines)
NGO	Non-governmental organisation
NSDP	National Strategic Development Plan 2014-2018 (Cambodia)
NSTDA	National Science and Technology Development Agency (Thailand)
NSTP	National Science and Technology Plan 2002-2020 (Philippines)
NUOL	National University of Laos
PCAARRD	Philippine Council for Agriculture, Aquatic and Natural Resources Research and Development
PNSA	Prek Leap National College of Agriculture (Cambodia)
PPP	Purchasing power parity
QUAIDS	Quadratic Almost Ideal Demand System
R&D	Research and development
RD&E	Research, development and extension
RUA	Royal University of Agriculture (Cambodia)
SCUs	State colleges and universities (Philippines)
SEA	Southeast Asia
SEARCA	Southeast Asian Regional Center for Graduate Study and Research in Agriculture
STI	Science, technology and innovation
TAC	Total allowable catch limit
TFP	Total factor productivity
TRIPS	WTO Agreement on Trade-Related Aspects of Intellectual Property Rights
USD	United States Dollars
VAAS	Vietnam Academy of Agricultural Sciences
VES	Village extension system (Lao PDR)
VHLSS	Viet Nam Household Living Standard Survey
VMS	Vessel monitoring system (Indonesia)
WIPO	World Intellectual Property Organization
YAU	Yezin Agricultural University (Myanmar)

Executive summary

Over the course of the last two decades, the countries that comprise the Association of Southeast Asian Nations (ASEAN) have transformed into a region described by *The Economist* in April 2016 as "at the forefront of the emerging markets success story". The features of this success – including strong gross domestic product (GDP) growth, rising agricultural productivity, output and agricultural incomes, aided by the increasing prominence of the region in global agro-food markets – have in turn spurred vast improvements in food security. These improvements are expected to continue over the 2015-24 period, with projections indicating that the number of undernourished people in the region as a whole will decline by almost 13 million.

As praiseworthy as these achievements may be, the challenges that lie ahead for regional food security should not be underestimated. In 2015, around 60 million people remained undernourished. Poverty rates remain worryingly high, surpassing 40% in a number of countries. Productivity growth in agricultural and fisheries sectors, necessary to meet future rises in demand for food, is likely to be hampered by land ownership rules and practices that have degraded the rural environment and marine ecosystems of the region. In addition, some populations in ASEAN, while *currently* considered to be food secure, remain at significant risk of undernourishment as a result of temporary income or price shocks.

In the lead-up to 2050 and beyond, climate change will pose a formidable challenge to food security in ASEAN, reducing yield growth and, in so doing, threatening farm incomes and generating rising commodity costs for consumers.

The response of ASEAN governments to challenges such as these has justifiably been the prioritisation of food security. Food security policy in the region is arguably "rice-centric". Although the dietary importance of rice has generally fallen over time, it remains a key production and consumption commodity, particularly for poor households, and has thus retained significant political importance.

Policies currently in place in the region include measures to promote food self-sufficiency and price stabilisation, together with often high-cost and untargeted public stockholding schemes. For some, these have proved ineffective and counterproductive to longer-term food security. While, for example, stockholding regimes and policies to incentivise domestic production in large importing ASEAN countries, underpinned by trade restrictions, have stabilised prices compared with ASEAN peers, this has been at a much higher price level. Given that undernourishment in the region tends to be concentrated in low-income households, this has had a negative impact on food security. Simulations suggest that removing current price support measures in Indonesia and the Philippines through ASEAN rice market integration would reduce the prevalence of undernourishment by 10% and 54% in these countries, respectively.

The artificially higher prices have also increased the vulnerability of households to food insecurity, certain farming households included. For example, in the Philippines, Indonesia and Myanmar, low-income farm households tend to purchase a significant proportion of the rice that they consume and, without improvements in their productive efficiency to capitalise on higher prices, are left on balance worse off.

In addition, the impacts of current policies are likely to *reinforce* expected price increases resulting from climate change. They also have the potential to exacerbate the negative impact of climate change on trade volumes. This amplification would have consequences, both for food security in ASEAN as a whole, and for the ability of households to manage risks to food security. Given the importance of regional and international trade for individual countries to manage climate-related production shocks that can increase food insecurity, falls in trade volumes can increase the vulnerability of countries to the price and production effects of climate change.

Finally, policies that constrain trade, such as import restrictions on specific agricultural and fishery products, can also negatively affect food security by increasing pressure on already constrained natural resources, jeopardising longer-term producer incomes and raising consumer prices.

Beyond domestic responses, ASEAN member states have laid the foundations for future progress by establishing regional frameworks to address key food security. There is much to be gained from capitalising on the valuable momentum generated by these collaborations and pursuing regional solutions to food security and risk management. Open markets, for example, can positively contribute to food security by providing new markets and opportunities for agricultural and fisheries income growth, while at the same time ensuring adequate supplies of diverse food products without the costs imposed by artificially higher prices. Indeed, analysis suggests that ASEAN rice market integration would reduce the undernourished population by 5% overall in the five ASEAN member states examined. In addition, open markets allow for the sharing of production risks across countries, aiding the general stability of access and availability: simulations of a number of alternative policies have revealed that ASEAN rice market integration would be particularly effective in mitigating the risk of undernourishment. Nevertheless, targeted support to vulnerable households that can be accessed both generally and more intensively at times of crisis is also required, and should be a part of future reforms aimed at improving food security.

Finally, the importance of additional measures to improve the enabling environment for sustainable productivity growth in the agricultural and fisheries sectors cannot be overemphasised. With some exceptions, common areas of weakness in ASEAN member countries include the lack of public investment in agricultural research and development (R&D), land market rights and access, farmer access to finance, the existence and quality of agricultural infrastructure, and the stringency and enforcement of environmental regulations.

Key policy recommendations

1. ***Provide targeted support to vulnerable households***
   - Improve access to food by poor households through conditional cash transfers or other targeted redistributive efforts such as food vouchers.
   - Provide training programmes to enable agricultural and fisheries producers to make better production and investment decisions, including through diversification to alternative activities.

2. ***Implement trade and domestic support reforms***
   - Gradually reduce trade barriers with a view to creating an open and competitive regional market, for rice in particular, and pursue more open markets with greater private sector involvement among a wider set of international trading partners.
   - Reduce distorting forms of domestic support to fisheries and agriculture.

3. ***Promote sustainable agricultural and fisheries productivity growth***
   - Strengthen the enabling environment through improving environmental governance; regulations on land, water and biodiversity resources; investments in infrastructure, agricultural R&D and agricultural innovation systems; improving rural land market rights and access, and increasing access to credit for farmers.
   - Improve sustainable resource management of fisheries through the adoption of inclusively-defined, science-based and measurable long-term management targets, for example.

Overview of challenges and opportunities for improving food security in Southeast Asia

ASEAN: A region of remarkable progress

The past 20 years have witnessed unprecedented changes in the economies and the agricultural and fisheries sectors of the countries that comprise the Association of Southeast Asian Nations (ASEAN).[1] Robust gross domestic product (GDP) growth, rising agricultural productivity and output, and strong growth in agricultural incomes have all contributed to significant improvements in the region's food security. The agricultural sector has undergone considerable structural adjustment, shedding a substantial amount of labour across all ASEAN member countries between the early 1990s and early 2010s, while simultaneously increasing production (World Bank, 2016). Despite these changes, the production mix of ASEAN as a whole has remained relatively stable, with rice continuing to represent the single largest production activity in value terms.

ASEAN member states are also playing an increasingly prominent role in world agro-food trade. Since the early 2000s, as a result of continued agricultural productivity growth, ASEAN has become a significant net agro-food exporter: net agro-food exports rose from around USD 6 billion in 1990 to over USD 50 billion in 2012 (WITS, 2016). This, together with rising imports – around a seven-fold increase over the same period – has meant that the region has forged strong links with international agricultural markets. The region has also become more interconnected internally, with intra-regional agro-food trade linkages strengthening over time. These new trade connections, both within the region and to global markets, are important for both consumers and producers. For consumers, they help to provide the domestic market with a more diverse range of food commodities, as well as alternative sources of food, contributing to more stable and counter-seasonal food supplies. For producers, these trade connections are becoming increasingly essential for incomes and livelihoods. Of the products traded on world markets, vegetable oils – palm oil in particular – are the most important agro-food export, accounting for the largest share of agro-food export value from the region as a whole, followed by fisheries and aquaculture (WITS, 2016).

Thanks to these changes, the majority of ASEAN member states have experienced strong improvements across a range of measures of the four Food and Agriculture Organization (FAO) dimensions of food security: (i) food availability, (ii) food accessibility, (iii) food utilisation, and (iv) the stability of the first three dimensions over time. One commonly used measure, the prevalence of undernourishment, reveals a general fall in rates since 1992 (FAO, 2016a). In comparison with other regions, Southeast Asia's reduction in undernourishment has been impressive: whereas in the early 1990s, the region's rates of undernourishment were the highest worldwide, they have since fallen below those observed in a number of other regions (Figure 1).[2]

Figure 1. Southeast Asia has experienced more dramatic declines in undernourishment than any other region

Prevalence (3-year moving average) as a percentage of population, 1992-2014

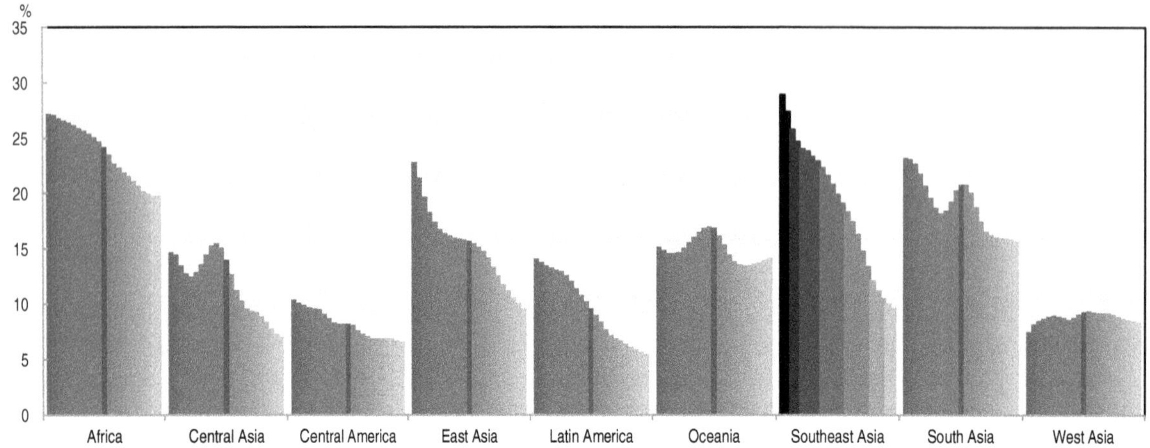

Source: FAO (2016a), *FAOSTAT*, http://faostat.fao.org/ (Chapter 1 of this report).

The medium-term outlook for food security is positive

The medium-term outlook for the region is also promising. Economic growth over the 2015-24 period is projected to equal or even exceed levels experienced over the past decade, and to surpass that of many other regions (OECD-FAO, 2015). Projected high levels of growth will in turn further diminish poverty and spur continued advances in food security in the region. Market outcomes projected for the medium term are expected to lower the prevalence of undernourishment in Southeast Asia by almost three percentage points – a greater reduction than that of other regions. Projected higher incomes across all income groups will, for example, enable greater access by ASEAN consumers to food, with the result that by 2024, the number and the share of undernourished individuals is projected to fall. For the region in total, the number of undernourished people is projected to decline by almost 13 million relative to 2015 (see Chapter 2 of this report).

Increased access to food will not only reduce the number of people facing undernourishment, but will also reduce its *depth*, thereby cutting the number of calories necessary to further reduce undernourishment. Overall, growth in agricultural and fisheries production and rising incomes across all income groups are expected to enable the share of undernourished people in ASEAN to fall to 6.8% of the population in 2024. For some of the larger ASEAN members (Indonesia and Thailand), the rate of undernourishment is projected to drop below the 5% threshold that the FAO uses to delineate food-insecure countries. Fisheries sector production, capture fisheries and aquaculture included, will play a particularly important role in food security in the ASEAN region over the medium term. Fish products already form an important part of the regional diet and represent an important source of income for many. Furthermore, they comprise the second-largest export item, with around USD 133 billion in exports in 2012 (WITS, 2016).

While development in the region will contribute to rising levels of food security, it will also place pressure on agricultural and fisheries markets to provide more and to deliver greater diversity of food products in a sustainable manner. Economic prosperity and population growth will both increase demand for a range of commodities and create shifts in the *nature* of demand. As incomes grow, there will be a general substitution away from crops such as rice to other products – particularly those derived from animal products, seafood included. Agricultural producers in ASEAN are expected to be able to rise to these challenges over the medium term, with regional production growth rates projected to exceed world averages for many products, including animal-derived products (OECD-FAO, 2015). This will also increase the importance of ASEAN for world supplies of agro-food commodities. Across individual countries, however, changes in the production of key commodities are more mixed. While the general patterns of increase remain, some changes in the production mix are observed, driven by land constraints and the changing relative returns from other production activities.

To meet these challenges, agricultural productivity improvements, part of which will be driven by improvements in yield, will be key. This will require a continuation of past trends in total factor productivity growth – trends underpinned by continued efforts at policy reform and additional investments by governments and producers. However, the limited availability of suitable land means that for the majority of agricultural activities, production growth over the medium term will be spurred by increases in intensification in particular, such as the greater use of improved seed varieties. For fisheries and aquaculture, medium-term continuation in productivity and production growth depends on the health of fish stocks and water ecosystems more generally. To ensure their sustainability, short-term reductions in production may be necessary.

Important challenges and risks remain

Notwithstanding the extraordinary development experienced by a number of member countries, ASEAN governments face a number of serious challenges. The impressive economic growth of the region as a whole hides differences in development and incomes among countries, which largely arise from differences in labour productivity growth in recent years. Farm size, which has fallen in Indonesia, the Philippines and Thailand, may also be a factor (Lowder et al., 2014). Agricultural total factor productivity growth may be curbed by falls in average farm sizes if these are also accompanied by further fragmentation of production activities. Poverty rates, measured as the percentage of the population living on less than USD 2 a day (in purchasing power terms) also remain high, exceeding 40% in a number of member states in 2013 (World Bank, 2016).

Furthermore, despite the significant improvements in food security, around 60 million people in the Southeast Asian region remain undernourished (FAO, 2015). Progress across countries has also been uneven (Figure 2): some (such as Lao People's Democratic Republic [PDR], Myanmar, the Philippines, Thailand and Viet Nam) have experienced continued declines in the prevalence of undernourishment, while for others, declines have been more erratic, due to macroeconomic conditions or domestic conflict. Moreover, in two of the countries that have experienced the greatest falls in undernourishment (Lao PDR and Myanmar), together with Cambodia, levels nevertheless remain so severe that extensive action will be required in order to better address food security. In addition, in the Philippines, there is a risk that current rapid population growth will lead to an *increase* in the absolute number of undernourished people, although this figure represents a smaller share of the total population than before. Medium-term projections for the region consequently point to the need for additional efforts by ASEAN governments to address food security.

Unsurprisingly, undernourishment in ASEAN tends to be concentrated in low-income households. In some countries, food crop-producing farm households have better access to food *despite* their lower average income levels, thanks to their own production. But this is not always the case: a significant number of farm households across ASEAN – low-income farm households in particular – tend to purchase a significant proportion of the rice that they consume. For example, the poorest 20% of farm households in the Philippines tend to be small-scale producers and landless workers whose rice production does not satisfy their own needs and who therefore purchase 77% of the rice that they consume on average (Chapter 4).

Other measures beyond undernourishment also point to remaining food security concerns. Regional levels of stunting among children under five years of age, for example, remain worryingly high (Figure 3).

There are also indications that current food security figures may be presenting an overly optimistic picture. While economic growth has enabled significant declines in populations suffering from chronic poverty, for some households, it may have only altered the *nature* of food insecurity from permanent to transitory, rather than eliminate it entirely. Some populations in ASEAN, albeit currently considered to be food secure, remain at significant risk of undernourishment as a result of temporary shocks that alter their incomes or the prices that they pay for food (Chapter 4). Such shocks can be related to economic events, natural disasters or domestic or international shocks to markets. Indeed, household-level data suggests that a great number of households may still be at risk of undernourishment. Low-income households are found to be particularly vulnerable to risks such as rice price increases and income loss, given their limited capacity to allocate additional expenditure to food. Of the three major risks that currently most concern regional agricultural policy makers – severe regional weather events (for example, an *El Niño*), domestic crop failures and economic downturns – domestic crop failure is expected to pose the greatest risk to undernourishment in Indonesia and Myanmar, a regional *El Niño* event is identified as the largest risk for the Philippines, and an economic downturn is expected to be the largest risk to food security in Thailand and Viet Nam.

Figure 2. Declines in the number of undernourished people have been uneven across ASEAN

Millions (3-year moving average) in selected years, selected ASEAN member states

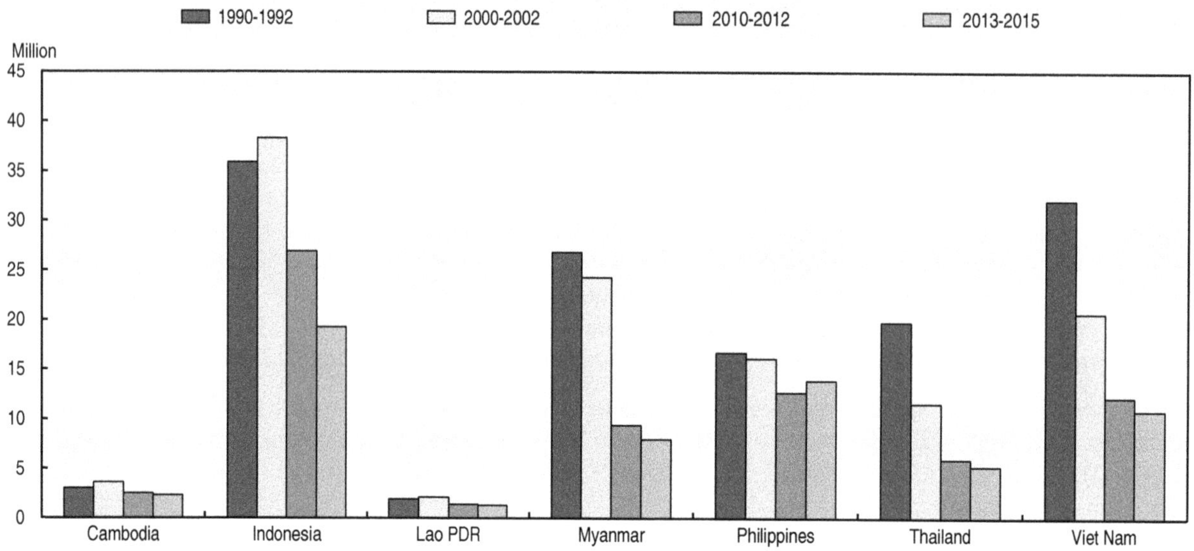

Source: FAO (2016a), *FAOSTAT*, http://faostat.fao.org/ (Chapter 1).

Figure 3. Rates of stunting among children under the age of five remain alarmingly high

Percentage of children under five years with stunted growth, various years, 2006-13

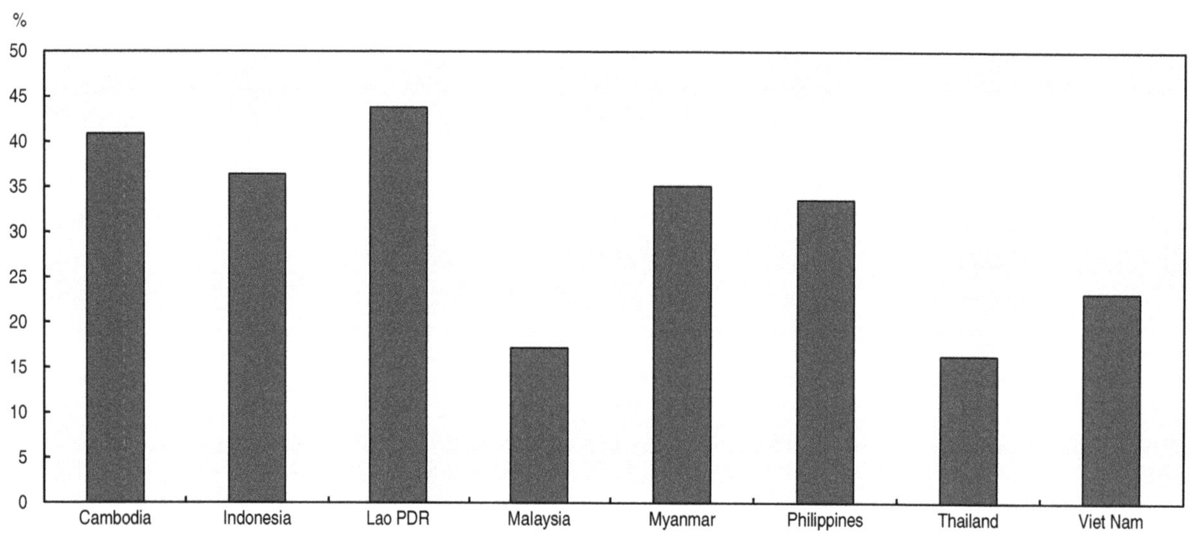

Source: FAO (2016a), *FAOSTAT*, http://faostat.fao.org/ (Chapter 1).

Climate change will pose challenges for food security

In addition to the abovementioned risks in the short term, food security in the ASEAN region will face a serious challenge in the lead-up to 2050 and beyond: climate change. While a number of uncertainties exist concerning the potential agricultural and socio-economic impacts of climate change, related to the extent of global action and to the potential role of contributing and countervailing natural influences, some of the negative effects of climate change and climate variability are already visible across the region. In recent decades, sea levels in the region have risen by 1–3 mm per year (ADB, 2009). Temperatures are also climbing, and are expected to increase by around 0.7–0.9°C between now and 2050 (Cruz et al., 2007). In addition, extreme weather events have not only been occurring more frequently, but are also increasing in intensity, leading to declines in water, soil and land resources (Yusuf and Francisco, 2009; Hijoika et al., 2014). Droughts – the climate feature that has in recent years had the greatest negative effect on annual production in the region – are expected to rise in intensity and frequency, while the intensity of tropical cyclones is also expected to increase. Over the longer term, these shocks are expected to not only lead to temporary disruptions to production, but to also affect the observed *trends*, which will in turn have a compounding effect both on farm household incomes and, ultimately, costs for consumers through rising prices of agricultural and food products (Figure 4).

Figure 4. Climate change is set to increase agricultural commodity prices in ASEAN and globally

Percentage change in prices in 2050 based on two climate models, compared with situation of no climate change

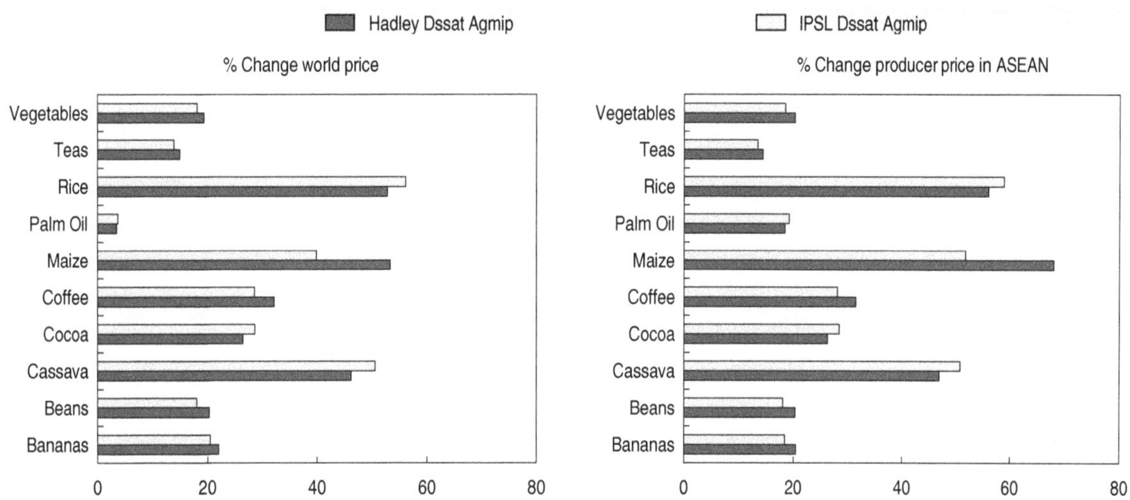

Source: OECD estimates (Chapter 2).

Projections for the region are calculated relative to a baseline with no climate change impact (Chapter 2). Like any analysis associated with projecting long-term trends, the projections make use of extrapolations that may not be representative of the reality that will exist in 2050. The precision of any particular number is therefore not meant to convey that this will inevitably be the result, when in fact over this period, a wide variety of unknown factors and adaptation actions will occur. But the existence of such factors is no reason to not consider the *trends*, which highlight the need for serious consideration of new policy measures. If acted upon, these measures should ensure that such projections never become reality. With this in mind, between now and 2050, climate change is expected to reduce yield growth for the majority of agricultural products in the ASEAN region – meaning that yields should still increase, but to a lesser extent than may have otherwise occurred – leading to falls in production relative to a situation of no climate change. As a large share of staple crop production in the region is rain-fed and hence more deeply affected by changes in precipitation patterns than are irrigated crops, the overall effect on production may be significant. Changes in yield, intermixed with rising and changing demand, will impact the relative returns to different crops, which are in turn projected to influence relative land use, as producers respond by adjusting the production mix. With strong demand for agricultural products and for staples in particular, lower yields in the major staple crops of rice, maize and cassava will spur increased allocation of land to these activities, thereby reducing available land for other crops. These effects will be further intensified by the reduction of land surface in coastal regions due to sea-level rises.

While real price declines are projected in the medium term (2015-24), in the longer term (beyond 2025), the effects of climate change on production, and on yields of staples in particular, are projected to increase prices for agricultural commodities worldwide compared with a situation of no climate change (Chapter 2). In Southeast Asia, moreover, prices of staple crops of rice, maize and cassava are projected to rise by *more* than average world prices, due to the influence of maintaining current agricultural domestic support and restrictive trade policies – particularly those related to rice – which compound some of the price effects of climate change. In addition, as a consequence of these production and price effects, if current policy settings are not reformed, agricultural trade between ASEAN and the rest of the world is also projected to be lower compared with a situation of no climate change. Finally, these price effects, and falls in trade volumes, can have implications for the ability of individual countries to respond to shocks that can disrupt production, and thus negatively influence food security.

Current policies may be working against food security objectives

Overall, regional policy frameworks recognise many challenges to food security, but current implementation and the trade and agricultural policy mix mean that the region's potential to further reduce food insecurity is not being realised.

In spite of overall regional improvements, current deficiencies and risks over the medium and long term mean that food security will remain a major concern for many ASEAN member states. This fact has been recognised by policy makers, who have developed regional policy architecture aimed at addressing food security. The ASEAN Integrated Food Security Framework (AIFS Framework) and the Strategic Plan for ASEAN Cooperation in Food, Agriculture and Forestry (ACFAF) provide a solid platform on which ASEAN member states are pursuing policies to tackle long-term food security. These regional frameworks include core policy areas and a number of "Strategic Thrusts", which set out actions for ASEAN member states to address food security. They are also supported by the ASEAN Plus Three Emergency Rice Reserve (APTERR), which seeks to provide food coverage across the region in times of severe, short-term need. However, while the regional frameworks identify a number of the key policy areas where action will be required – such as those related to the development of regional agro-food markets, sustainable production and increased investment in the sector – reform has been slow and, in some cases, current policy choices are actually *undermining* food security (Chapter 3).

Moreover, notwithstanding these regional frameworks, the current policy approach to food security in ASEAN member states tends to be domestically-focused and largely oriented towards the management of international market risks through food self-sufficiency, price stabilisation and social safety nets, rather than responding to domestic supply risks – which, as for many other regions, are actually more frequent and higher-impact events – or addressing long-term issues that underpin future food security in the region. Self-sufficiency in particular has often been strongly promoted by a number of ASEAN governments as a strategy to reduce vulnerability to world price movements, such as those observed during the food price crisis of 2007/08.

Generally speaking, food security policy in ASEAN members can be characterised as "rice-centric" (Chapters 3 and 4), focused on ensuring that rice production is sufficient to meet domestic demand, often irrespective of countries' production capacity. Rice remains the dominant crop in the region, representing the single largest production activity in value terms, a significant net export commodity – even if some countries are net importers – and a large part of the energy and nutrition intake of regional consumers. Although the dietary importance of rice has generally fallen over time, particularly as incomes have increased and diets have diversified, its status as a key production and consumption commodity, particularly for poor households, means that it has retained significant political importance for food security.

While food security policies vary across the region, there are some commonalities, depending on countries' importing or export status. In general, policies in *importing* countries tend to focus on the domestic production of rice in particular – and also of other commodities, such as soybeans, maize, sugar and beef – through the use of price support, trade barriers and input subsidies. By contrast, in *exporting* countries, interventions in export markets (taxes, bans and licencing arrangements), together with attempts to set aside land for rice production, are favoured. Indonesia has implemented the most ambitious production targets in the region, aiming for self-sufficiency across all main staple products (Chapter 3). While other substantial investments to support agriculture have been made – for example, in Viet Nam and Indonesia, there have been significant investments in irrigation and other agricultural infrastructure – these tend to be much smaller in value terms than the other forms of support. On the consumer side, meanwhile, some countries, such as Indonesia, Malaysia and the Philippines, have combined production and trade-based policies with public stockholding regimes aimed at stabilising rice prices and providing subsidised public food distribution.

Many of these current food security policies that are aimed at achieving self-sufficiency through production-oriented and trade-restrictive policies have important drawbacks. For the economy as a whole, they can create inefficiencies in resource allocation, generate greater uncertainties that discourage private investment that would otherwise help to improve producer incomes by better linking producers to markets, and crowd out foreign direct investment (FDI). They can also delay the structural adjustment that is necessary to promote sector productivity and resilience over the longer term, by locking in expensive forms of production, which in turn diminish competitiveness. For governments, they impose a significant drain on public finances which could otherwise be more effectively put to use elsewhere (Deuss, 2015) – input subsidies, stockholding and price stabilisation policies all require significant public resources to maintain.

But critically, the effectiveness of production-oriented and trade-restrictive policies at improving food security for consumers is also questionable. Export restrictions have exacerbated global and domestic price volatility, undermining food security both within the region and from a global perspective. Meanwhile, stockholding regimes and policies to incentivise domestic production in large importing ASEAN countries, both supported by trade restrictions, have stabilised prices compared with ASEAN peers, but at a much *higher* level of prices (Figures 5 and 6). As food insecurity is concentrated among lower-income households, the net impact of these policies on food security is in fact negative. In Indonesia, for example, the gap between domestic and world prices has been widening in recent years, and in 2012-14, domestic prices rose to 70% above comparable world prices (Figure 6 and OECD, 2015a). As poorer households tend to spend a greater proportion of their income on food – in the case of Indonesia, Myanmar, the Philippines, Thailand and Viet Nam, at least – the overall price increases have serious consequences for their food security, albeit to a lesser extent in Thailand. Simulations suggest, for example, that removing price support measures through rice market integration would improve access to rice and reduce undernourished populations in Indonesia and the Philippines by 10% and 54% respectively (Chapter 4). Moreover, the higher prices resulting from these policies increase not only the levels of undernourishment overall, but also the vulnerability of households, leaving them less able to cope with the impacts of more frequently occurring domestic production disruptions. Outside the sector itself, other policies such as restrictions on inward foreign investment have had similar effects on prices.

These policies have impacts on both non-farming *and* farming households. For example, in the Philippines, Indonesia and Myanmar, where small-scale subsistence farms or landless farm workers dominate the sector, the rice production of many farm households – low-income farm households in particular – does not always satisfy their consumption needs. As these households tend to purchase a significant proportion of the rice that they consume, higher rice prices worsen their food security (Chapter 4). Policies need to take into consideration the fact that vulnerable farm households tend to be small-scale producers that source a large part of their rice consumption from the market. Furthermore, price support is most often a highly inefficient and ineffective way of addressing low farm incomes of the poorest farmers, as it is larger, more competitive producers, with significant marketable surpluses, who capture most of the benefits from higher domestic prices. As a result, the benefits of support often accrue to otherwise food-secure households – to both larger and more food-secure crop producers, as in the case of Indonesia and the Philippines (Dawe, Moya and Casiwan, 2006).

Figure 5. Countries with policies that limit trade have higher rice prices than ASEAN peers

Real rice prices in USD, April 2001 to March 2015

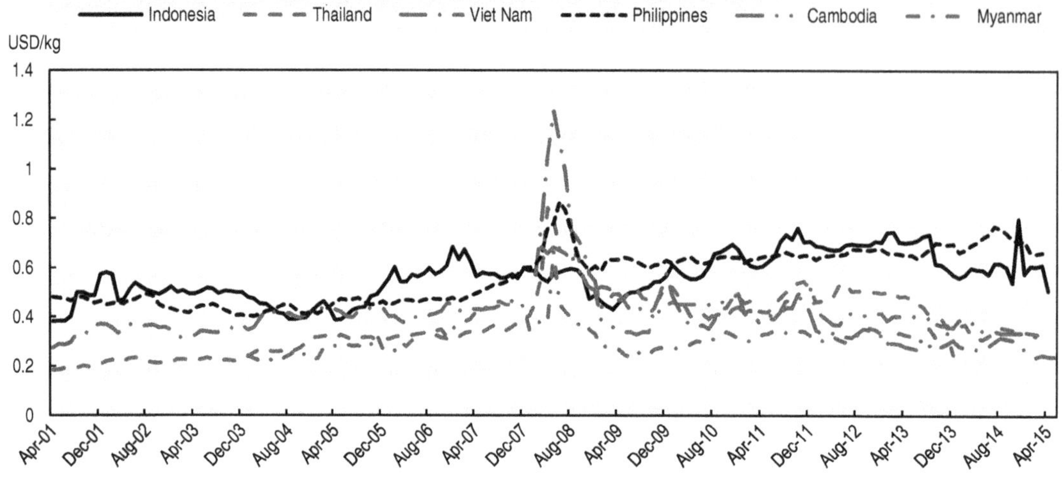

Sources: Indonesian Ministry of Agriculture (2015); Bank of Thailand (2016), *Wholesale Price of Certain Commodities in Bangkok Metropolis*, www2.bot.or.th/statistics/BOTWEBSTAT.aspx?reportID=89&language=ENG; International Grains Council (2016), *International Grains Council*, www.igc.int/en/markets/marketinfo-prices.aspx; Philippines Bureau of Agricultural Statistics (2016); FAO (2016b), *Food Security Portal*, www.foodsecurityportal.org/api/rice; and World Bank (2014) (Chapter 1).

Figure 6. Price gaps for these countries are also increasing

Percentage difference to Thai wholesale price of like quality (proxy for world price), real prices USD, April 2001 to March 2015

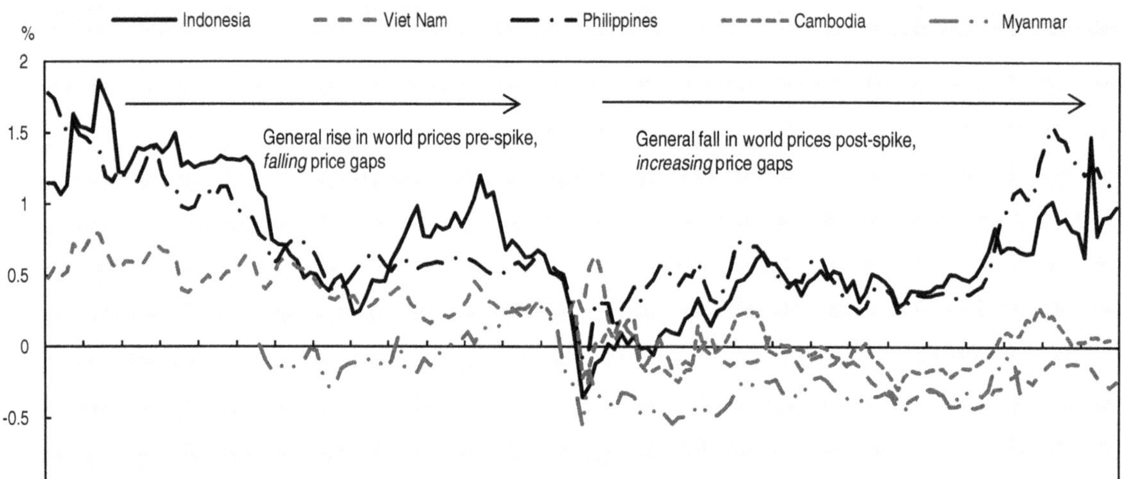

Notes: The price gaps depicted are simple price gaps and do not account for transport margins. Prices would be expected to be higher in import markets compared with export markets due to these margins. However, for Indonesia and the Philippines, the gaps depicted are significant, and when transport margins and differences in domestic processing and wholesale costs are corrected for, significant price gaps between domestic and international markets remain. For details on these calculations, see OECD (2012) for Indonesia and OECD (2017a) for the Philippines. Latest information on reference prices and price gaps, as measured in estimates of market price support for rice, can be found at OECD (2017b).

Sources: Indonesian Ministry of Agriculture (2015); Bank of Thailand (2016), *Wholesale Price of Certain Commodities in Bangkok Metropolis*, www2.bot.or.th/statistics/BOTWEBSTAT.aspx?reportID=89&language=ENG; International Grains Council (2016), *International Grains Council*, www.igc.int/en/markets/marketinfo-prices.aspx; Philippines Bureau of Agricultural Statistics (2016); FAO (2016b) *Food Security Portal*, www.foodsecurityportal.org/api/rice; and World Bank (2014) (Chapter 3).

A similar problem occurs with existing efforts at redistribution to address the food insecurity of poor households. While welcome in principle, a number of the larger food distribution programmes, such as those in Indonesia and the Philippines, have been burdened by difficulties in effective targeting and considerable leakages to non-needy households (Chapters 3 and 4). There are a number of options, beyond the physical distribution of food to provide targeted assistance for poor households, which do not involve the higher prices and increased vulnerability associated with current policies (these options are discussed further in the following section).

Moreover, the impacts of current policies are likely to be greater in the context of – and are indeed likely to reinforce – expected price increases resulting from climate change (Chapter 2). They also have the potential to amplify the negative effects of climate change on trade volumes. This amplification would have consequences, both for food security in the region as a whole, and for the ability of individual households to manage any transitory risks to food security. For ASEAN, both regional and international trading links will be important for individual countries to manage climate-related production shocks that can increase food insecurity. Falls in trade volumes in already thin markets[3] can therefore increase the vulnerability of countries to the price and production effects of climate change, which can in turn bring about further decreases in trade and therefore incomes, GDP and food security.

In addition, policies that limit trade can also increase pressure on natural resources. In the case of the Indonesian fisheries sector, for example, import restrictions are in place on a number of seafood species that can be caught or produced domestically, and direct support in the form of fuel tax concessions is provided to fishers. These policies encourage greater fishing pressure on stocks in a context of insufficient fisheries stock management and thus pose a particular risk to overfished and endangered fish stocks. As a result, such policies jeopardise longer-term producer incomes, consumer prices and therefore food security (Chapter 7, see also Box 3).

Despite the drawbacks of many of these approaches, it is often argued that current policies help to insure against the risks of fast price movements on international markets and combat domestic risks related to macroeconomic and environmental events (OECD, 2015b). While current policies have reduced volatility in domestic markets in Indonesia and the Philippines, this review finds evidence that they have not been successful in mitigating risks more holistically – that is to say, they manage a risk but at the cost of exacerbating this risk and others in *future*. Indeed, from an international market perspective, policy measures such as these, along with export restrictions and import measures, combined with panic purchases by some governments, largely contributed to the food price spike of 2007/08 (Alavi et al., 2012; OECD, 2008; Piesse and Thirtle, 2009; Naylor and Falcon, 2010; Headey, 2011). For rice in particular, changes in global demand, supply and stocks were not sufficient to explain the price spike that occurred. Ultimately, those actions that were intended to keep downward pressure on rice prices actually caused them to soar higher in domestic markets than would otherwise have likely occurred (Anderson, Ivanic and Martin, 2013). As such, these policies actually undermined food security.

Alternative policies can have significant benefits

A range of alternative policies, or policy focuses, have been explored to help inform the improved design and implementation of policies to manage risk and reduce food insecurity in ASEAN (Chapter 4). These have revealed that alternative approaches and a shift in policy focus would be more effective than current measures that rely on market interventions to improve food security. In some cases, these alternatives would reduce not only current levels of food insecurity, but also the exposure of households to food insecurity risks.

Three key policy areas are identified below. First, to squarely address the objective of food security, a fundamental action is to set in place mechanisms that target food-insecure households in a way that is effective, efficient and minimises negative impacts on the agricultural and fisheries sectors or costs to the economy as a whole. A second set of policies focuses on the role of market integration within ASEAN in boosting food security by promoting stable access to food, including through the management of weather and economic risks. A final key set of policies to promote food security addresses the need to ensure that the agricultural and fisheries sectors are able to supply needed food over the longer term, by focusing on improving sustainable productivity.

i) Targeted support to vulnerable households will be critical

Targeted support to vulnerable households that can be accessed both generally and more intensively during times of crisis should be a priority for all governments in the region. While higher incomes and greater productivity of agricultural producers will all contribute to faster falls in undernourishment, improved *access* to food by poor households through non-market distortionary means, namely social safety nets or other targeted redistributive efforts such as food vouchers, are projected to have the greatest impact on undernourishment and therefore food security in the region in the short term. That is, it is not the lack of available food that is the fundamental obstacle to regional – and worldwide – food security, but rather effective access to that food by poorer households. In Indonesia and the Philippines in particular, replacing the current public rice distribution programme – which is underpinned by trade and other market interventions – with targeted cash transfers or food vouchers would have greater potential to mitigate against food insecurity risks arising from a range of sources, and to improve food security overall (Figure 7). Of the two alternative policies (food vouchers or cash transfers), the optimal choice will depend on individual country circumstances: both have their advantages, but each has greater potential to be more effective in certain scenarios in one country than in another.

Targeted support programmes are important in all countries, including those with significant rice exports, such as Viet Nam, Thailand and Myanmar. While overall, producers and countries will benefit, reforms that break down trade barriers may increase world prices and so too domestic rice prices in exporting countries, with implications for poorer urban households, especially in Myanmar and Viet Nam (Chapter 4). While the positive income effects for rice farmers, in Myanmar in particular, should stimulate economic growth, increasing employment opportunities more generally, it will not happen immediately, and reforms will benefit some households and have adverse impacts on others. An important part of the policy tool kit in dealing with this, as in net importing countries, will be the provision of targeted support to vulnerable households to mitigate the initial negative impact on urban consumers.

Figure 7. Targeted support programmes would perform significantly better than existing rice distribution programmes in Indonesia and the Philippines

Percentage point change in the share of population that is undernourished

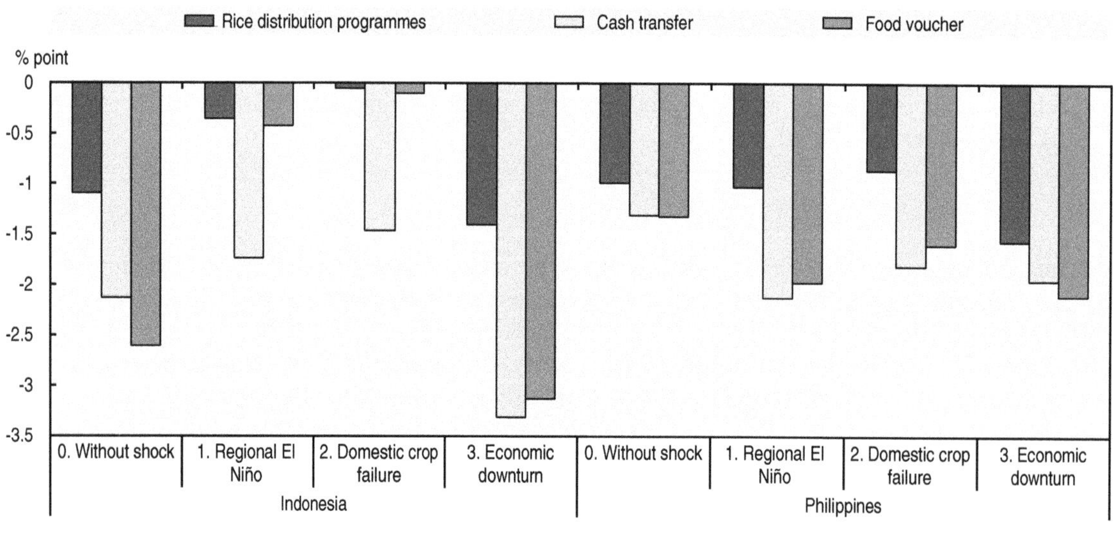

Source: OECD estimates (Chapter 4).

Beyond redistributive approaches, direct support for livelihood diversification may also be needed in certain instances. In both the agricultural and fisheries sectors, for example, the transition to more remunerative alternative activities, where necessary, could be facilitated through training programmes and the provision of credit facilities (discussed further in Box 3).

The extent to which price rises will occur will depend on the responses of producers to the price effects of integration and the time frame over which reforms are implemented. For exporting countries in ASEAN, greater access to regional rice markets and higher returns for producers will induce new productivity-enhancing investments in the sector and encourage greater production. In the case of Myanmar, for example, the large productivity gaps that already exist with other ASEAN countries indicate considerable scope for supply responses that will mitigate the domestic price rises experienced through regional rice market integration (Box 1).

ii) Trade reforms can underpin risk management to improve food security

Implementing trade reforms that enable an open regional market, particularly for rice, will be an important step in improving regional food security. While these are already key elements of the ASEAN Integrated Food Security Framework, and the Strategic Plan for ASEAN Cooperation in Food, Agriculture and Forestry, more efforts are needed. As ASEAN members now constitute a larger share of world agricultural trade, there is even greater need for a transparent, less-distorted regional – and indeed international – agricultural market that will enable ASEAN producers to exploit their growing competitive advantages in the agricultural and food sectors, fisheries and aquaculture included. The current push towards the closer integration of ASEAN economies, including free trade in agricultural products, aided by improved trade facilitation measures and the harmonisation of food regulations, could significantly increase growth opportunities and incomes in the region and thereby improve food security (Chapter 4). The importance of open markets is highlighted by the case of rice. Of the alternative policies examined, targeted cash transfers and food vouchers included, ASEAN rice market integration has the *greatest* potential to reduce the risk of food insecurity in the region (Box 1), particularly in its two largest rice-importing countries: Indonesia and the Philippines (Figure 8). The region, already self-sufficient in rice, can exploit its production base and natural comparative advantages for the benefit of all through the integration of the ASEAN rice market. Taking this step would be more effective at reducing the risk of undernourishment than the current policy emphasis of some countries on domestic market solutions.

Figure 8. ASEAN rice market integration can reduce undernourishment

Percentage point change in the share of population that is undernourished

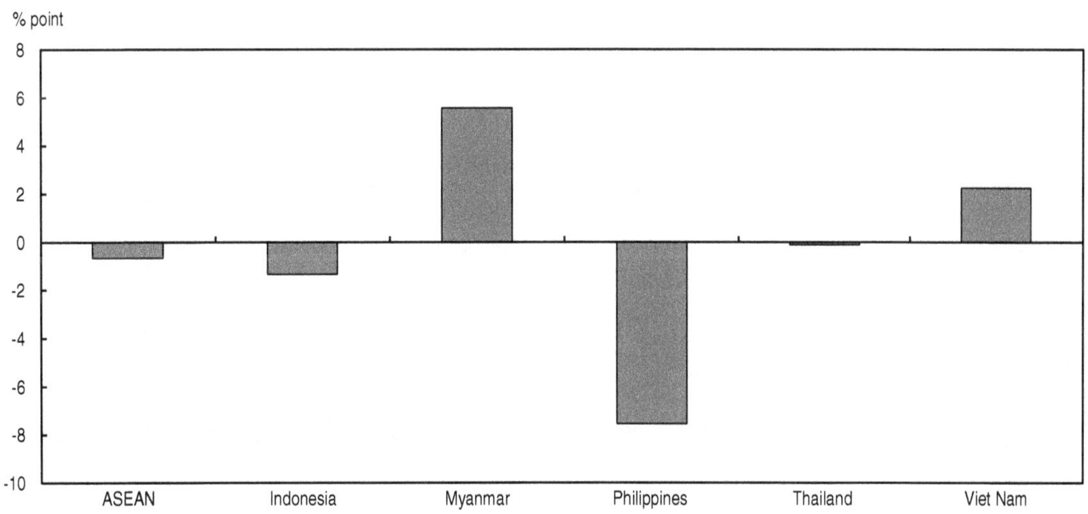

Note: "ASEAN" in this instance refers to the five ASEAN countries included within the study.
Source: OECD estimates (Chapter 4).

By contrast, increases in restrictive trade policies – and indeed in agricultural protection in general – both within ASEAN and worldwide, can have significant negative effects on rice and agriculture more broadly in the region. Increasing trade restrictions impacts not only trade but also GDP, rural incomes and, ultimately, food security. The pursuit of more open markets among a wider set of trading partners, both regionally and internationally, together with reductions in distorting forms of domestic support to agriculture, should therefore significantly improve food security in the ASEAN region.

Box 1. ASEAN rice market integration is vital for food security

ASEAN rice market integration, involving both the removal of all trade barriers – including export restrictions – and other actions that will enhance trust in regional markets and improve producers' abilities to access markets, represents a step towards achieving the goals of the ASEAN Economic Community Blueprint 2025, and is also a component of the ASEAN Integrated Food Security Framework and the Strategic Plan for ASEAN Cooperation in Food, Agriculture and Forestry. However, current policy settings in a number of countries actually work against market integration. The analysis conducted in this report shows that there is much to be gained – in terms of managing risk and improving food security – from moving towards regionally integrated rice markets.

Overall, ASEAN rice market integration would reduce the undernourished population by 5% in the five countries examined. Of these five, undernourishment in two rice-importing countries (Indonesia and the Philippines) would fall the most as a result of the resulting decreases in domestic prices. The integration of regional rice markets also mitigates the otherwise large impact of weather risks in the region. In particular, increased consumer access in both Indonesia and the Philippines could offset the food insecurity impact of a regional *El Niño* or of domestic crop failure, which are identified as the largest risks to food security, respectively speaking, for these two countries. While the regional *El Niño* scenario increases the undernourished population in five ASEAN member states by 49% under the current rice trade regime, integrating the regional rice market could mitigate the impact to an 11% increase. Furthermore, for both Indonesia and the Philippines, despite a shift to larger import volumes, 89% and 73% of their respective domestic consumption would still be provided by local production (compared with current levels of 99% to 86% respectively). This highlights that in both countries, regional integration and a vibrant and internationally competitive rice sector can indeed co-exist.

However, integration may also increase consumer rice prices in exporting countries, such as Myanmar, Viet Nam, and – to a lesser extent – Thailand, depending on regional interactions with the world market and the extent to which domestic prices are currently below world prices (due to a lack of market access). While in the case of Myanmar, in particular, the positive income effects generated by rising producer prices should increase aggregate demand and thus economic growth, safety nets can help to mitigate the potential negative effects of such increases on poor urban households. In any case, the extent to which price rises *will* occur will depend on the responses of producers to more favourable market conditions and the scope for production expansion either through closing productivity gaps, increased intensification or area expansion – that is, the size of the supply response. For some countries in the region, such as Myanmar, the large productivity gaps that already exist with other ASEAN countries, current lack of market access and low levels of mechanisation indicate considerable scope for improvement in this respect. It is also likely that the gradual integration of the regional rice market would actually prevent a sharp increase in rice prices in exporting countries as producers in all countries will have time to adjust to the new rice market conditions.

In addition, greater involvement of the private sector in regional rice trade could help to facilitate the necessary market integration, as well as providing benefits in terms of greater efficiency, reduced distortions and greater potential for growth. Viet Nam could, for example, allow its private exporters to play a significant role in the export market by reducing the role of state-owned enterprises in rice exports, while in the Philippines and Indonesia, the role of state agencies could be restricted to the neutral management of emergency stocks to enable the greater involvement of private traders in rice imports.

Source: Chapter 4.

iii) Promoting sustainable productivity growth is essential for food security over the longer term

Undernourishment in some countries is so prevalent and persistent that additional efforts across all dimensions of food security will be needed. In particular, in addition to targeted transfers and removing barriers to regional integration, investments will be needed to ensure the sustainable supply of food over the longer term.

Maintaining current strong rates of agricultural total factor productivity growth is critical for ASEAN member states to achieve food security. This will require additional measures to address the enabling environment for the agricultural and fisheries sectors as a whole (Box 2). These include: ensuring competitive markets; transparent policy making; and investments in research and development (R&D) and in improving agricultural innovation systems (AIS), together with efforts to ensure that R&D outputs are adopted by farmers and fishers. That said, while agricultural output in the region has significantly increased in recent decades, spurred by productivity improvements across the majority of countries, previous growth levels may not be sustainable, due to the heavy reliance on natural resource exploitation and land expansion which has resulted in the degradation of land and water resources in some areas (OECD, 2012, 2014, 2015c). In future, ASEAN countries will also need to increasingly adopt agricultural and fisheries development strategies that prioritise total factor productivity growth while protecting the natural resource base on which these sectors rely. These sustainable investments will contribute to more stable food supply, less volatile prices and increased investment, and create an environment that is conducive to improving food security across all its dimensions.

Box 2. OECD Agricultural Growth Enabling Index (AGEI)

The enabling environment broadly comprises the settings within which the agricultural sector and wider economy operate, including non-distorting and stable policies, the adequate provision of public goods, good governance through laws and regulations that are conducive to private sector economic activity while at the same time addressing market failures, and robust and effective institutions. As such, it is multifaceted, including aspects of agricultural and environmental policies, private investment, and broader policies relating to infrastructure, macroeconomic settings, innovation, and research and development (R&D). These elements combine to shape incentives for investment by farms and other businesses in agricultural value chains and the building of economic capacities. A positive agricultural enabling environment is therefore important to maximise the returns to agricultural policies and investments in the sector.

The OECD Agricultural Growth Enabling Index (AGEI) (Diaz-Bonilla, Orden and Kwieciński, 2014) compares the performance of selected countries (Cambodia, Indonesia, Lao PDR, Malaysia, Myanmar, the Philippines, Thailand and Viet Nam, together with a wider set of relevant non-ASEAN comparators) across the various components of the enabling environment. As such, it provides an overview of government measures and activities that potentially aid or hinder agricultural growth.

Figure 9. There is scope to improve a number of areas of the enabling environment

AGEI normalised scores for each country relative to average for the 32 countries included within the AGEI

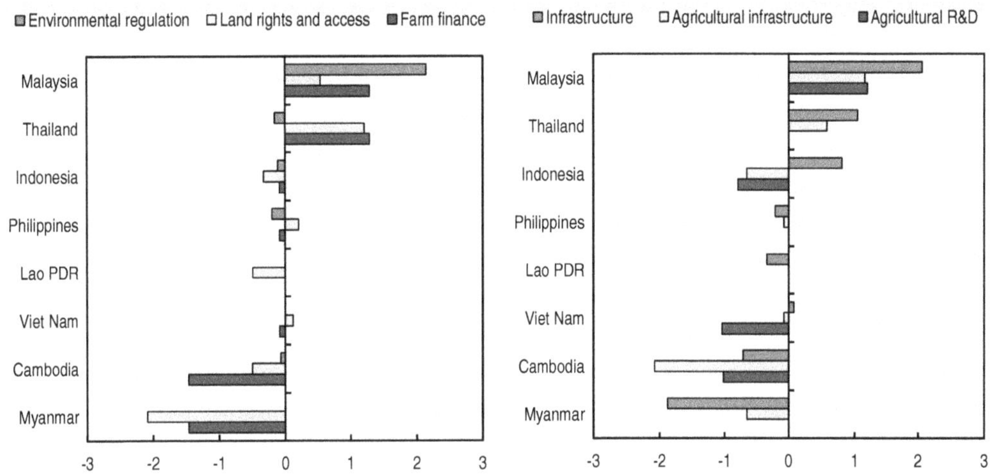

Source: Chapter 5.

Although the performance of the ASEAN countries analysed varies significantly across the AGEI, the results reveal some common relative strengths and weaknesses. Relative strengths include aspects of economy-wide policy settings – such as the macroeconomic environment, labour market operations and human capital – and relatively abundant water resources, while common areas of relative weakness include agricultural and sustainability aspects of the enabling environment. With the exception of Malaysia, ASEAN member states tend to score relatively poorly with regard to public investments in agricultural R&D, land market rights and access, farmer access to finance, the existence and quality of agricultural infrastructure – although Thailand also scores above average in this respect – and for the stringency and enforcement of environmental regulations. Indeed, the results suggest that, compared with other sectors, agriculture in ASEAN member countries may actually be underprovided with public goods and other economic services. Some countries, for example, score highly for infrastructure and financial market development, yet receive below-average scores for agricultural infrastructure and/or farmer access to finance. In the future, therefore, sustainable total factor productivity growth in ASEAN may be hampered by underinvestment in agricultural sector capacities, as well as by a lack of policies that protect the natural resources upon which agricultural production relies.

Furthermore, while common strengths and weaknesses can be identified, the performance of countries often varies across the various subcomponents of the AGEI, indicating significant opportunities for ASEAN member states to learn from each other's experiences with different policy approaches and reforms.

Source: Chapter 5.

a) Increased public investment in agricultural R&D and government efforts to improve agricultural innovation systems more broadly are needed

Public investment in agricultural R&D is essential for sustainable agricultural total factor productivity growth (Sheng et al., 2011; Smeets Kristkova, Van Dijk and Van Meijl, 2016; Chapter 6). By ensuring that farmers have access to innovations that meet their diverse and complex needs, public spending on agricultural R&D is proven to be more effective at raising sustainable agricultural productivity than other public expenditures in agriculture, such as irrigation and fertiliser subsidies (Diaz-Bonilla, Orden and Kwieciński, 2014). For example, the development of a new rice variety that is better adapted to the projected changes in climate could not only mitigate a number of the expected negative effects from climate change, but would also help address the impacts on the production of other crops, agricultural prices and farmer incomes (Chapter 2). By increasing the rice supply, the new rice variety could enable reductions in the volume of land needed for rice cultivation, thereby increasing the land available for other crops, and therefore also increasing the production of these crops, limiting the pressure on commodity prices overall. The aquaculture sector is another example that confirms the importance of technology for sustainable productivity growth: technological innovation has played an important role in growth in every aspect of aquaculture operations, particularly in the development of efficient feed, and will also contribute significantly to the response of the aquaculture sector to climate change, including through research on the domestication of species that are resistant to higher salinity and water temperatures (Chapter 7).

Notwithstanding the extensive evidence of the benefits of agricultural research, and indeed despite recent increases in expenditure in a number of ASEAN countries, public investment in agricultural research in the region is generally low. Although the share of private sector R&D is rising in some countries, mainly in higher-value, market-oriented commodities, this remains a minor share of total agricultural R&D funding, and is insufficient to compensate for the lack of public support (Chapter 6). Moreover, research intensities (public expenditure on agricultural R&D as a percentage of agricultural GDP) have decreased in some countries (ASTI, 2016; IFPRI, 2015; Stads, 2015), implying that public funding for agricultural research, which is already low, is struggling to keep pace with agricultural sector growth in these countries.

As a result of these shortfalls in funding, AIS in some ASEAN countries lack the necessary research capacities to develop and adapt innovations that address the challenges facing their agricultural sectors. If current levels of underinvestment in agricultural R&D continue, long-term sustainable agricultural total factor productivity growth in ASEAN – and, consequently, the capacity of countries to eliminate food insecurity – will be severely constrained. ASEAN member states should address this by, for example, reallocating funding to R&D from other, more distorting forms of sectoral support, such as input subsidies, and by exploring strategies to encourage and facilitate increased private investment in agricultural innovation. These strategies include the strengthening of intellectual property rights and the facilitation of public private partnerships (PPPs) for innovation. Other potential sources of funding, such as commodity levies, could also be explored in consultation with industry.

Beyond agricultural R&D funding, the capacity of agricultural innovation systems to contribute to sustainable agricultural productivity growth depends on the active engagement of governments in a number of key aspects of innovation systems. These include governance arrangements that support efficient and effective AIS which are both demand-driven and responsive to the needs of farmers, and which encourage collaboration between AIS actors; effective training, extension and advisory services that facilitate farmers' access to and adoption of technology and knowledge; and networks and capacity for regional and international research collaboration to maximise the gains from domestic resources so that countries can better leverage their domestic research resources and benefit from specialisation and international research spillovers, and more efficiently address mutual challenges, food security included.

Government involvement to date has been more effective in some aspects of AIS than others. With regard to research collaboration, for example, ASEAN member state research agencies are successfully forging strong international research links with a large number of international research institutes and development agencies, and higher education institutes are actively co-operating with a view to building the capacity of agricultural education in the region. Efforts in this area should be continued. Where governance is concerned, however, there is room for further improvement (Chapter 6). Stronger AIS governance is vital to maximise the payoffs to investments in agricultural R&D. This includes mechanisms to increase farmer engagement in priority-setting and the development of strategies for AIS that establish agricultural research priorities and co-ordinate key AIS actors, the private sector included.

Long-term sustainable agricultural total factor productivity growth depends not only on the capacity of AIS to meet the innovation needs of farmers, but also the capacity and willingness of the farmers themselves to adopt and implement these innovations. The provision of extension services can be a highly effective means to engage with producers and facilitate their access to technology and knowledge. This is of particular importance in countries with large numbers of small-scale producers, who usually lack the resources to innovate on their own. Although extension services are undergoing reforms in a number of ASEAN countries with a view to better addressing the needs of farmers, they have nevertheless been weakened by funding constraints, poor co-ordination of extension providers and, in some decentralised systems, insufficient capacity at the local government level (IFAD, 2014; OECD, 2015c; OECD, 2017a). ASEAN governments can strengthen agricultural extension systems by increasing funding for extension services; continuing reforms to ensure that extension and advisory systems are demand-driven and encompass a wide range of relevant issues, including natural resource management and marketing skills; engaging farmers in the setting of priorities; and co-ordinating the roles of extension providers, in order to minimise duplication of activities and ensure that extension is made available to all farmers. In countries with decentralised extension systems, governments should invest in increasing the training capacity of local governments and other extension service providers.

While effective and efficient agricultural innovation systems are essential elements of total factor productivity growth, without further efforts to improve other agriculture-related elements of the broader enabling environment, the benefits of investments in R&D will not be maximised. Without the necessary incentives for co-investment by producers, R&D outputs risk remaining "on the shelf", with limited uptake by producers. If this occurs, long-term, sustainable productivity growth will not be achieved. Those elements where scope for further improvement in ASEAN countries remain, namely infrastructure, agricultural infrastructure in particular; land tenure and access to credit by farmers; and the enforcement of environmental regulations, are set out below.

b) Greater investment in agricultural infrastructure would bring multiple benefits

Investments in infrastructure can not only enable the dissemination of innovations, via communication infrastructure, for example – they can also promote food security by making production more profitable by cutting the transaction costs of getting product to market and responding to changing consumer demands. Well-developed transport and communication infrastructure can enable all producers to be connected to markets (OECD, 2015d), and can allow them to access knowledge and extension services (Waite et al., 2014; Chapters 5 and 7). In doing so, these connections can improve total factor productivity, reduce operational costs and stimulate value creation, all of which can also increase the competitiveness of domestic producers in international markets.

The AGEI performance of ASEAN countries varies with respect to infrastructure (Box 2). While Malaysia scores the highest of all countries included within the index, non-ASEAN included, Myanmar receives the lowest score. Roads, electricity and telecommunications infrastructure are particularly poor in rural areas of Myanmar, constraining the modernisation of the agricultural sector. But while investments in general infrastructure are important for the overall economic development of ASEAN countries, it is important that agricultural infrastructure is not left behind. Underinvestment can negatively impact agricultural total factor productivity in a number of respects: projections indicate significant potential for agricultural infrastructure to help offset, or at least moderate, the negative effects of climate change, for example. In general, ASEAN member states perform poorly with respect to some aspects of agricultural infrastructure, such as storage facilities (Malaysia and Thailand excluded). This pattern of prioritising general infrastructure is also evident in those ASEAN countries that perform slightly better than average compared with countries – non-ASEAN included – at a similar level of development (Box 2, Figure 9). It should be noted, however, that the index only measures the ability to store and transport crops to market, thus overlooking the often necessary heavy investments in irrigation infrastructure, in Viet Nam and Indonesia, for example. Indeed, in some ASEAN countries, the majority of investments in agricultural infrastructure tend to be focused on irrigation.

Governments have a key role to play in supporting infrastructure. Nevertheless, the burden on public funds could be alleviated through the formation of alternative financing models beyond reliance on tax revenues. One such example relates to public-private partnerships, which have the potential to offer advantages to financially constrained governments. However, the decision by governments to adopt a PPP approach or any other financing method should be guided by the careful balance of costs and benefits, and there remains a need for governments to complete necessary due diligence before committing to any project.

c) Reforms to land tenure are also key

Secure land market rights and access support sustainable agricultural total factor productivity growth by increasing incentives for landholders to make long-term investments and to also preserve the natural resource base (FAO, 2011). Secure land market rights also enable the use of land as collateral for loans, which can be essential for the funding of productivity and sustainability-enhancing innovations (MCC, 2014). Clear and secure land rights and access are all the more important in regions where limited agricultural land is available, such as ASEAN. Yet, as mentioned earlier (Box 2), ASEAN member states tend to score relatively poorly in these respects, with Myanmar, Lao PDR, Cambodia and Indonesia all performing below average in the AGEI, compared with a wider set of countries at a similar level of development. In Myanmar, for example, state ownership of land and the underdeveloped rule of law have led to a lack of clarity regarding land rights in many cases (OECD, 2016), while in Indonesia, the slow registration of land rights is a significant constraint for smallholders in particular (OECD, 2012). Even in Viet Nam, which scores slightly above average, land use rights are both restricted and insecure (OECD, 2015c), resulting in limited access to credit by small-scale farmers in particular.

ASEAN governments should continue to pursue reforms to improve regulatory and institutional frameworks that govern rural land market operations. This can involve the acceleration of land registration, the harmonisation of land legislation and the recognition and protection of customary land tenure.

d) Access to credit for farmers – small-scale farmers in particular – needs to be improved

Closely linked to land tenure is the issue of access to credit. Access to a range of finance options enables farmers to manage price shocks, invest in innovations, and take advantage of market opportunities, export and domestic. ASEAN member country performance in this area is mixed, with Malaysia and Thailand featuring among the top performers of all countries considered by the AGEI, while Cambodia and Myanmar are among the worst performers. In Myanmar, for example, of all sectors of the economy, farm households are the most underserved by the formal financial system (OECD, 2016). Farm households in Myanmar have considerably less access to formal financial services than non-farm households, and some segments, such as farmers without land rights, are effectively excluded due to a lack of collateral. In Viet Nam, which scores slightly below average in the AGEI, financial markets in rural areas tend to be highly concentrated, limiting farmer access to formal banking services (OECD, 2015c). In East Java in Indonesia, a country which also scores just below average, 95% of farmers have never obtained credit from banks (OECD, 2012).

In addition, governments should focus on increasing the range of financing options available to farmers, small-scale farmers included. Effective rural financial systems encompass a variety of institutions, including commercial banks, development finance institutions, microfinance facilities, and specialised financial services such as leasing and finance companies. In Viet Nam, for example, credit bureaus can reduce the dependence on conventional collateral, as the reputation of borrowers can replace collateral (OECD, 2015c). It can also lower operational costs and speed up the time required to obtain loan approval. In Myanmar, commercial banks should be allowed to operate in the agricultural sector, and the ceiling on bank lending rates should be raised or even eliminated (OECD, 2015d).

e) Environmental governance should be strengthened

In a number of ASEAN countries, the development of agriculture and fisheries – and broader economic growth – has relied on natural resource exploitation. Continued exploitation in the absence of necessary environmental controls has the potential to threaten the natural resource base for agricultural production. Between 1990-2010, for example, 17% of new palm oil plantations in Malaysia and 63% of those in Indonesia came at the expense of lost biodiversity-rich tropical forest (Gunarso et al., 2013; Koh et al., 2011; Pirker et al., 2016), and in addition contributed to increased carbon emissions from the sector (Carlson et al., 2012; Miettinen et al., 2012; Omar et al., 2010, in Pirker et al., 2016), threatening the future productive capacity of the region. In the case of fisheries, meanwhile, the future of the sector as a source of livelihoods and food security increasingly rests on the sustainable management of resources to avoid pressure on over-fished stocks and degradation of ecosystems resulting from poorly managed aquaculture.

The agricultural and fisheries sectors also face growing competition from other domestic and industrial users for water and land resources. Land is an important input for aquaculture production in the region, for example. Although the intensity of land use is dependent on the species, production increases in the aquaculture sector have generally depended on greater access to land. Increasing land use will place greater pressure on an already constrained regional resource. The constraints vary across countries, but in some regions, competition exists between agricultural crops, such as rice, and aquaculture (Tran et al, 2017). Finally, as mentioned earlier, climate change poses a significant risk to agricultural and fisheries production.

Given these challenges, ASEAN member states urgently need to prioritise environmentally sustainable agricultural total factor productivity growth. Nevertheless, with the exception of Malaysia, environmental governance and regulations on land, water and biodiversity resources are generally weak across the region (Chapter 5). Governments should therefore both strengthen regulations on access to and use of natural resources, and ensure that existing regulations are enforced. Improved environmental regulation should be seen as part of an overall strategy to promote sustainable food security (Box 3).

Box 3. Case study: Fishing for food security in Indonesia

Food security remains an important concern in Indonesia, despite its rapid economic development. As in most ASEAN countries, poverty and undernourishment rates have both significantly improved in recent decades, and indications are that this trend will continue. Nevertheless, over 19 million people remain undernourished (FAO, 2015), the largest absolute number in the Southeast Asian region. Indicators of malnutrition also remain at alarming levels – in 2013, more than a third of children under the age of five were stunted.

The fisheries and aquaculture sector plays an important dual role in food security in Indonesia in terms of diet and income generation. Capture fisheries and aquaculture not only contribute significantly to the diet and nutrition of the Indonesian population, but also sustain livelihoods for about 20 million people (Chapter 7). Seafood consumption in Indonesia has grown significantly since the early 1990s, reaching one of the world's highest rates in 2015. The country has also recently become the second-largest producer of seafood worldwide, after China.

Indonesia has responded to the need for food security primarily by encouraging increases in domestic production with a view to increasing the availability of domestically-produced food. Recent measures that have been put in place in the fisheries and aquaculture sector to assist this include efforts to combat illegal fishing; the provision of direct support to artisanal fishers and aqua-farmers through subsidised fuel, vessel, gear, equipment and inputs distribution; public investments in infrastructure; and restrictions on trade and foreign investment to protect domestic fishers and aqua-farmers from global competition.

Policy recommendations

To enable the Indonesian fisheries and aquaculture sector to further contribute to the improvement of food security and nutrition, government policy should focus on three key objectives: (i) improving sustainable resource management, and (ii) increasing economic opportunities for fishers and aqua-farmers while (iii) improving consumer access to seafood.

Improving sustainable resource management: Continued seafood production growth under a business as usual scenario is not an option in Indonesia – the majority of fish stocks cannot absorb further increases in catch effort, and the expansion of aquaculture production will only be sustainable if it consumes less land and water resources, and if externalities are better controlled to avoid the degradation of ecosystems. Government efforts to combat illegal fishing are a welcome step towards reducing pressure on overfished stocks. Nevertheless, they should be complemented by better management of domestic fisheries, notably the adoption of inclusively-defined, science-based and measurable long-term management targets. There is also potential for the improved regulation of the aquaculture sub-sector, with better monitoring of its impact on ecosystems, notably through the clearer allocation of responsibilities and sharing of information across government structures.

Increasing economic opportunities for fishers and aqua-farmers while improving consumer access to seafood: Sustainable increases in the incomes of fishers and aqua-farmers should be predominantly sought through increases in the value of seafood produced as well as reductions in production cost, without interventions in markets. Catch limits and other measures aimed at improving the sustainability of fish resources should contribute to increasing producer incomes in the long run by increasing the value of the catch per unit of effort, leading to a reduction in production costs, particularly if complemented by an efficient catch allocation system. Better regulation of environmental externalities linked to aquaculture production and the adoption of sustainable production practices also have the potential to increase the returns from aquaculture farms in the medium to long run. Transitional costs are however to be expected, and careful anticipation of impacts and the design of adaptive strategies will be necessary to avoid income disruptions. The direct support currently provided to individual fishers and aqua-farmers – mainly in the form of subsidies for fuel and vessel or productive equipment – with a view to increasing their incomes or reducing their costs nevertheless biases production incentives and increases pressure on resources. To prevent this, it is recommended, particularly during periods of transition to more sustainable production practices, that the government instead directly support livelihoods via the provision of social safety nets that are targeted at vulnerable households, and help the conversion to more remunerative activities by means of education and training. Directing public support to R&D and infrastructure would also favour greater value creation, while at the same time reducing prices for consumers – thus improving access to food, the third key objective – and increasing food availability. Reliance on restrictive trade and investment barriers, on the other hand, supports incomes at the expense of reduced consumer access to seafood products, and therefore has the potential to be counterproductive to food security in the long run.

Source: Chapter 7.

Summary of recommendations

The recommendations suggested below, which are not listed in order of importance, are derived from the analysis undertaken in this report. These recommendations are not exhaustive, and should be interpreted as a starting point for government consideration. In particular, choices will need to be made as to which policy actions should and can be implemented quickly, and which might be acted upon more gradually.

- ASEAN member states have put in place a sound regional architecture to address many of the key food security challenges facing the region. The ASEAN Integrated Food Security Framework and the Strategic Plan for ASEAN Cooperation in Food, Agriculture and Forestry provide a solid platform on which ASEAN member states are pursuing policies to address long-term food security. These regional policy frameworks are underpinned by core policy areas and a number of "Strategic Thrusts", which set out actions for ASEAN member states. This regional framework is also supported by the ASEAN Plus Three Emergency Rice Reserve (APTERR) that seeks to provide food coverage across the region in times of severe short-term need.
- A key finding of this review is that ASEAN member states can significantly benefit from additional efforts and policy choices that are consistent with the core policy areas identified in the regional frameworks. Despite the strong regional links, the current policy approach to food security in ASEAN member states remains by and large domestically-focused and oriented towards the management of international market risks rather than responding to the – more important and more frequent – risks from domestic supply disruptions, or addressing long-term issues that will help to underpin future regional food security. Policies that are used include measures to promote food self-sufficiency and price stabilisation, along with often poorly-targeted social safety nets underpinned by public stockholding schemes, which have proved costly and have likely been counterproductive to food security in the longer run.
- ASEAN governments could benefit from adopting a broader understanding of food security and further embracing the move towards regional solutions to food security and risk management. Open markets can positively contribute to food security. They can improve access, by increasing incomes of poorer households; the availability of food, by providing opportunities for food to be imported by food-deficit countries and by increasing the diversity of products; and the general *stability* of access and availability, by the sharing of production risks across countries. With respect to the rice sector, regional rice trade integration has been found to positively improve regional food security to a significant degree. The functioning of the rice market could also be improved by the greater involvement of the private sector in regional rice trade.

- In order to avail of some the benefits that further regional – and even global – integration can bring, there needs to be reductions in trade barriers to make food more accessible to consumers, together with more targeted social safety net measures, such as conditional cash transfers and food voucher programmes for vulnerable households. Aligned with these shifts, ASEAN governments need to make investments in promoting sustainable total factor productivity growth.

- Sustainable agricultural total factor productivity growth for food security will require investments in and reforms to the enabling environment in the ASEAN region. These include the improvement of environmental governance; regulations on land, water and biodiversity resources; together with investments in infrastructure and agricultural R&D. Investments in transport and communications infrastructure, for example, would help to connect farms to market opportunities, knowledge and other services, extension included. Governments should also persevere with reforms to improve regulatory and institutional frameworks that govern rural land market rights and access, and should consider opportunities to increase farmer access to credit, small-scale farmers included. In general, governments will need to at least maintain the past intensity of policy reforms that address the remaining constraints to productivity growth and continue with the intensity of investments if *past* trends in regional productivity are to be maintained. If sustainable productivity growth is to contribute further to improving regional food security, however, this intensity needs to be increased.

  – Current underinvestment in agricultural R&D threatens long-term total factor productivity growth in the region. There is a need for additional investment above and beyond current levels. A well-functioning national and regional agricultural innovation system will play an important role in this respect. The involvement of producer and industry organisations and the development of strategies for the diffusion of the outputs of R&D are important.

  – Public investment in infrastructure can help to connect producers to markets, allow them to take advantage of new opportunities, and provide access to knowledge and other services, such as extension. In doing so, it helps to improve productivity, reduce operational costs and thereby stimulate value creation. Ensuring that investments in agricultural infrastructure are not sidelined is of particular importance for ASEAN members.

  – Secure land market rights and access increase incentives for long-term investments and technology adoption by farmers. These can be achieved via the acceleration of land registration, the harmonisation of land legislation, and the recognition and protection of customary land tenure, among other measures.

  – Facilitating access by investors to long-term financing and enabling small-scale producers to access credit increases investment and thus productivity. Opportunities to increase the range of financing options available to farmers through reforms to the financial services sector include allowing greater involvement by the commercial banking sector or the adoption of micro-financing models, as well as efforts to facilitate the process of applying for loans.

  – More robust environmental governance is necessary to preserve natural resources and provide incentives for environmentally-sustainable production. This requires the strengthening of regulations on access to and use of natural resources, together with better enforcement of existing regulations.

In the case of the fisheries sector more specifically, sustainable resource management is a first and essential step to sustainably improve producer incomes in the long run while also increasing the availability of fisheries products and reducing prices for consumers, thereby improving food security. Lessons from Indonesia suggest that to enable efficiency improvements without increasing consumer prices or pressure on resources, production-distorting policies should be reduced. Government support should instead take the form of investment in R&D and transport and energy network infrastructure, together with the provision of targeted social safety nets and training to enable the transition to more remunerative activities. Finally, good governance of fisheries management frameworks is critical to ensure that any reforms are socially acceptable and durable.

Notes

1. The ten ASEAN member states are Brunei Darussalam, Cambodia, Indonesia, Lao People's Democratic Republic (PDR), Malaysia, Myanmar, Philippines, Singapore, Thailand and Viet Nam.
2. Undernourishment is defined as chronic calorie deficiency. It should be noted, however, that food insecurity can also result from insufficient intake of protein and micronutrients.
3. Thin markets are markets in which there are low imports and exports relative to production and consumption (OECD, 2009). For example, rice markets are often considered thin, with an average of 8% of total production traded over the period 2005-16, compared with other staple crops such as soybeans, wheat and maize which had 37%, 21% and 12% of production traded respectively (USDA, 2017).

References

ADB (2009), *The Economics of Climate Change in Southeast Asia: A Regional Review*, Asian Development Bank, Manila.

Alavi, H., A. Htenas, R. Kopicki, A. Shepherd and R. Clarete (2012), *Trusting Trade and the Private Sector for Food Security in Southeast Asia*, World Bank, Washington, DC.

Anderson, K., M. Ivanic and W. Martin (2013), "Food price spikes, price insulation, and poverty", *Policy Research Working Paper*, No. 6535, World Bank, Washington, DC.

ASTI (2016), *Agricultural Science and Technology Indicators* (database), International Food Policy Research Institute (IFPRI), www.asti.cgiar.org/data (accessed on 13 July 2016).

Bank of Thailand (2016), *Wholesale Price of Certain Commodities in Bangkok Metropolis* (dataset), Bank of Thailand, Bangkok, www2.bot.or.th/statistics/BOTWEBSTAT.aspx?reportID=89&language=ENG (accessed on 15 June 2015).

BPS-Statistics Indonesia (2010), *National Socioeconomic Survey,* Central Bureau of Statistics of Indonesia.

Carlson, K.M., L.M. Curran, G.P. Asner, A.M. Pittman, S.N. Trigg and J.M. Adeney (2012), "Carbon emissions from forest conversion by Kalimantan oil palm plantations", *Nature Climate Change,* Vol. 3/3, pp. 283–287.

Cruz, R.V., H. Harasawa, M. Lal, S. Wu, Y. Anokhin, B. Punsalmaa, Y. Honda, M. Jafari, C. Li and N. Huu Ninh (2007), "Asia" in *Climate Change 2007: Impacts, Adaptation and Vulnerability. Contribution of Working Group II to the Fourth Assessment Report of the Intergovernmental Panel on Climate Change,* M.L. Parry, O.F. Canziani, J.P. Palutikof, P.J. van der Linden and C.E. Hanson (eds.), Cambridge University Press, Cambridge, United Kingdom, pp. 469-506.

Dawe, D., P. Moya and C. Casiwan, eds. (2006), *Why Does the Philippines Import Rice? Meeting the Challenge of Trade Liberalization,* International Rice Research Institute and Philippine Rice Research Institute, Manila.

Deuss, A. (2015), "Review of the performance and impacts of recent stockholding policies", in *Issues in Agricultural Trade Policy: Proceedings of the 2014 OECD Global Forum on Agriculture*, OECD Publishing, Paris, http://dx.doi.org/10.1787/9789264233911-5-en.

Diaz-Bonilla, E., D. Orden and A. Kwieciński (2014), "Enabling environment for agricultural growth and competitiveness: Evaluation, indicators and indices", *OECD Food, Agriculture and Fisheries Papers*, No. 67, OECD Publishing, Paris, http://dx.doi.org/10.1787/5jz48305h4vd-en.

FAO (2016a), *FAOSTAT* (database), Food and Agriculture Organization of the United Nations Statistics Division, Rome, http://faostat.fao.org/ (accessed on 22 February 2016).

FAO (2016b), "Rice Prices", *Food Security Portal* (database), Food and Agriculture Organization of the United Nations, Rome, www.foodsecurityportal.org/api/rice (accessed on 27 April 2015).

FAO (2015), *Regional Overview of Food Insecurity Asia and the Pacific: Towards a Food Secure Asia and the Pacific*, Food and Agriculture Organization, Bangkok.

FAO (2011), *The State of the World's Land and Water Resources for Food and Agriculture (SOLAW): Managing Systems at Risk*, Food and Agriculture Organization, Rome, and Earthscan, London.

Gunarso, P., M.E. Hartoyo, F. Agus and T.J. Killeen (2013), *Oil Palm and Land Use Change in Indonesia, Malaysia and Papua New Guinea*, Round Table on Sustainable Palm Oil (RSPO), Singapore.

Headey, D. (2011), "Rethinking the global food crisis: The role of trade shocks", *Food Policy*, Vol. 36, Elsevier, Amsterdam, pp. 136-146.

Hijioka, Y., E. Lin, J.J. Pereira, R.T. Corlett, X. Cui, G.E. Insarov, R.D. Lasco, E. Lindgren and A. Surjan (2014), "Asia", in *Climate Change 2014: Impacts, Adaptation, and Vulnerability. Part B: Regional Aspects, Contribution of Working Group II to the Fifth Assessment Report of the Intergovernmental Panel on Climate Change*, Barros, V.R., C.B. Field, D.J. Dokken, M.D. Mastrandrea, K.J. Mach, T.E. Bilir, M. Chatterjee, K.L. Ebi, Y.O. Estrada, R.C. Genova, B. Girma, E.S. Kissel, A.N. Levy, S. MacCracken, P.R. Mastrandrea and L.L. White (eds.), Cambridge University Press, Cambridge, United Kingdom and New York, United States, pp. 1327-1370.

IFAD (2014), *Fostering Agricultural Revitalization in Myanmar (FARM): Final Project Design Report*, International Fund for Agricultural Development.

IFPRI (2015), "Food policy indicators: Tracking change", in *2014–2015 Global Food Policy Report*, International Food Policy Research Institute, Washington, DC.

Indonesian Ministry of Agriculture (2015), data received from Indonesian Ministry, Bangkok.

International Grains Council (2016), "Export Prices", *International Grains Council* (database), London, www.igc.int/en/markets/marketinfo-prices.aspx (accessed on 7 May 2015).

Koh, L.P., J. Miettinen, S.C. Liew and J. Ghazoul (2011), "Remotely sensed evidence of tropical peatland conversion to oil palm", *Proceedings of the National Academy of Sciences of the United States of America – PNAS*, Vol. 108, pp. 5127–5132.

Lowder, S.K., J. Skoet and S. Singh (2014), "What do we really know about the number and distribution of farms and family farms worldwide?", Background paper for *The State of Food and Agriculture 2014*, *ESA Working Paper* No. 14-02, Food and Agriculture Organization of the United Nations, Rome.

MCC (2014), "Guide to the Indicators and the Selection Process, FY 2015" (webpage), Millennium Challenge Corporation, www.mcc.gov/resources/doc/report-guide-to-the-indicators-and-the-selection-process-fy-2015 (accessed on 30 April 2015).

Miettinen, J., A. Hooijer, C. Shi, D. Tollenaar, R. Vernimmen, S.C. Liew, C. Malins and S.E. Page (2012), "Extent of industrial plantations on Southeast Asian peatlands in 2010 with analysis of historical expansion and future projections", *GCB Bioenergy*, Vol. 4/6, pp. 908–918.

MNPED (2010), *Integrated Household Living Conditions Survey in Myanmar (2009-2010),* Myanmar Ministry of National Planning and Economic Development, United Nations Development Programme, United Nations Children's Fund, and the Swedish International Development Cooperation Agency, Yangon, Myanmar.

Naylor, R.L. and W.P. Falcon (2010), "Food security in an era of economic volatility", *Population and Development Review*, Vol. 36/4, New Jersey, United States, pp. 693-723.

OECD (2017a), *Agricultural Policies in the Philippines*, OECD Publishing, Paris, http://dx.doi.org/10.1787/9789264269088-en.

OECD (2017b), *Agricultural Policy Monitoring and Evaluation* (website), OECD Publishing, Paris, http://www.oecd.org/tad/agricultural-policies/monitoring-and-evaluation.htm (accessed on 21 February 2017).

OECD (2016), *Multi-Dimensional Review of Myanmar: Volume 3. From Analysis to Action*, OECD Development Pathways, OECD Publishing, Paris, http://dx.doi.org/10.1787/9789264256545-en.

OECD (2015a), *Agricultural Policy Monitoring and Evaluation 2015*, OECD Publishing, Paris, http://dx.doi.org/10.1787/agr_pol-2015-en.

OECD (2015b), *Managing Food Insecurity Risk: Analytical Framework and Application to Indonesia*, OECD Publishing, Paris, http://dx.doi.org/10.1787/9789264233874-en.

OECD (2015c), *Agricultural Policies in Viet Nam 2015*, OECD Publishing, Paris, http://dx.doi.org/10.1787/9789264235151-en.

OECD (2015d), *Multi-dimensional Review of Myanmar: Volume 2. In-depth Analysis and Recommendations*, OECD Publishing, Paris, http://dx.doi.org/10.1787/9789264220577-en.

OECD (2014), *Towards Green Growth in Southeast Asia*, OECD Green Growth Studies, OECD Publishing, Paris, http://dx.doi.org/10.1787/9789264224100-en.

OECD (2012), *OECD Review of Agricultural Policies: Indonesia 2012*, OECD Publishing, Paris, http://dx.doi.org/10.1787/9789264179011-en.

OECD (2009), *OECD Annual Report 2009*, OECD Publishing, Paris, http://dx.doi.org/10.1787/annrep-2009-en.

OECD (2008), "Rising food prices: Causes and consequences", *OECD Policy Brief*, OECD Publishing, Paris, www.oecd.org/trade/agricultural-trade/40847088.pdf.

OECD-FAO (2015), *OECD-FAO Agricultural Outlook 2015-2024*, OECD Publishing, Paris, http://dx.doi.org/10.1787/agr_outlook-2015-en.

Omar, W., N. Abd Aziz, A. Tarmizi, M.H. Harun and A. Kushairim (2010), *Mapping of Oil Palm Cultivation on Peatland in Malaysia*, MPOB Information Series No. 529, Kuala Lumpur.

Piesse, J. and C. Thirtle (2009), "Three bubbles and a panic: An explanatory review of recent food commodity price events", *Food Policy*, Vol. 34, Elsevier, Amsterdam, pp. 119-129.

Philippine Statistics Authority (2014), *Survey of Food Demand for Agricultural Commodities in the Philippines,* Department of Agriculture Bureau of Agricultural Statistics, Quezon City, Philippines.

Philippines Bureau of Agricultural Statistics (2016), "Rice", *Bureau of Agricultural Statistics* (database), Manila.

Pirker, J., A. Mosnier, F. Kraxner, P. Havlík and M. Obersteiner (2016), "What are the limits to oil palm expansion?", *Global Environmental Change*, Vol. 40, September, pp. 73-81.

Sheng, Y., E.M. Gray, J.D. Mullen and A. Davidson (2011), *Public Investment in Agricultural R&D and Extension: An Analysis of the Static and Dynamic Effects on Australian Broadacre Productivity*, Australian Bureau of Agricultural and Resource Economics and Sciences Report to the Grains Research and Development Corporation, Canberra.

Smeets Kristkova, Z., M. Van Dijk and H. Van Meijl (2016), "Projections of long-term food security with R&D driven technical change – A CGE analysis", *NJAS – Wageningen Journal of Life Sciences*, Vol. 77, Elsevier, Amsterdam, pp. 39-51.

Stads, G.-J. (2015), *A Snapshot of Agricultural Research Investment and Capacity in Asia*, Asia-Pacific Association of Agricultural Research Institutions (APAARI) and International Food Policy Research Institute (IFPRI), Washington, DC and Bangkok.

Tran, N., U-P. Rodriguez, C.Y. Chan, M.J. Phillips, C.V. Mohan, P. J.G. Henriksson, S. Koeshendrajana, S. Suri and S. Hall (2017), "Indonesian Aquaculture Futures. Part 1: An analysis of fish supply and demand in Indonesia to 2010 and role of aquaculture using the AisaFish Model", *Marine Policy*, Vol. 79, pp.25-32.

USDA (2017), *Production, Supply and Distribution Online* (database), United States Department of Agriculture, Washington, DC, http://apps.fas.usda.gov/psdonline/psdquery.aspx (accessed on 22 March 2017).

Waite, R., M. Beveridge, R. Brummett, S. Castine, N. Chaiyawannakarn, S. Kaushik, R. Mungkung, S. Nawapakpilai and M. Phillips (2014), "Improving productivity and environmental performance of aquaculture performance", *World Resources Institute Working Paper*, World Resources Institute, Washington, DC.

WITS (2016), *World Integrated Trade Solution* (database), World Bank, Washington, DC, https://wits.worldbank.org/WITS/WITS/Restricted/Login.aspx (accessed on 15 November 2016).

World Bank (2016), *World Bank World Development Indicators* (database), http://databank.worldbank.org/data/ (accessed on 25 November 2016).

World Bank (2014), *Myanmar: Rice Price Volatility and Poverty Reduction*, Economic and Sector Work, Report No. 89687-MM, World Bank, Washington, DC.

Yusuf, A.A. and H.A. Francisco (2009), *Climate Change Vulnerability Mapping for Southeast Asia*, Economy and Environment Program for Southeast Asia, Singapore.

Chapter 1

Agriculture and food security in ASEAN

This introductory chapter presents the structure of the overall report, followed by an overview of agriculture and food security across the ten Association of Southeast Asian Nations (ASEAN) members. It summarises developments in agricultural production in ASEAN, focusing on production and trade of key agro-food products including rice, after which it provides a snapshot of food security within ASEAN across a range of indicators, including examining the changes that have taken place over the past 20 years.

Key points

- Over the past 20 years, the economies of Association of Southeast Asian Nations (ASEAN) members have experienced strong gross domestic product (GDP) growth, rising household incomes and structural changes that have benefited food security and development.
- Agriculture has undergone significant structural changes, shedding a significant amount of labour from the sector in all member countries between the early 1990s and early 2010s while at the same time increasing production.
- Despite the structural changes, the production mix of ASEAN as a whole has remained relatively stable, with rice the dominant crop produced.
- Since the early 2000s, on the back of continued agricultural productivity growth, the region has become a significant net agro-food exporter, with both growing regional and world trading links.
- While food security in the region has experienced the most significant improvements compared with other regions worldwide, it nevertheless remains an issue, with high rates of undernourishment, stunting and at risk households in some countries.

1.1 Introduction

Food insecurity remains a global problem. Across the world, in 2014-16, around 793 million people were undernourished, with a greater number consuming food at levels considered inadequate for a normal healthy life (FAO, 2016a). For Southeast Asia, despite a number of countries within the region having undergone significant economic development and transformation, leading to higher incomes, the region still contains around 60 million of the world's undernourished (FAO, 2016a).

Understandably, therefore, food security remains high on the policy agenda of most countries that comprise the Association of Southeast Asian Nations (ASEAN).[1] Going forward, while a number of positive developments are likely to take place in these economies, there will also be developments – such as those related to climate change – that will create challenges and risks for the elimination of food insecurity in the region.

Food insecurity is a complex, multidimensional problem related to food availability, access to affordable food, the effective use by people of the food that they consume, and the stability of these elements over time. Given this, governments generally apply a range of policy responses to help tackle issues of food insecurity within their countries. However, within the set of policies used, policy makers in Southeast Asia and across the globe often turn to agriculture and, to a lesser extent, fisheries,[2] as a central pillar in their policy mix. The two sectors' primacy in policy responses to tackle food insecurity stems from their links to both food *availability* through production and its *accessibility* through incomes and prices. Agricultural and food markets more generally have a dual role in providing income for poor producers and supplying food to consumers.

It is nevertheless important that agricultural policies are both appropriately designed and implemented, as the dual role played by agricultural and food markets can mean that interventions have a number of drawbacks. If interventions are poorly designed, they create inefficiencies in production systems that inevitably impose a drag on economic growth and income generation – elements critical for improving food security. They also increase food prices – an effective tax on poor consumers – and can expose households to greater price risks. Under such situations, it is often the poorest and most food-insecure households, as net buyers of food, that are worst affected.

Given the importance of agriculture and fisheries within the suite of food security policies, this report explores the range of agriculture- and fisheries-related policies that have been used within Southeast Asia in pursuit of addressing food security concerns. Specifically, the report focuses on the ten countries that comprise ASEAN.

Within ASEAN, countries differ in a number of ways that indicate potential for both closer co-operation and opportunities to learn from policy approaches applied in each member state. This potential has long been recognised by regional leaders, resulting in the development of a number of significant regional policy frameworks to set the direction of food security policies and economic development more broadly. In light of this, the report seeks to set out policy recommendations from its findings in relation to these regional frameworks, and to develop an evidence base for regional policy makers to enable them to continue to address issues of food insecurity within the region.

1.2 Structure of the report

This report explores a range of agriculture and fisheries-related issues and policies that are important for the region in its efforts to achieve food security and respond to the risk of food insecurity. Through the analysis provided, findings and policy recommendations are presented to help guide the development of food security policy. The report first explores the background to the region and the future outlook for agricultural markets and food security, after which it reviews and analyses the current policy approaches, exploring issues of effectiveness and efficiency and presenting recommendations for change. Finally, solutions to address food insecurity in the longer term are considered. These include options for investments and reforms in the agricultural and fisheries sectors that will help these sectors to sustainably contribute to improvements in food security. A brief summary of each chapter in the report is provided below.

This chapter presents a brief overview of the changes that have occurred in the ASEAN economies and agricultural sectors over the past 20 years. It describes some of the key developments that have occurred, and shows how the region has dramatically improved food security. Despite this, it highlights the considerable diversity that exists across the ten ASEAN members, and points to a number of remaining food security challenges that regional policy makers continue to face.

Chapter 2 presents information on the agricultural markets and food security outlook for ASEAN. It explores both the possible medium-term changes and the implications that these will have for future levels of food security in the region, along with the longer-term impacts of climate change. The chapter highlights that while continued development and income growth will sustain the fall observed in levels of food security in the region over the medium term, it will not be enough to fully eliminate food insecurity in the region. Furthermore, in the longer term, the risks created by climate change could have a number of negative implications for food security in the region. The chapter emphasises that additional policy action will be required if food insecurity is to be eliminated, pointing to the importance of additional investments in research and development to foster innovation, improving access to food, and continued investments in the other aspects of the enabling environment for the agricultural and fisheries sectors.

Chapter 3 presents a stocktake of agriculture-related food security policies applied by ASEAN member states. The report highlights that in a number of countries, food security objectives pursued through agricultural policy have relied on market interventions and can generally be considered "rice-centric". Policy approaches that have relied on market interventions have generally been found to have a number of unintended impacts on food security and, in some cases, have even worked *against* improvements in food security. Furthermore, market interventions have generally created inefficiencies in the agricultural sector, limiting its ability to improve food security in the longer term through sustainable improvements in producer incomes.

Chapter 4 analyses the effectiveness of a number of current policy approaches in dealing with the risk of food insecurity. Chapter 3 highlighted a number of the efficiency and indirect costs created by market interventions; however, the analysis conducted in Chapter 4 also shows that market interventions and distortions to trade in agricultural products are also not effective in managing food insecurity risks. In contrast, of the options explored, open markets and targeted household support provide the most effective means to manage such risks. The chapter provides evidence on the effectiveness of ASEAN's push for open markets, not only for continued regional development but as a key policy tool to enhance regional food security.

Chapter 5 shifts the focus from current policies to an assessment of factors that will be important for the region to achieve food security in the longer term. A critical element of this is the quality of the agricultural enabling environment. A country's enabling environment is the multifaceted settings within which the agricultural sector and economy more broadly operate, comprising a wide range of factors from policies and the provision of public goods, to the institutions through which government programmes and services are delivered. This chapter presents an analysis of the enabling environment in ASEAN with reference to a number of other countries at a similar level of development. The analysis shows that while some aspects of the enabling environment are relatively strong – such as stable macroeconomic environments, labour market operations, along with human and natural capital – others are relatively weak. In particular, levels of public investment in agricultural research and development and the quality and presence of agricultural infrastructure indicate that agriculture in ASEAN may be underprovided with public goods and other economic services relative to other sectors. Sustainability aspects of the enabling environment are also relatively weak in ASEAN. Agricultural land is scarce, and environmental governance and regulations are weak and/or poorly enforced.

Chapter 6 explores opportunities to enhance food security through improvements to agricultural innovation systems in ASEAN member states. Well-functioning agricultural innovation systems will be important for the region to continue to achieve productivity growth in the medium term, and in dealing with the risks created by climate change in the longer term. Through a stocktake of agricultural innovation systems for ASEAN member states, the chapter highlights a number of areas for improvement, including correcting underinvestment in agricultural research and extension to enable countries to respond to the challenges facing their agricultural sectors, and developing stronger governance such as through mechanisms to increase farmer engagement in priority setting. Governments also need to improve the co-ordination of extension systems to ensure that all farmers are reached.

Chapter 7 explores the role of fisheries policies in helping to deliver food security within ASEAN in the long term. For a number of ASEAN member states, fisheries and aquaculture are important for the diets and livelihoods of a significant share of the population, and thus directly affect food security. Through a case study of the policy settings in Indonesia, this chapter highlights that policies that provide for the sustainable exploitation of resources and ecosystems can both reduce the risk associated with the sector's contribution to food security and increase incomes for fishers and aqua-farmers. The chapter also highlights that combating illegal fishing, as pursued actively by Indonesia, is an important part of the policy framework for the sector. Similar to agriculture, directing public support to research and development and infrastructure has the potential to improve the incomes of fishers and aqua-farmers, while also reducing prices for consumers and increasing food availability. Reliance on restrictive trade and investment barriers, on the other hand, supports incomes at the cost of *reduced* access to seafood products for consumers. Furthermore, direct support to individual fishers and aqua-farmers should not bias production incentives or increase pressure on resources, but rather directly support livelihoods and help the transition to more remunerative activities where relevant.

1.3 Agriculture and trade in ASEAN

Economies in ASEAN have experienced significant levels of economic growth over the past 20 years. Most ASEAN countries have been closing the gap with advanced economies over this period, experiencing income growth and increases in their levels of development. However, growth has been uneven, with differences largely explained by differences in labour productivity growth over the past 20 years (OECD, 2013). Overall, trends in agricultural production and trade also point to increasing food availability within the region which, coupled with increased incomes, suggests that food security has indeed improved.

ASEAN countries vary in their size, level of development and incomes (Table 1.1). Across ASEAN members, Singapore has the highest level of gross domestic product (GDP) per capita, followed by Brunei Darussalam and Malaysia. Indonesia is the region's most populous nation, with more than double the population of the second-largest country, the Philippines. Indonesia is also the largest country in terms of physical size, both in total and in terms of agricultural land. In Thailand, the Philippines, Viet Nam and Cambodia, however, the proportion of total land that is deemed agricultural is greater than that of Indonesia.

Table 1.1. General statistics for ASEAN members, 2013

	General					Agricultural		
	GDP per capita (USD)	Total land (sq. km)	Population (millions)	Trade (%GDP)	Poverty (% pop.)	Land size (sq. km)	GVA per worker	Land (% total)
Brunei Darussalam	38 563	5 270	0.4	109	n.a.	134	n.a.	2.5
Cambodia	1 006	176 520	15.1	140	41.3	57 550	527	32.6
Indonesia	3 644	1 811 570	249.9	49	43.3	565 000	1 034	31.2
Lao PDR	1 653	230 800	6.8	83	62.0	24 690	495	10.7
Malaysia	10 538	328 550	29.7	154	n.a.	77 495	9 674	23.6
Myanmar	1 101	653 290	53.3	n.a.	n.a.	125 930	n.a.	19.3
Philippines	2 765	298 170	98.4	60	41.7	123 950	1 136	41.6
Singapore	55 980	700	5.4	360	n.a.	7	36 625	1.0
Thailand	5 779	510 890	67.0	144	3.5	218 600	1 166	42.8
Viet Nam	1 909	310 070	89.7	165	12.5	108 420	476	35.0

Notes: n.a.: Not available. GDP per capita measured in 2013 USD. Poverty rate statistics represent the percentage of the population living on less than USD 2.00 a day at 2005 international prices (PPP). Gross value added (GVA) per worker in agriculture comprises value added from forestry, hunting, and fishing as well as crop cultivation and livestock production measured in constant 2005 USD.

Source: World Bank (2016), *World Development Indicators*, http://databank.worldbank.org/data/.

Value added by agricultural workers also varies considerably. Of ASEAN's major agricultural producing countries, excluding Brunei Darussalam and Singapore, Malaysian workers generate the highest value added in constant USD terms (2005) – more than nine times that of the next ranked country, Thailand.

ASEAN members also vary in terms of their relative openness to other economies across the world. In 2013, measured as total trade as a percentage of GDP, Singapore was the most engaged with the rest of the world, while Indonesia was the least.

Poverty, measured as the percentage of the population living on less than USD 2 a day (in purchasing power terms), continues to be a sizeable problem for a number of countries in ASEAN. Poverty rates are the highest in Lao PDR, but are also above 40% in Cambodia, Indonesia, and the Philippines.

Agricultural production in ASEAN

The relative size of the agricultural sector in ASEAN countries has changed significantly over time (Figure 1.1). Agriculture's relative importance in GDP and employment declined in most countries between 1996 and 2014 (earliest and latest years for which data are available). Productivity improvements and opportunities outside agriculture have led to significant labour-shedding in a number of countries, but most notably in Cambodia and Viet Nam. Interestingly, the agricultural share of employment fell in Thailand even as its share of GDP rose over the period, representing a shift to relatively higher-value production along with relative changes in other sectors of the economy.

The significant structural adjustment that has occurred within the agricultural sectors of ASEAN countries has contributed to strong production growth. Since the 1960s, annual compound production growth in the Southeast Asian region as a whole has been strong (Figure 1.2).[3] However, agricultural production growth has only exceeded population growth since the 1980s, driven by both slowing population growth rates and increases in agricultural growth rates during the 1980s and 2000s.

For ASEAN as a whole, the structure of production has remained relatively stable across broad commodity groupings. Rice cultivation is the main agricultural production activity, accounting for a greater share of gross production value than any other commodity. Despite the general stability of production, some trends do appear. Most notably, the contribution of rice to total gross agricultural production value has fallen since the early 1990s – from around 40% to close to 30% in 2013 (Figure 1.3). Much of the change has been driven by the increasing contribution of palm oil to total agricultural production value in the region. Within categories, there have also been changes, such as increasing poultry production within the meat sector.

Figure 1.1. Agricultural sector share of employment and gross domestic product (GDP) (%)

1996 and 2014

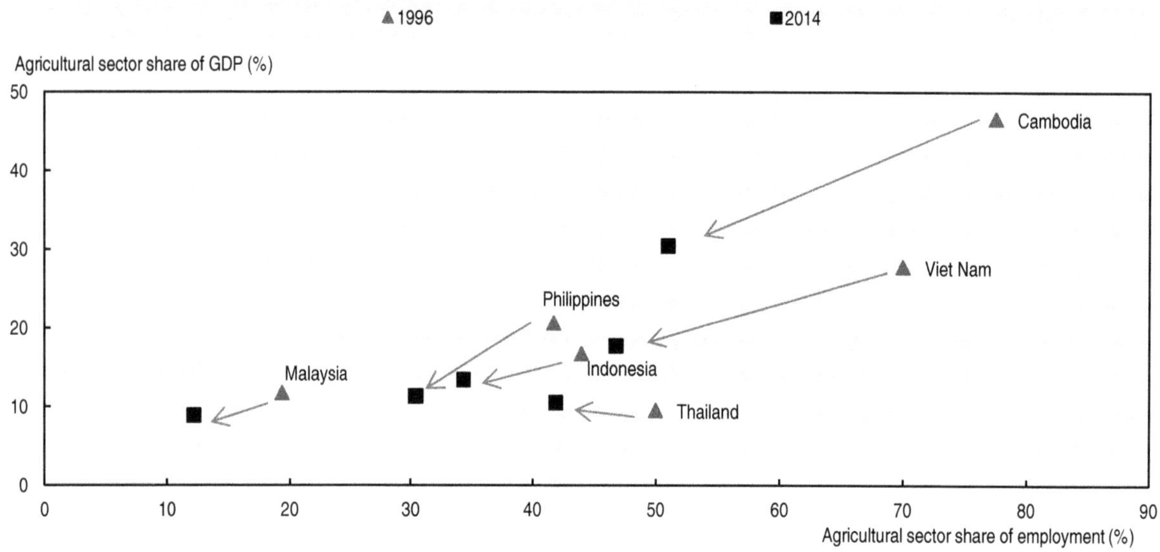

Notes: Data for Cambodia are for 1998 and 2012, and data for employment share are for 1996 and 2013 for Thailand and Viet Nam.
Source: World Bank (2016), *World Development Indicators,* http://databank.worldbank.org/data/.

Figure 1.2. Production growth in Southeast Asia

Decadal annual compound growth rates (%) 1960-2012

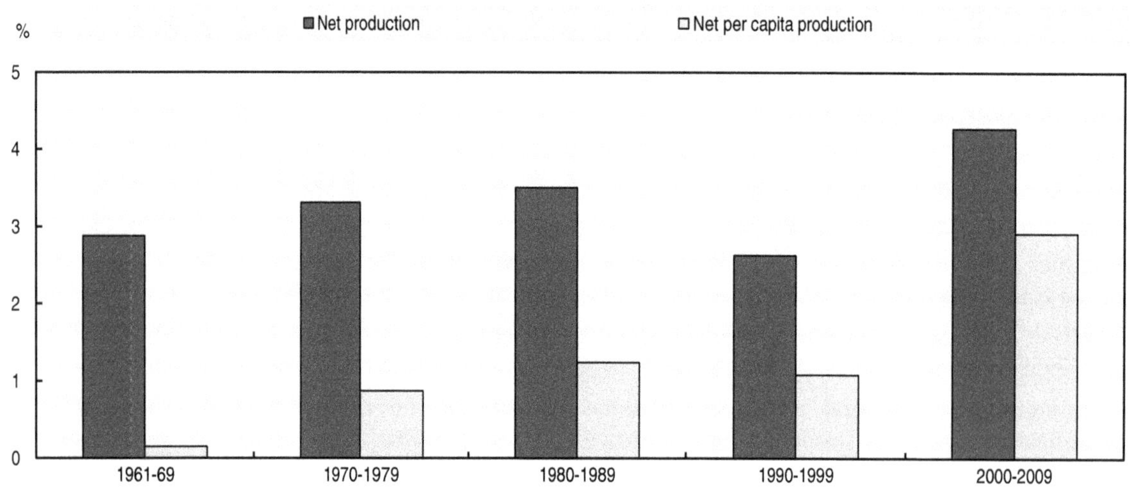

Notes: Net production refers to total production less cereal use for livestock feed. The FAO (2016a) calculates net production in the form of an index – see www.fao.org/faostat/en/#data/QV for further details.
Source: FAO (2016a), *FAOSTAT,* http://faostat.fao.org/.

Figure 1.3. ASEAN agricultural production shares

Commodity shares of gross value of production in international prices, 1963-2013

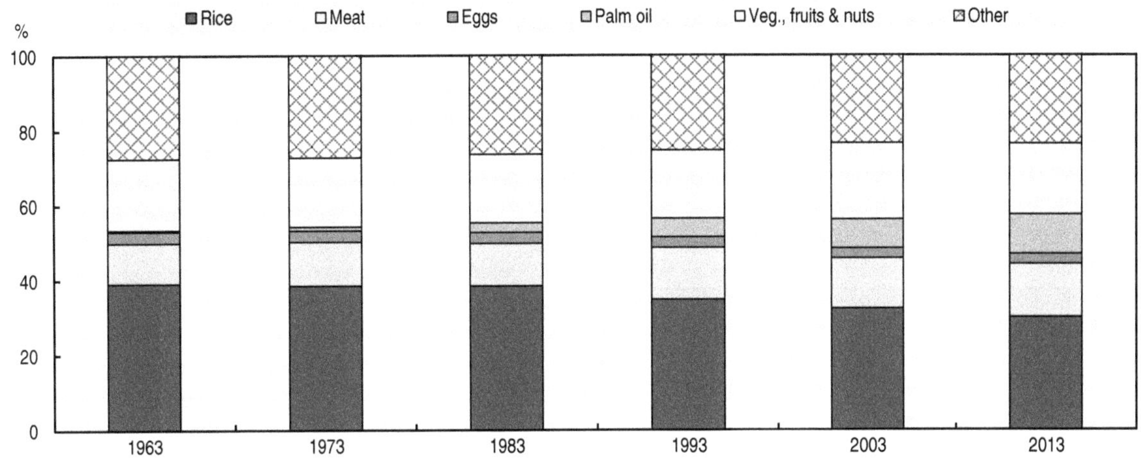

Notes: International prices are used to overcome issues in the aggregation of commodities that cannot be added up with their physical weights. The FAO (2016a) uses international prices in determining gross production value so that production trends can be seen without the influence of changes in exchange rates – see www.fao.org/faostat/en/#data/QV for further details.
Source: FAO (2016a), *FAOSTAT*, http://faostat.fao.org/.

Across individual ASEAN members, there are clear variations in the extent to which the structure of production has changed over time (Figure 1.4). In Brunei Darussalam, production has become centred on meat products, almost entirely poultry; Malaysia has shifted to palm oil production; and Singapore has shifted its production away from poultry to eggs. Myanmar has also seen changes to its production mix, increasing both meat and fruit and vegetable production. The share of Myanmar's production value coming from rice has also fallen – by around 20 percentage points between 1963 and 2013.

Of all ASEAN members, in 2013, the agricultural sectors of Brunei Darussalam, Cambodia, Malaysia and Singapore were most reliant on one production activity or sector: meat, rice, palm oil and eggs respectively. Others are more diversified – however, rice remains a dominant crop for most. Brunei Darussalam, Malaysia and Singapore have experienced the most change within their agricultural sectors. In contrast, the Philippines has become more concentrated, with the share of rice in its total agricultural production value increasing.

Little comparative information is available on the characteristics of agricultural producers across ASEAN countries. This is due to both a lack of information at the country level and the inherent difficulties in building a consistent and coherent database. Some information is provided by the Food and Agriculture Organization of the United Nations (FAO) on the basis of various agricultural censuses and other similar data collected at the country level (Figure 1.5). While data are partial and dated for ASEAN members (the most recent data available relates to estimates collected during the 2000s), they indicate that, in those countries for which data are available, average farm size is small. Estimates range between 0.8 ha per agricultural holding in Indonesia to 3.2 ha in Thailand (Lowder et al., 2014).

In terms of total holdings, Indonesia has the largest number, in line with its larger population, with close to an estimated 25 million agricultural holdings, based on 2003 data. For those countries for which time series data exist (Indonesia, the Philippines and Thailand), patterns indicate a generally falling average farm size. In some of these countries, changes in average land holdings can be traced towards policy moves that have redistributed land, for example, such as those in the Philippines (OECD, 2017). The trend of falling farm size may have broader long-term implications for agricultural productivity growth if it is also accompanied by further fragmentation of production activities. In contrast, two countries appear to have exhibited trends of *increasing* farm size: Myanmar and Viet Nam. In the case of Viet Nam, this has been in the area of livestock (OECD, 2015).

Data on the distribution of farm size is equally sparse. Again, while limited and dated, Lowder et al. (2014) report that farms of less than 1 hectare of land dominate (Figure 1.6). Indonesia and Viet Nam have the largest share of total producers who have less than 1 hectare of land. Thailand and Myanmar also stand out as countries with different patterns of ownership – both have a relatively higher number of producers who farm between 2-5 hectares, compared with other countries.

Figure 1.4. Agricultural production within ASEAN members

Share of production value in international prices

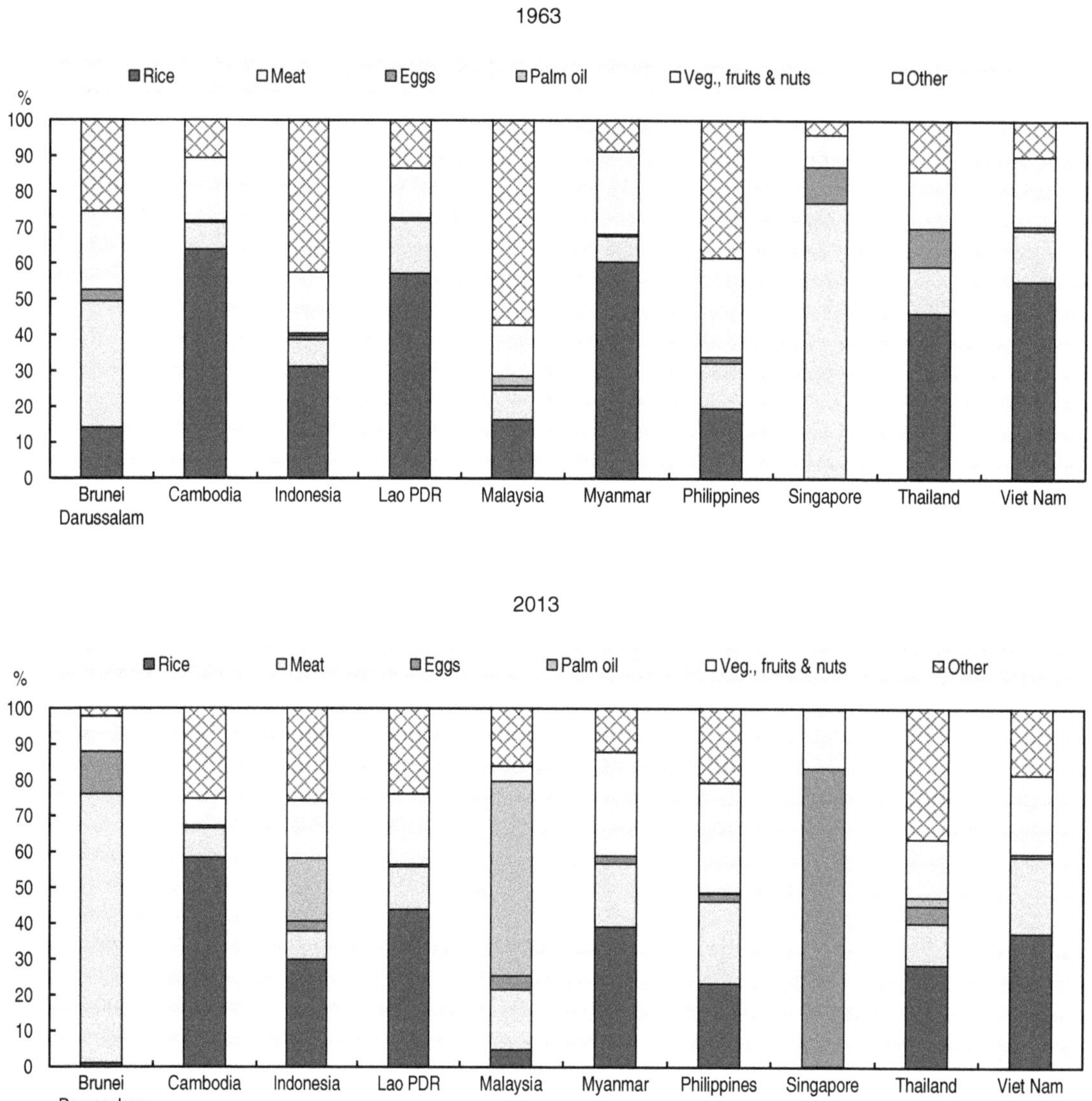

Notes: International prices are used to overcome issues in the aggregation of commodities that cannot be added up with their physical weights. The FAO uses international prices in determining gross production value so that production trends can be seen without the influence of changes in exchange rates – see www.fao.org/faostat/en/#data/QV for further details.
Source: FAO (2016a), *FAOSTAT*, http://faostat.fao.org/.

Figure 1.5. Number of agricultural holdings and average size

Estimates during the 2000s

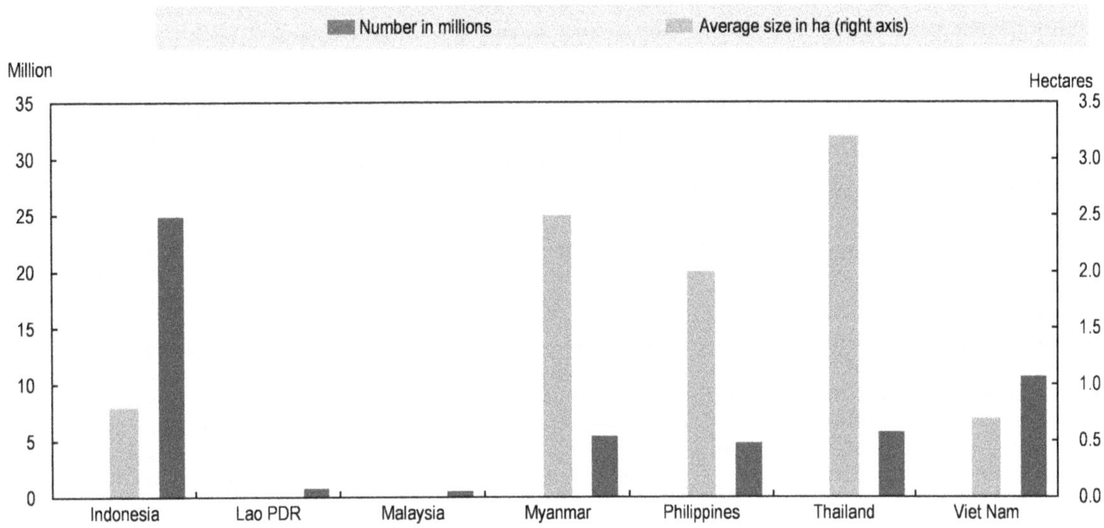

Notes: Estimates for each country relate to data collected during the 2000s. Specifically, Indonesia (2003), Malaysia (2005), Myanmar (2010), the Philippines (2002), Thailand (2003) and Viet Nam (2001).
Source: Lowder et al. (2014).

Figure 1.6. Distribution of farm size

Percentage of farm holdings by size, estimates during the 2000s

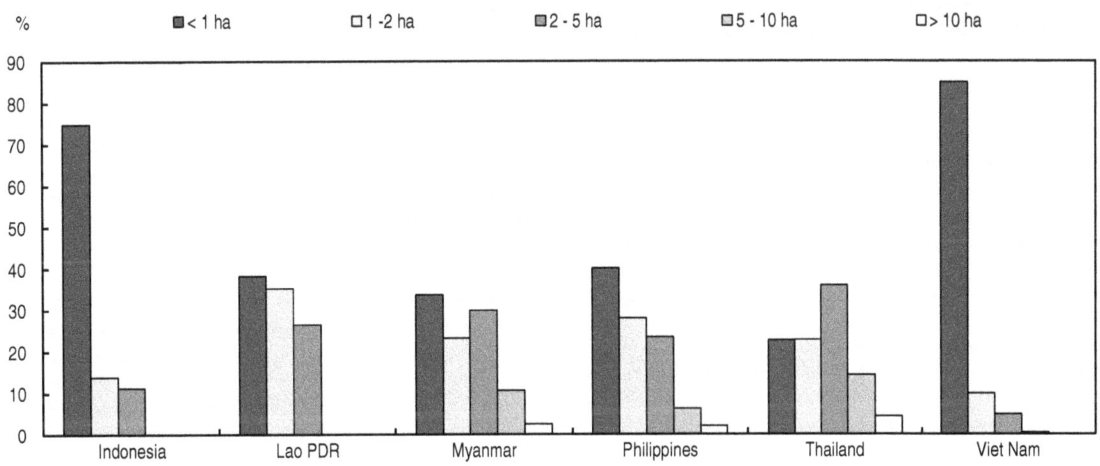

Notes: Estimates for each country relate to data collected during the 2000s. Specifically, Indonesia (2003), Lao PDR (1998-99), Myanmar (2003), the Philippines (2002), Thailand (2003) and Viet Nam (2001).
Source: Lowder et al. (2014).

Agricultural total factor productivity in ASEAN

Improvements in agricultural productivity have played a key role in driving the significant growth in agricultural output realised by Southeast Asia in recent decades. As estimated by the USDA, total factor productivity (TFP)[4] for the region as a whole has increased at an average annual rate of 2.2% a year since 1991 (1.4% a year on average for the period 1961 to 2012). Moreover, agricultural productivity growth has accounted for an increasing share of output growth over time (Figure 1.7). Between 2001 and 2012, productivity growth accounted for over 60% of output growth, compared with 13% in the 1980s, when increasing input use of 2.8% a year drove agricultural output growth.

Figure 1.7. Composition of agricultural output growth in Southeast Asia, by period (%)

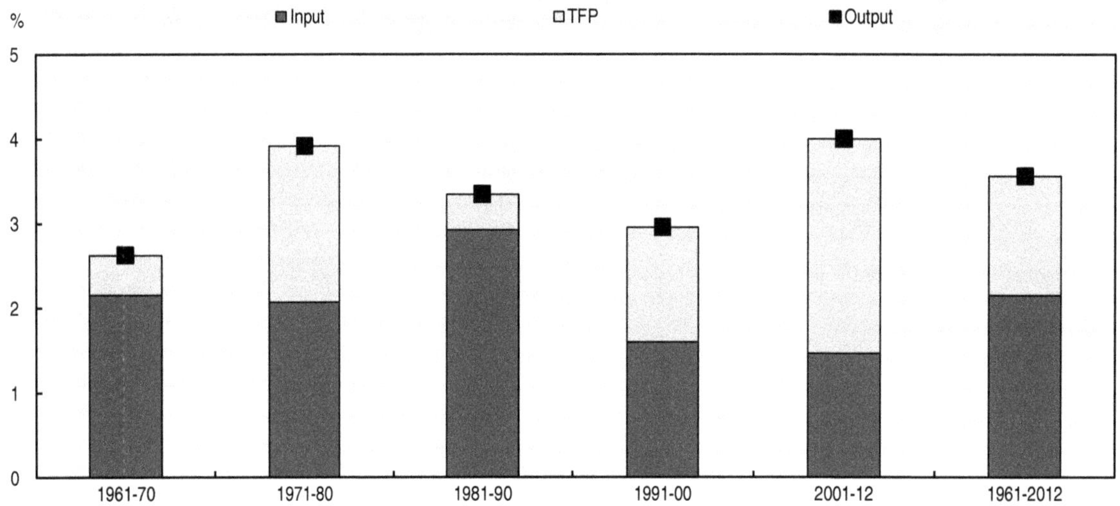

Notes: Southeast Asia, as defined in this instance, includes Brunei Darussalam and Timor Leste, but excludes Singapore. Output is the gross agricultural output for Southeast Asia. Inputs include labour, land, livestock, machinery and intermediate inputs. The USDA Economic Research Service's methodology for measuring international agricultural TFP growth is available at www.ers.usda.gov/data-products/international-agricultural-productivity/documentation-and-methods/.

Source: USDA (2016a), *International Agricultural Productivity*, www.ers.usda.gov/data-products/international-agricultural-productivity.aspx.

Productivity growth rates in Southeast Asia compare favourably with those achieved in other world regions (Figure 1.8). For the period 2001-12, agricultural productivity growth in Southeast Asia exceeded growth realised in all other regions except the rest of Asia.[5] Moreover, for the same period, productivity growth accounted for a similar share of agricultural output growth (63%) to the world average.

Comparing TFP trends across ASEAN member states, productivity grew strongest in Cambodia and Myanmar in recent decades, with most growth taking place since 2000. Productivity growth was weakest in the Philippines, Lao PDR and Indonesia (Figure 1.9). However, viewed over a longer period (1961 to 2012), Malaysia's TFP growth has exceeded that achieved in the other ASEAN member countries, reflecting the significant increase in the share of palm oil in the gross value of production (Figure 1.4 and Table 1.A1.1 in the Annex).

The importance of productivity growth in driving agricultural output growth varies across ASEAN member states. For the period 2001-12, productivity growth accounted for 90% of output growth in Malaysia, and 82% and 72% of output growth in Thailand and the Philippines, respectively. For the same period, productivity growth accounted for around 60% of output growth in Cambodia, Indonesia, Myanmar and Viet Nam. In contrast, increasing input use drove output growth in Lao PDR, accounting for over 70% of output growth for the period 2001-12 (Table 1.A1.1 in the Annex).

Figure 1.8. Trends in agricultural total factor productivity (TFP) growth, Southeast Asia and selected regions

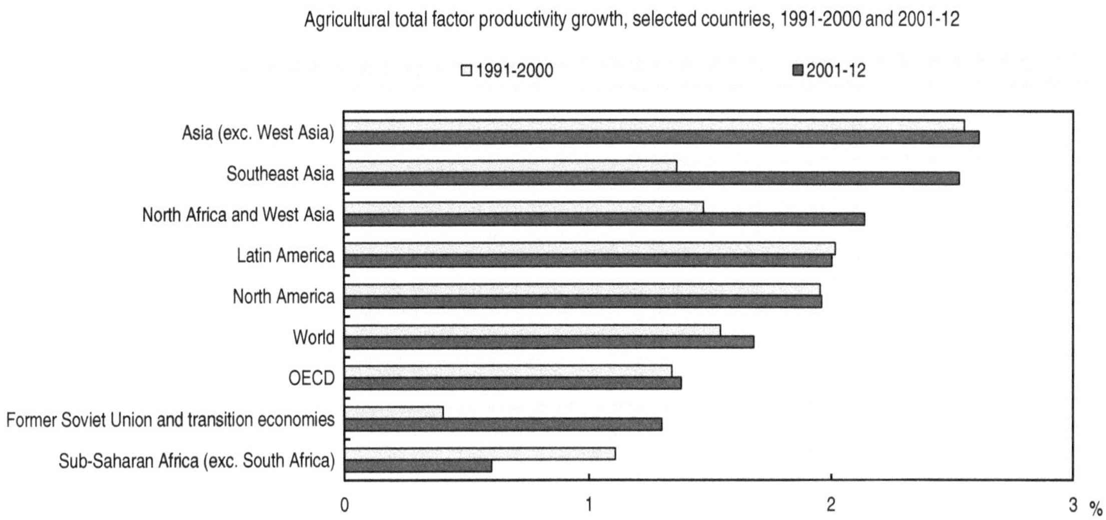

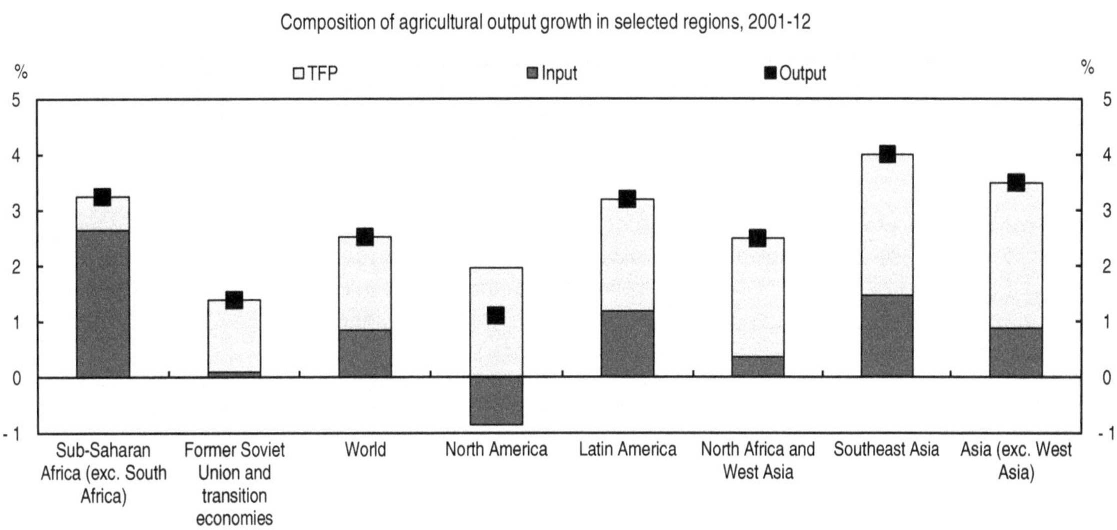

Notes: Southeast Asia, as defined in this instance, includes Brunei Darussalam and Timor Leste, but excludes Singapore. Output is the gross agricultural output for Southeast Asia. Inputs include labour, land, livestock, machinery and intermediate inputs. The USDA Economic Research Service's methodology for measuring international agricultural TFP growth is available at www.ers.usda.gov/data-products/international-agricultural-productivity/documentation-and-methods/.

Source: USDA (2016a), *International Agricultural Productivity*, www.ers.usda.gov/data-products/international-agricultural-productivity.aspx.

Figure 1.9. Trends in total factor productivity (TFP), by country, 1991-2012

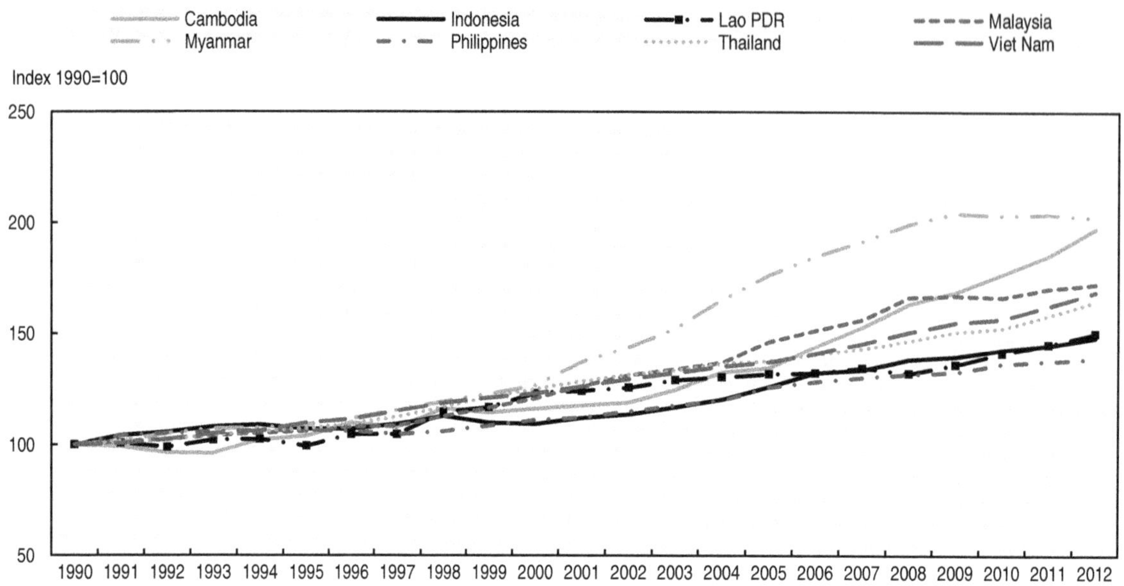

Source: USDA (2016a), *International Agricultural Productivity*, www.ers.usda.gov/data-products/international-agricultural-productivity.aspx.

Sustainability trends

Southeast Asia has abundant natural resources, including oil, gas and minerals, and forests and fisheries, which make a significant contribution to the wealth and economic growth of member countries (OECD, 2014). The region is endowed with abundant freshwater resources. However, land is scarce. Agricultural land per capita ranges from 0.12 ha per capita in Viet Nam to 0.38 ha per capita in Cambodia, significantly lower than the world average of 0.7 ha per capita (World Bank, 2016).

Growth in agricultural production in Southeast Asia has increased pressure on natural resources. The agricultural area has increased in all countries over the last two decades, as non-agricultural areas (forest, grasslands, and wetlands or peat lands) were converted to agricultural production – mainly cash crops (rubber and palm oil in particular, and also coffee in Viet Nam), but also subsistence farming (ADB, 2009). The region has experienced a net loss in forested area and, in recent years, the expansion of agricultural land has been a more important driver of deforestation than logging (FAO, 2011a).

The intensification of agricultural production has also contributed to natural resource degradation, land and water included. In general, Southeast Asia has low levels of natural soil nutrient availability, and soil quality constraints affect more than half of the cultivated land base in the region. However, natural soil constraints have been compounded by excessive use of agro-chemical inputs such as fertilisers, pesticides and other chemicals; unsustainable cultivation practices; and erosion, including as a result of deforestation, resulting in a decline in soil quality and structure (for example, see OECD, 2012, 2015). Agricultural intensification has also contributed to deteriorating water quality, as a result of nutrient runoff from the excessive application of nutrients (particularly nitrate and phosphate fertilisers) and the growing livestock sector. More broadly, agrochemical pollution is a serious problem in much of Southeast Asia (FAO, 2011b).

Greenhouse gas emissions (methane and nitrous oxide) from agriculture are increasing (Table 1.2). Measured in carbon dioxide (CO_2) equivalents, the increase in greenhouse gas emissions over the period 1990 to 2010 ranged from 24% in Indonesia to 62% in Myanmar. In 2010, the share of agriculture in total emissions ranged from 5.1% of emissions in Malaysia to 39% of emissions in Cambodia (FAO, 2016a).

Table 1.2. Emissions from agriculture in CO_2 equivalents (gigagrams per year)

	1990	2000	2010	Change 1990-2010 (%)
Cambodia	13 115.3	14 274.9	18 762.5	43.1
Indonesia	126 343.6	135 554.3	156 226.5	23.7
Lao PDR	5 136.0	5 775.3	7 369.4	43.5
Malaysia	11 250.9	12 164.7	14 967.9	33.0
Myanmar	39 841.8	50 095.8	64 635.5	62.2
Philippines	38 165.5	46 678.2	50 783.6	33.1
Thailand	55 795.4	55 679.6	70 795.6	26.9
Viet Nam	41 147.9	57 642.7	61 098.8	48.5

Source: FAO (2016a), *FAOSTAT,* http://faostat.fao.org/.

Rice cultivation accounts for the largest share of greenhouse gas emissions from agriculture for the region as a whole, followed by livestock production and the use of synthetic fertilisers. Emissions from the use of synthetic fertilisers more than doubled between 1990 and 2010, the fastest growing source of emissions from agriculture. There have also been significant increases in emissions from agricultural activities such as burning and decomposition of crop residues, and various practices related to manure management, albeit from a low base. Emissions from rice cultivation increased by 34% over the period 1990-2010 (FAO, 2016a). Land use change, including the conversion of non-agricultural land to cropland, is also a major source of emissions from the region (ADB, 2009).

Trade in agricultural products

ASEAN is playing an increasingly important role in world agro-food trade. ASEAN as a whole has increasingly become a net agro-food exporter, with around USD 133 billion in exports in 2012, compared with 83 billion worth of agro-food imports (Figure 1.10).

Intra-regional agro-food trade is also an important component of food supply. The share of agro-food imports sourced from within the ASEAN group has trended up over time, and accounted for close to 30% of the region's total imports in 2012 (depicted in Figure 1.10). Given significant agro-food export growth to non-ASEAN countries, particularly in the 2000s, regional trade markets are less important for agricultural exporters – around 20% in 2012. However, the relative importance of regional markets has trended up over time.

Of the products traded, vegetable and animal fats and oils (in this case, palm oil) are the most important agro-food export, accounting for the largest share of agro-food ASEAN export value – a share that has grown over time (Figure 1.11, top panel). Fisheries and aquaculture are also important to the region, with exports of fish and seafood products ("Fish" in Figure 1.11) the second-largest export earner. Exports are also quite concentrated, with the top ten products accounting for around 80% of total export value.

On the import side, there is more diversity. The top ten imported products account for around 60% of total imports. Flours, brans and other food industry preparations and residues, dairy products, fish and seafood and wheat are all major import products within the region (Figure 1.11, bottom panel).

Figure 1.10. Agro-food trade in ASEAN

Total trade (USD billion) and share of intra-regional trade (%), 1996-2012

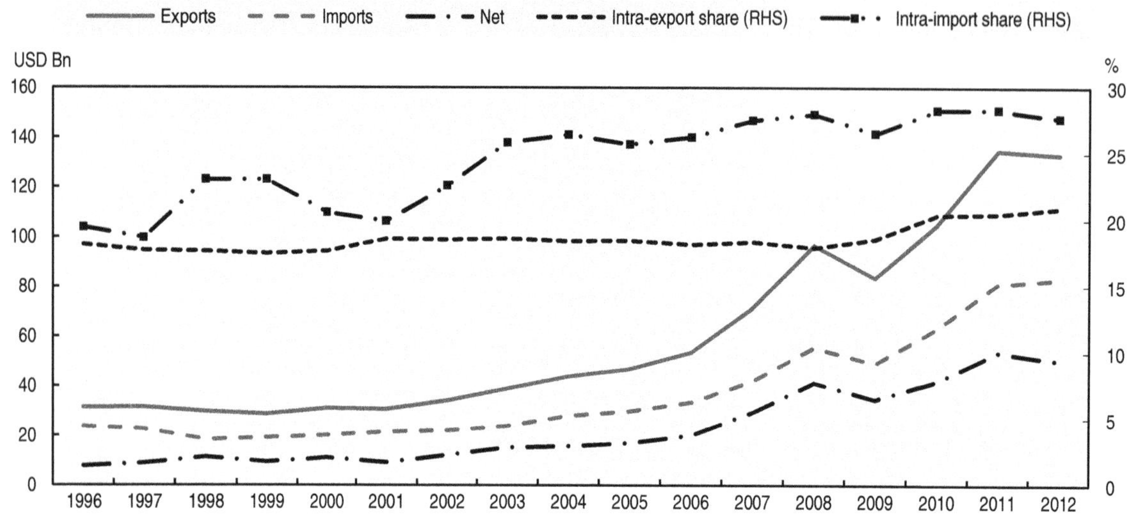

Source: WITS (2016), *World Integrated Trade Solution*, https://wits.worldbank.org/WITS/WITS/Restricted/Login.aspx.

The rice production and trade landscape

ASEAN is comprised of a mix of mainland countries and island or narrow peninsula states. With this comes a mix of geographic and climatic conditions that fundamentally influence agricultural production systems and inherent comparative advantages. With respect to food production, the geography of the region determines natural rice production capabilities.

Over the long run, the production and trade positions of ASEAN member countries show a consistent pattern of high production and net exports from mainland countries, and net imports by island and peninsula states (Dawe, 2013). This pattern is primarily driven by the fact that mainland states are dominated by large river systems that provide ample water and flat land that is well suited to rice production. In general, these conditions mean that Viet Nam, Cambodia, Lao PDR, Thailand and Myanmar have historically been net exporters of rice, producing more than domestic consumption levels. On the other hand, Indonesia, Malaysia, the Philippines, Singapore and Brunei Darussalam have traditionally been net importers.

Historical trends in net export/import status appear to persist today. The region as a whole is a significant net exporter of rice – however, this position is driven primarily by exports from Thailand and Viet Nam (Figure 1.12).[6] Recently, Cambodia also became a net exporter, while the remaining countries for which data exists are net importers.

The influence of geography on rice production is also represented in production per capita statistics (Figure 1.13). All mainland producers have high per capita production rates compared with other ASEAN members. In terms of yield, differences across countries are less marked than differences observed in production (Figure 1.14). A number of net importing – or traditional net importing – countries have higher yields than mainland producers. Indeed, Viet Nam is the only exception, due to a sustained period of strong yield growth since the early 1990s. Higher yields in importing countries are likely to be partly explained by pressures for "induced innovation" (Hayami and Ruttan, 1985). With relatively limited suitable land resources, competition from mainland rice producers who have a natural comparative advantage has provided a strong incentive for producers in these countries to seek productivity improvements. Results are also likely to be influenced by government policy, including the provision of input subsidies by some (discussed in Chapter 3).

Figure 1.11. Major ASEAN agro-food exports and imports

Share of total export and import value, selected years

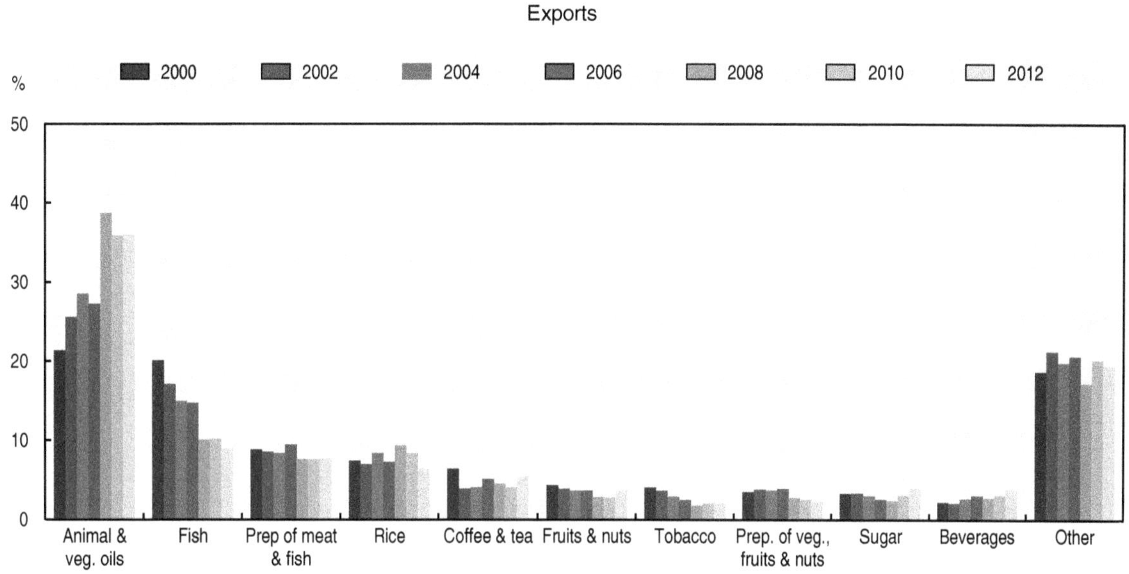

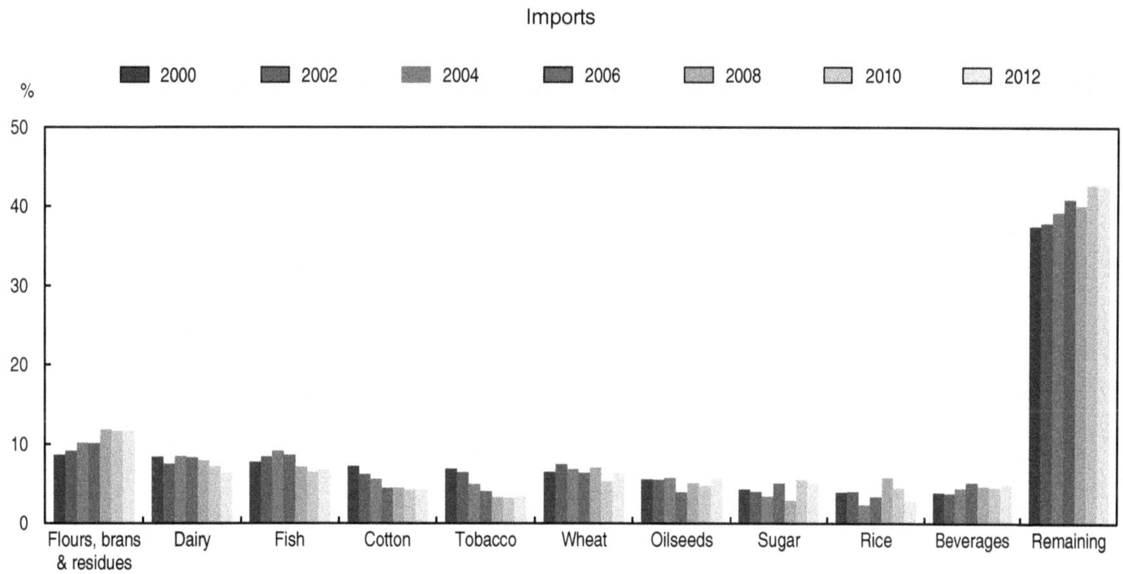

Source: WITS (2016), *World Integrated Trade Solution*, https://wits.worldbank.org/WITS/WITS/Restricted/Login.aspx.

Figure 1.12. Net exports of rice

USD million, 2000-13

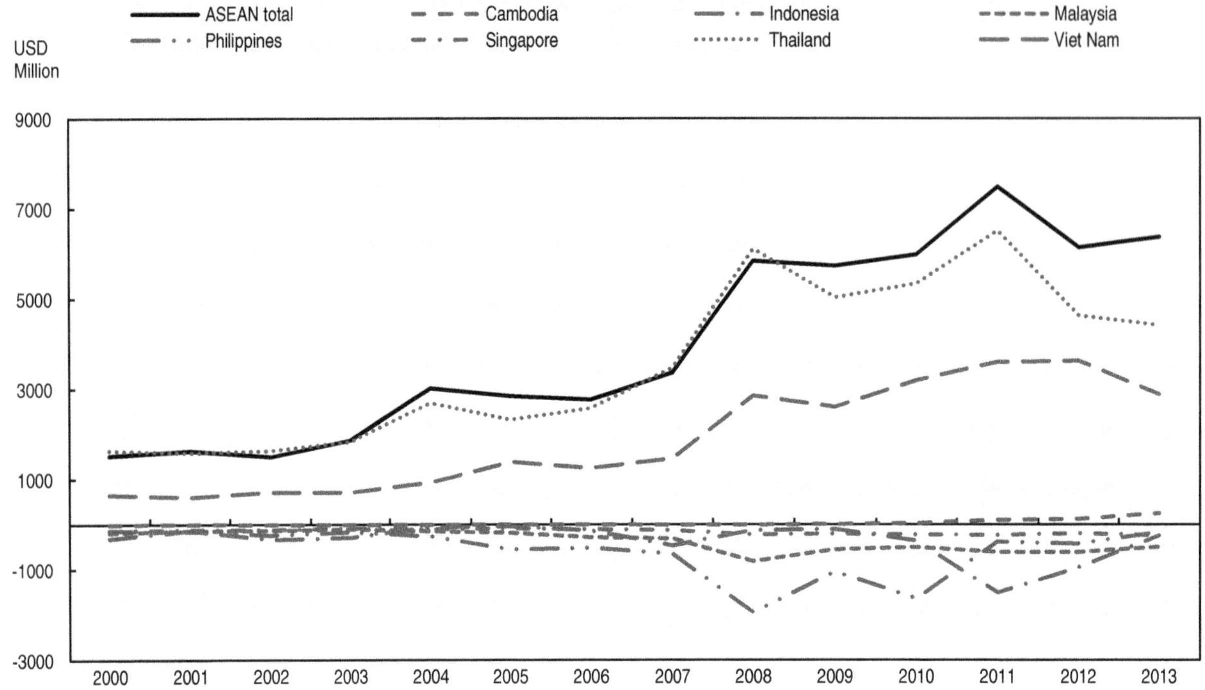

Source: WITS (2016), *World Integrated Trade Solution*, https://wits.worldbank.org/WITS/WITS/Restricted/Login.aspx.

Figure 1.13. Rice production per capita in ASEAN

Average decadal kg/person

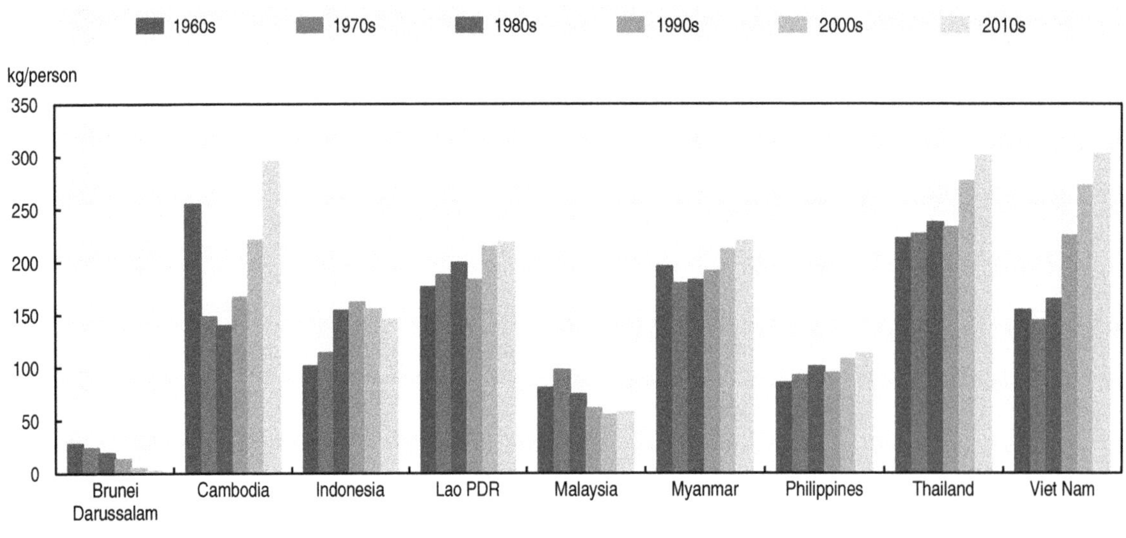

Sources: USDA (2016b), *Production, Supply and Distribution Online*, http://apps.fas.usda.gov/psdonline/psdquery.aspx, and World Bank (2016), *World Development Indicators*, http://databank.worldbank.org/data/.

Figure 1.14. Rice yields across ASEAN

Tonnes per hectare, 1961-2015

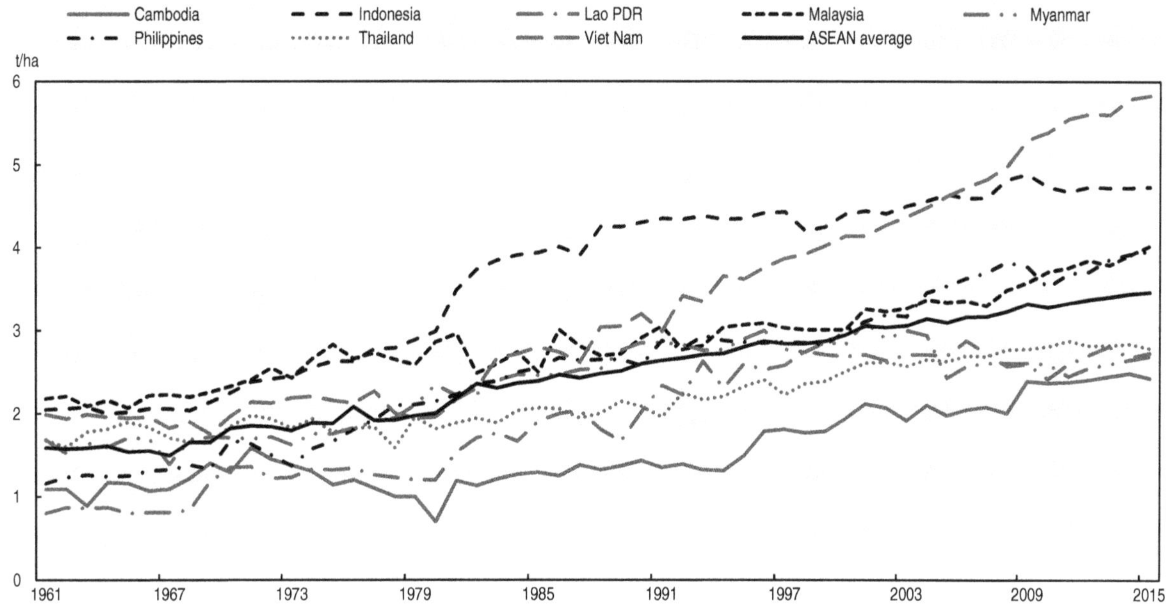

Source: USDA (2016b), *Production, Supply and Distribution Online*, http://apps.fas.usda.gov/psdonline/psdquery.aspx.

Trends in per capita rice for food consumption are also mixed across the region (Figure 1.15). With the exception of Indonesia, net exporting countries generally have higher consumption rates. For some countries, per capita consumption has fallen over time – a trend which is particularly evident for Malaysia and Thailand. For others, such as Indonesia and the Philippines, the opposite is true. However, rice for food consumption is only part of the story. In certain countries such as Myanmar, some published statistics report significant consumption of rice for animal feed. United States Department of Agriculture (USDA) data, for example, suggests total domestic consumption of rice is higher than other estimates, mainly due to the amounts accounted for by feed consumption. Furthermore, USDA estimates suggest that the recent increases in consumption in Viet Nam and Myanmar are also larger than that shown in Figure 1.15.

But while aggregate consumption of rice reveals a mixed picture, the importance of rice in meeting total *caloric* consumption has fallen in almost all ASEAN countries since the 1960s (Figure 1.16). The rates of change have been most significant for both Thailand and Malaysia, indicating that diets have diversified most in these countries compared with the 1960s. In contrast, the importance of rice in the average diet has increased in the Philippines – and indeed in Brunei Darussalam more recently, albeit to a lesser extent and from a much smaller base.

Irrespective of the production capacities of various countries in the region, a number have sought to become self-sufficient in rice as a means to improve food security (self-sufficiency as a policy approach is discussed in Chapter 3). However, despite strong yields in countries such as Indonesia, some research has suggested that achieving self-sufficiency in island and peninsula states will be difficult. Making use of agricultural production information on yields and available resources, such as land and water, Clarete (2013) suggests that – even in the presence of expected yield improvements, based on a continuation of historical trends – there is only a small probability that countries such as Indonesia and the Philippines could achieve self-sufficiency over the long run if historical shocks to yield are replicated in the future. Over the long term, Clarete (2013) suggests that there is only a small likelihood that Indonesia will become self-sufficient in rice production, and only a 5% likelihood for the Philippines. It was noted, however, that if sustained yield improvements were achieved, self-sufficiency over the long run may be possible, but this would come at a significant cost, as high domestic prices would be needed to encourage production – and to discourage consumption.

Figure 1.15. Rice consumption per capita across ASEAN

Average decadal kg/person

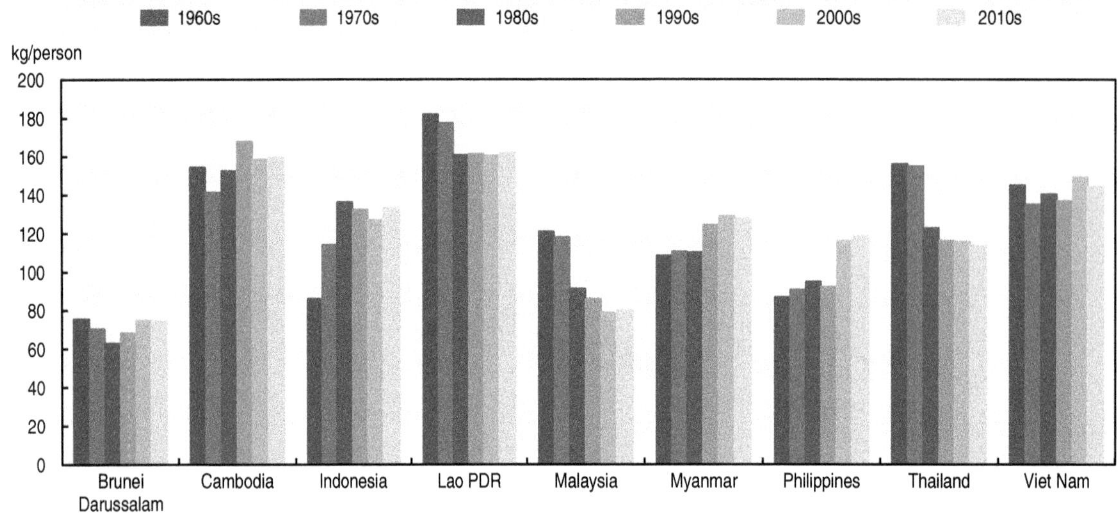

Notes: Consumption drawn from FAO food balance sheets, and represents rice used for food, expressed as kg per capita per year. Decade averages taken as the simple average over each decade, with 2010s being incomplete.
Source: FAO (2016a), *FAOSTAT*, http://faostat.fao.org/.

Figure 1.16. Share of rice in total caloric intake

Average percentage of total caloric intake per person per day

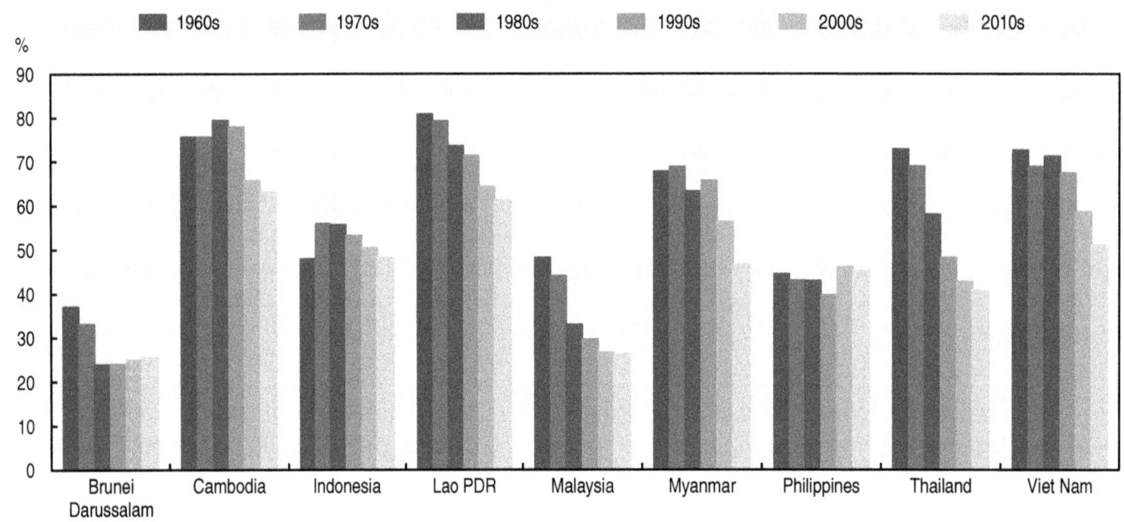

Source: FAO (2016a), *FAOSTAT*, http://faostat.fao.org/.

Figure 1.17. Domestic rice price volatility

Coefficient of variation in real prices in USD, April 2001-March 2015

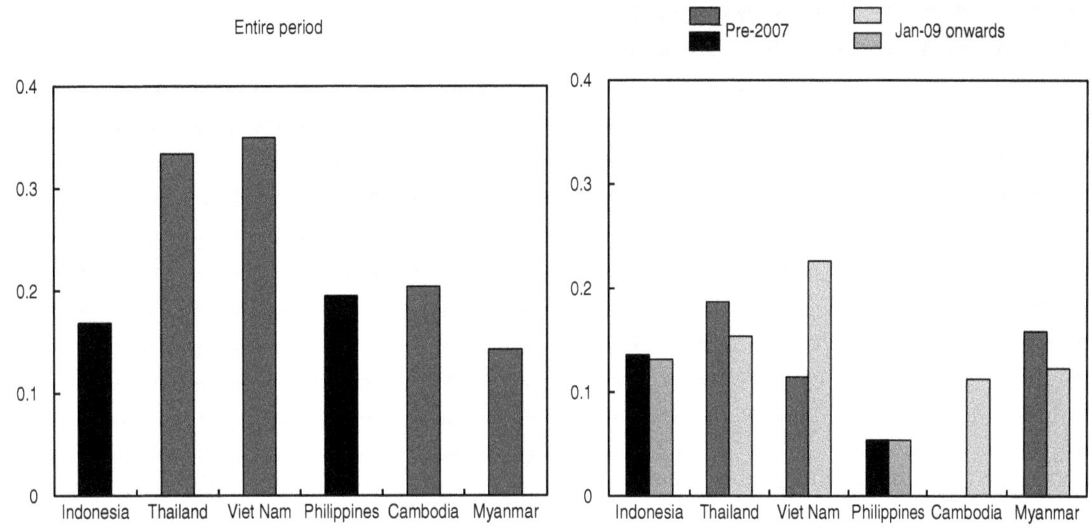

Notes: Volatility measured as the coefficient of variation of real prices expressed in United States Dollars for the entire period. Black bars indicate net importers, blue net exporters.

Sources: Various sources: Indonesia: Indonesian Ministry of Agriculture (2015) data on rice (15% broken); Thailand: Bank of Thailand (2016) data on wholesale price (5% broken), www2.bot.or.th/statistics/BOTWEBSTAT.aspx?reportID=89&language=ENG; Viet Nam: International Grains Council (2016) data on 5% broken export rice price, www.igc.int/en/markets/marketinfo-prices.aspx; Philippines: Philippines Bureau of Agricultural Statistics (2015), regular milled rice wholesale price (25% broken based on Intal, Cu and Illescas [2012]); Cambodia: FAO (2016b) *Food Security Portal*, www.foodsecurityportal.org/api/rice for rice (mix), which the FAO reports as low-quality rice (Maltsoglou, Dawe and Tasciotti [2010]); and Myanmar: Staple domestic market variety (5% broken), from World Bank (2014).

Figure 1.18. Domestic rice prices across ASEAN

Real prices in USD, April 2001-March 2015

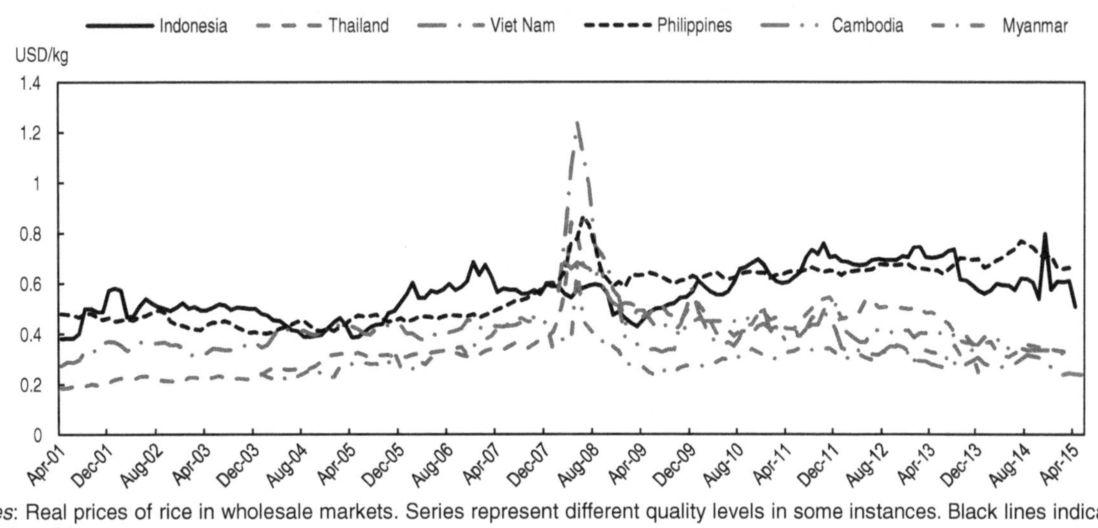

Notes: Real prices of rice in wholesale markets. Series represent different quality levels in some instances. Black lines indicate net importers, blue net exporters.

Sources: Various sources: Indonesia: Indonesian Ministry of Agriculture (2015) rice data (15% broken); Thailand: Bank of Thailand (2016) data on wholesale price (5% broken), www2.bot.or.th/statistics/BOTWEBSTAT.aspx?reportID=89&language=ENG; Viet Nam: International Grains Council (2016) data on 5% broken export rice price, www.igc.int/en/markets/marketinfo-prices.aspx; Philippines: Philippines Bureau of Agricultural Statistics (2015), regular milled rice wholesale price (25% broken based on Intal, Cu and Illescas [2012]); Cambodia: FAO (2016b), *Food Security Portal*, www.foodsecurityportal.org/api/rice for rice (mix), which the FAO reports as low-quality rice (Maltsoglou, Dawe and Tasciotti [2010]); and Myanmar: Staple domestic market variety (5% broken), from World Bank (2014).

Another influence on rice production relates to market prices. Many importing countries in the region have taken steps to reduce price volatility, concerned that it harms both consumers and producers. For producers, excessive volatility can make production and incomes uncertain, lessening their ability to make investments that would otherwise increase their productivity. This is particularly so when there are large numbers of small producers. For consumers, as many spend a large share of their budget on food, excessive price movements can push households into poverty and food insecurity. Such reasons have spurred attempts by governments to reduce rice price volatility. Price stabilisation policies have most often been implemented in historically importing countries. In importing countries for which data are available, price volatility appears to be lower than in exporting countries as a result of government policy, but prices are significantly *higher* (Figures 1.17 and 1.18). In contrast, for countries like Viet Nam, where producers have faced lower and more volatile prices, significant increases in production have still been achieved. For all countries, with the exception of Viet Nam, prices pre-2007 were more volatile than prices post. This trend is in line with changes in price volatility observed on world markets for a number of commodities in recent years (see OECD, 2016 for more details).

1.4 Food security in ASEAN

Food security is a multi-dimensional concept. According to the FAO definition agreed at the 1996 World Food Summit and expanded upon at the 2001 Summit, food security exists when all people, at all times, have physical, social and economic access to sufficient, safe and nutritious food to meet their dietary needs and food preferences for an active and healthy life.

The FAO definition encompasses four dimensions of food security, suggesting that people will only be food secure when sufficient food is available, they have access to it, and it is well utilised. The fourth requirement is that those three dimensions are stable over time. The multi-dimensional nature of food security means that no single indicator captures current and past levels of food security. Despite this, most ASEAN members have shown significant increases in food security since the early 1990s across a range of different measures related to the four dimensions.

One commonly used measure – the prevalence of undernourishment[7] – shows a general fall in rates for the region since 1992 (Figure 1.19). Despite this, trends across individual countries are not uniform: Brunei Darussalam, Malaysia and Singapore are all considered to be food secure based on the undernourishment indicators, as less than 5% of their populations are deemed to be undernourished. Some countries, such as Lao PDR, Myanmar, the Philippines, Thailand and Viet Nam, have experienced continued and persistent declines in the prevalence of undernourishment. For others, the declines have not been consistent, due to a range of factors, including macroeconomic conditions, such as the Asian Financial Crisis in the case of Indonesia, and domestic conflict in the case of Cambodia.

For the Southeast Asia region as a whole, reductions in undernourishment have been impressive (Figure 1.20). In the early 1990s, undernourishment rates for the region were the world's highest, but by 2014, these rates had fallen *below* those seen in a number of other regions.

In terms of absolute numbers, Indonesia has the largest population of undernourished people, with just under 20 million people in 2013-15 (Figure 1.21). Trends in absolute numbers largely follow those seen in the rates – with the exception of the Philippines, where numbers of undernourished people have been fairly stable since the early 1990s.

Figure 1.19. Undernourishment in ASEAN

Prevalence (% of population, 3-year moving average) 1990-2016

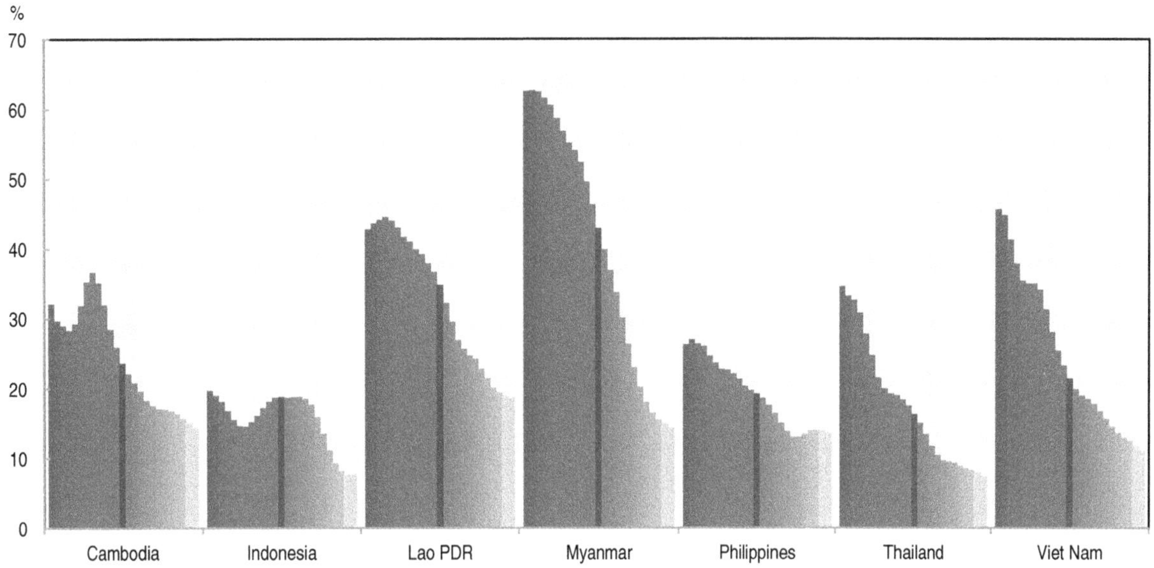

Source: FAO (2016a), *FAOSTAT*, http://faostat.fao.org/.

Figure 1.20. Undernourishment across the world

Prevalence (% of population, 3-year moving average) 1992-2014

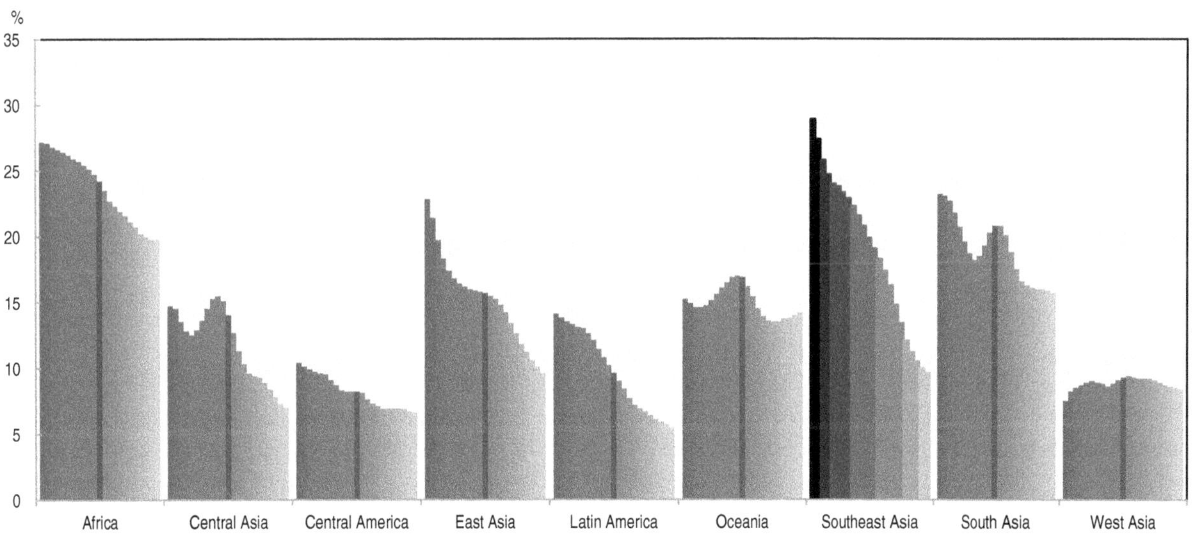

Source: FAO (2016a), *FAOSTAT*, http://faostat.fao.org/.

Figure 1.21. Number of undernourished people in ASEAN

Millions (3-year moving average) in selected years

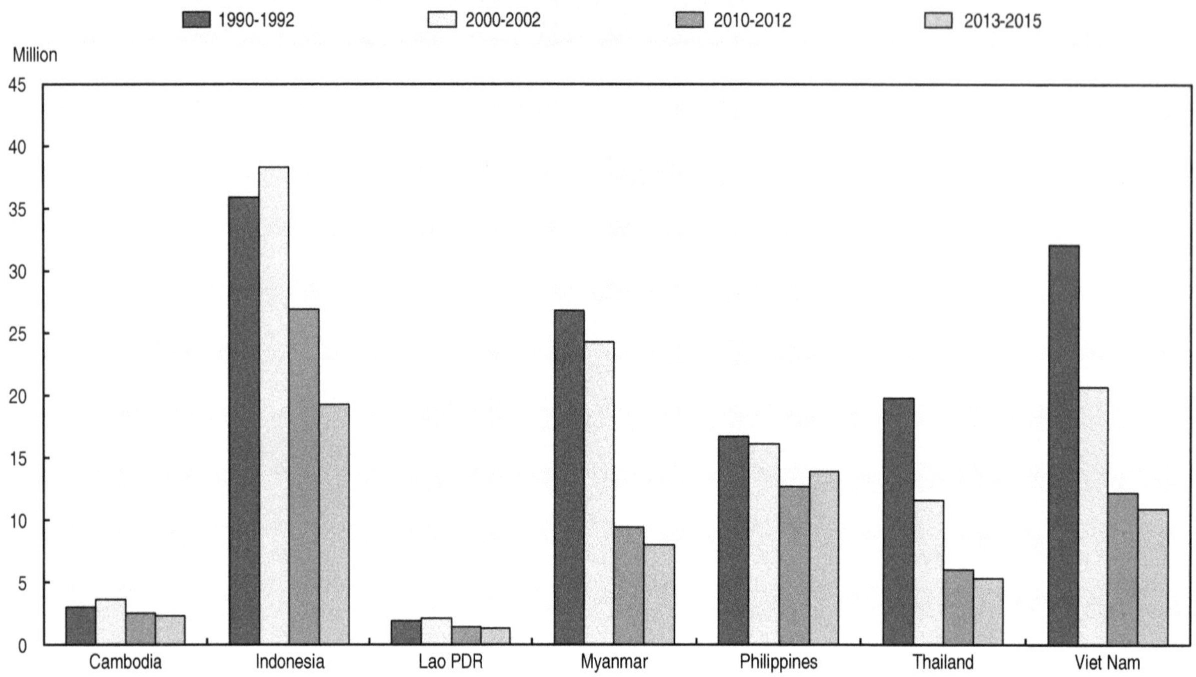

Source: FAO (2016a), *FAOSTAT*, http://faostat.fao.org/.

The undernourishment indicator shows the number of food insecure people at a given point in time. However, a greater number of people may be at *risk* of food insecurity due to the inadequacy of their current dietary intake.[8] The at risk population can be broadly thought of as those who have diets that are inadequate but not to the extent that they mean they would be classified as undernourished – essentially the difference between the two indicators. FAO statistics show that on average, a further 10% of the population across ASEAN countries are at risk of food insecurity. That is, food inadequacy rates shown in Figure 1.22 are around 10 percentage points higher than rates of undernourishment shown in Figure 1.19. Even in the case of more developed members, such as Malaysia and Brunei Darussalam – but not so for Singapore – food inadequacy rates are above the 5% threshold usually used to deem a country food secure. These statistics suggest that risks to food security remain for most members across the region.

Beyond the significant number of at risk people, other measures still point to significant challenges in achieving food security in the region. A measure used to better depict the stability of food security, that of stunting among children under five years of age, points to high rates across ASEAN. Stunting occurs when children do not receive enough food consistently – or do not fully utilise the food that they receive – to grow to reference heights for their age determined with respect to the demographic characteristics of their particular population. As such, while rates of undernourishment provide a snapshot of individuals at any given point of time, rates of stunting partially indicate how common it is for families to experience some level of food insecurity to such an extent that it influences health outcomes. The levels of stunting in ASEAN indicate that there remains persistent long-term food insecurity, a lack of health care and hygiene, or a combination of these in parts of the region. Stunting rates are highest in Lao PDR and Cambodia, and lowest in Thailand and Malaysia, although the figure for Malaysia is dated (Figure 1.23). Rates have also reportedly not changed much over the past two decades (FAO, 2015).

Figure 1.22. Food inadequacy in ASEAN

Prevalence (% of population, three-year moving average) 1990-2016

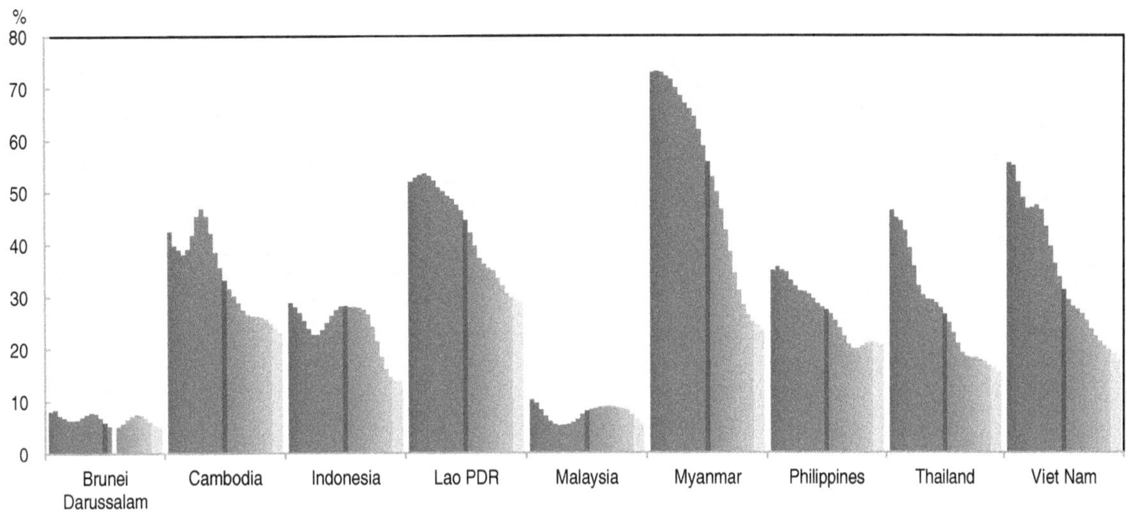

Source: FAO (2016a), *FAOSTAT*, http://faostat.fao.org/.

Figure 1.23. Rates of stunting among children under the age of five

Percentage of children under five years with stunted growth

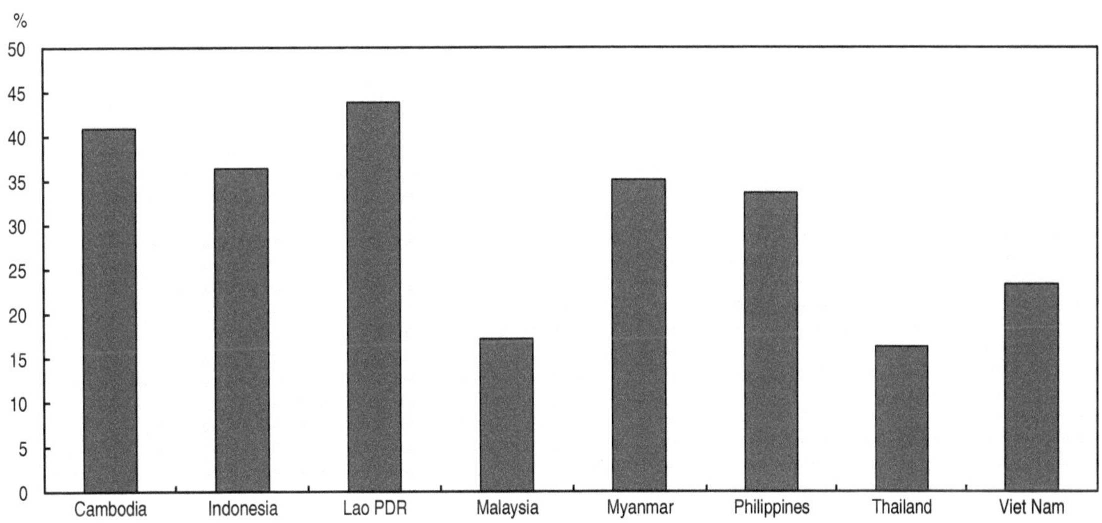

Notes: Data is for latest year available for each country, specifically: Cambodia (2010), Indonesia (2013), Lao PDR (2011), Malaysia (2006), Myanmar (2009), Philippines (2011), Thailand (2012) and Viet Nam (2010). Given a lack of data, a consistent time series for this indicator is not available and, as such, it is not possible to show trends over a longer time period as depicted for rates of undernourishment.
Source: FAO (2016a), *FAOSTAT*, http://faostat.fao.org/.

Figure 1.24. Household food consumption

Share of total energy intake (%), selected ASEAN members

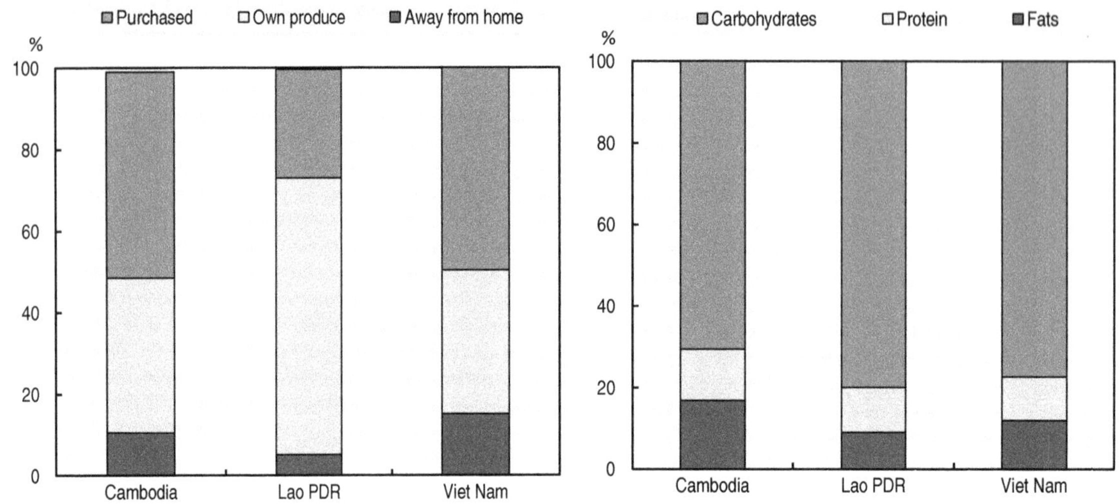

Note: Survey years are Cambodia (2009), Lao PDR (2008) and Viet Nam (2006).
Source: FAO (2016a), *FAOSTAT*, http://faostat.fao.org/.

Figure 1.25. Daily per capita energy availability

Total energy intake (kcal), 2007-09

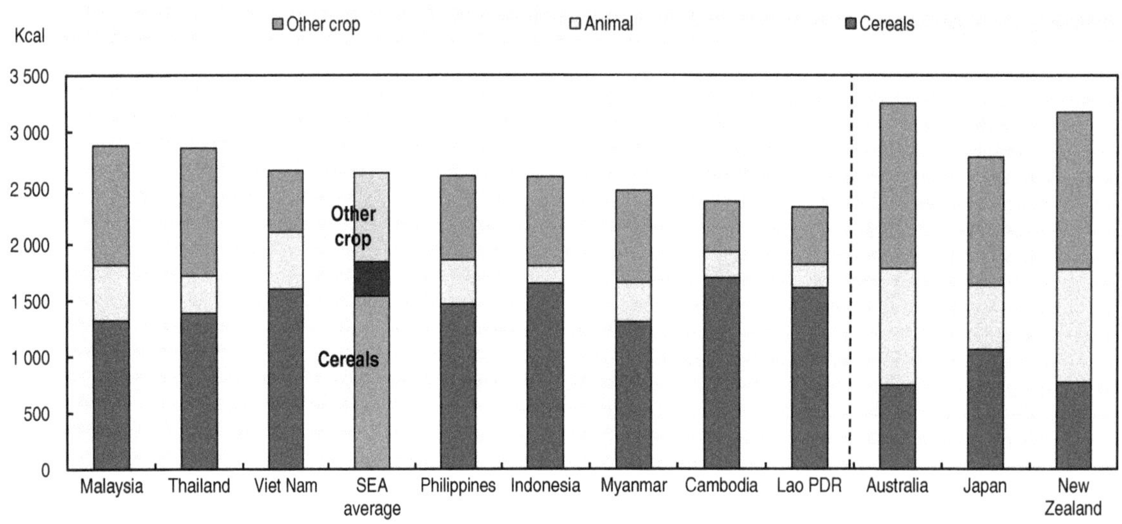

Note: SEA: Southeast Asia.
Source: FAO (2012).

Nevertheless, headline food security indicators, stunting included, mask a number of differences between and within populations that are important for policy makers to consider when designing efficient and effective responses to tackle food insecurity. The socio-economic characteristics of food-insecure households are particularly important, taking into account, for example, an understanding of their location (rural versus urban); whether they are food producers, and if so, the extent to which they are net sellers or buyers; and the demographic characteristics of these households. Some information is available from FAO-collected household surveys for Cambodia, Lao PDR and Viet Nam (Figure 1.24). Looking at food consumption, on average, households in these three countries rely most heavily on carbohydrates for their energy needs – between 70% and 80% of total energy consumption. For many households, much of the food consumed comes from own production, particularly for households in Lao PDR. Although dated, data based on the availability of different types of food in all ASEAN member countries confirm the importance of cereals in total caloric intake (Figure 1.25). When compared with developed countries in the Asia-Pacific region, the importance of cereals for ASEAN countries is noticeably high. In 2007-09, cereals accounted for close to 59% of food availability in Southeast Asia as a whole, with animal (fish included) products comprising around 11% and other crop products the remaining 30%. Cambodia and Lao PDR had the highest reliance on cereals for food availability (around 70%) while Viet Nam and Malaysia had the highest rates of food available from animal and fish products: 19% and 17% respectively.

Household-level data of the sort used to estimate indicators shown in Figure 1.24 can also be used to shed light on the greatest risks to food security faced by households, along with the impacts of policy interventions. A more in-depth depiction of the socio-economic characteristics of food insecurity is provided in Chapter 4.

1.5 Concluding comments

ASEAN is comprised of countries at different levels of development, incomes and size. In terms of agriculture, some in the region have experienced significant levels of structural adjustment, with employment in agriculture falling rapidly over the past 20 years. For others, much is yet to come. Rice remains the dominant crop in the region, representing the single largest production activity in value terms. The region as a whole is also a significant net exporter of this commodity. However, the primacy of rice in agriculture does not hold for all countries, with palm oil, poultry and eggs also of critical importance for some.

Rice also remains a critical component of diets in the region. And while its importance has generally fallen over time, its status as a key production and consumption commodity has meant that it has remained of significant political importance in terms of food security. Rice production is also inherently divided by geographic factors that have, over a long history, governed levels of comparative advantage and subsequent trading patterns. With self-sufficiency seen by many as the tool to improve food security, the mix of political importance and natural comparative advantages in rice production have fundamentally shaped food security policy in the region.

Notes

1. The ten ASEAN member states are Brunei Darussalam, Cambodia, Indonesia, Lao People's Democratic Republic (PDR), Malaysia, Myanmar, Philippines, Singapore, Thailand and Viet Nam.
2. Throughout this report, the term "fisheries" is used to refer to both wild harvest fishery activities (inland and marine), together with aquaculture production.
3. Unless indicated otherwise, Southeast Asia is defined as the ten ASEAN members and Timor-Leste.

4. Total factor productivity (TFP) is an indicator for measuring agricultural productivity. It takes into account all of the market inputs used in agricultural production (labour, land, livestock, machinery and intermediate inputs) and compares these with the total market outputs produced (crop and livestock commodities).
5. Excluding West Asia.
6. Net exports of rice are calculated as total exports to all partners less total imports from all partners.
7. The FAO's prevalence of undernourishment indicator is based on a comparison of usual food consumption expressed in terms of dietary energy (kcal) with certain energy requirement norms.
8. Like measures of undernourishment, inadequacy of dietary intake is based on the demographic composition of a country and with reference to a minimum intake requirement. Dietary intake is classified as inadequacy by using a higher caloric need threshold than undernourishment rates in order to measure the percentage of the population at risk of not meeting their food requirements.

References

ADB (2009), *The Economics of Climate Change in Southeast Asia: A Regional Review*, Asian Development Bank, Mandaluyong City, Philippines.

Bank of Thailand (2016), *Wholesale Price of Certain Commodities in Bangkok Metropolis* (database), Bank of Thailand, Bangkok, www2.bot.or.th/statistics/BOTWEBSTAT.aspx?reportID=89&language=ENG (accessed 15 June 2015).

Clarete, R. (2013), *Rice Self-Sufficiency and Rice Trade: Options for Food Security in ASEAN*, Contributed Paper to the 2nd ASEAN Rice Trade Forum, 4-5 June 2013, Yogyakarta, Indonesia.

Dawe, D. (2013), "Geographic determinants of rice self-sufficiency in Southeast Asia", *ESA Working Paper* No. 13-03, Food and Agriculture Organization of the United Nations, Rome.

FAO (2016a), *FAOSTAT* (database), Food and Agriculture Organization of the United Nations Statistics Division, Rome, http://faostat3.fao.org/home/E (accessed on 22 February 2016).

FAO (2016b), "Rice Prices", *Food Security Portal* (database), Food and Agriculture Organization of the United Nations, Rome, www.foodsecurityportal.org/api/rice (accessed on 27 April 2015).

FAO (2015), *Regional Overview of Food Insecurity Asia and the Pacific: Towards a Food Secure Asia and the Pacific*, Food and Agriculture Organization of the United Nations, Bangkok.

FAO (2012), *Selected Indicators of Food and Agricultural Development in the Asia-Pacific Region 2001-2011*, FAO Regional Office for Asia and the Pacific, Bangkok.

FAO (2011a), *Southeast Asian Forests and Forestry to 2020*, Food and Agriculture Organization, Bangkok.

FAO (2011b), *The State of the World's Land and Water Resources for Food and Agriculture (SOLAW): Managing Systems at Risk*, Food and Agriculture Organization, Rome, and Earthscan, London.

Hayami Y. and V. Ruttan (1985), *Agricultural Development: An International Perspective*, Johns Hopkins University Press, Baltimore.

Indonesian Ministry of Agriculture (2015), data received from Indonesian Ministry, Bangkok.

Intal, P.S., L.F. Cu and J.A. Illescas (2012), "Rice prices and the National Food Authority", *Philippine Institute for Development Studies Discussion Paper Series*, No. 2012-27, Makati City, Philippines.

International Grains Council (2016), "Export Prices", *International Grains Council* (database), London, www.igc.int/en/markets/marketinfo-prices.aspx (accessed on 7 May 2015).

Lowder, S.K., J. Skoet and S. Singh (2014), "What do we really know about the number and distribution of farms and family farms worldwide?", Background paper for *The State of Food and Agriculture 2014*, *ESA Working Paper* No. 14-02, Food and Agriculture Organization of the United Nations, Rome.

Maltsoglou, I., D. Dawe and L. Tasciotti (2010), "Household level impacts of increasing food prices in Cambodia", *Environment and Natural Resources Management Working Paper*, No. 37, Food and Agriculture Organization of the United Nations, Rome.

OECD (2017), *Agricultural Policies in the Philippines*, OECD Publishing, Paris, http://dx.doi.org/10.1787/9789264269088-en.

OECD (2016), *Evolving Agricultural Policies and Markets: Implications for Multilateral Trade Reform*, OECD Publishing, Paris, http://dx.doi.org/10.1787/9789264264991-en.

OECD (2015), *Agricultural Policies in Viet Nam 2015*, OECD Publishing, Paris, http://dx.doi.org/10.1787/9789264235151-en.

OECD (2014), *Towards Green Growth in Southeast Asia,* OECD Green Growth Studies, OECD Publishing, Paris, http://dx.doi.org/10.1787/9789264224100-en.

OECD (2013), *Innovation in Southeast Asia*, OECD Publishing, Paris, http://dx.doi.org/10.1787/9789264128712-en.

OECD (2012), *OECD Review of Agricultural Policies: Indonesia 2012*, OECD Publishing, Paris, http://dx.doi.org/10.1787/9789264179011-en.

Philippines Bureau of Agricultural Statistics (2015), "Rice", *Bureau of Agricultural Statistics* (database), Manila.

USDA (2016a), *International Agricultural Productivity* (database), United States Department of Agriculture Economic Research Service, www.ers.usda.gov/data-products/international-agricultural-productivity.aspx (accessed on 17 August 2016).

USDA (2016b), *Production, Supply and Distribution Online* (database), United States Department of Agriculture, Washington, DC, http://apps.fas.usda.gov/psdonline/psdquery.aspx (accessed on 1 September 2015).

WITS (2016), *World Integrated Trade Solution* (database), World Bank, Washington, DC, http://wits.worldbank.org/default.aspx (accessed on 15 November 2016).

World Bank (2016), *World Development Indicators* (database), World Bank, Washington, DC, http://databank.worldbank.org/data/ (accessed on 25 November 2016).

World Bank (2014), *Myanmar: Rice Price Volatility and Poverty Reduction*, Economic and Sector Work, Report No. 89687-MM, World Bank, Washington, DC.

Annex 1.A1.

Background table on the agricultural situation in ASEAN

Table 1.A1.1. Productivity, output and input growth in Southeast Asia, by period (%)

	Cambodia	Indonesia	Lao PDR	Malaysia	Myanmar	Philippines	Thailand	Viet Nam	Southeast Asia
1961-70									
TFP	-1.0	1.7	0.7	3.0	-1.9	-0.2	0.3	-0.6	0.5
Output	2.7	2.7	5.7	5.4	1.4	2.6	3.4	0.5	2.6
Input	3.6	1.0	5.0	2.4	3.3	2.9	3.1	1.1	2.2
1971-80									
TFP	-4.4	1.3	-0.9	2.1	1.9	3.2	2.3	1.3	1.8
Output	-7.0	3.3	1.2	4.4	4.2	5.1	5.0	2.9	3.9
Input	-2.6	2.1	2.1	2.3	2.3	1.8	2.7	1.6	2.1
1981-90									
TFP	3.3	0.3	1.0	3.0	-0.4	0.1	0.0	1.1	0.4
Output	6.1	4.6	3.0	4.6	0.4	1.8	2.7	4.0	3.3
Input	2.8	4.2	2.0	1.6	0.8	1.7	2.7	2.9	2.9
1991-00									
TFP	2.3	0.6	2.2	1.8	2.3	0.5	2.3	2.3	1.4
Output	4.7	2.1	5.3	2.5	4.9	2.1	2.4	5.7	3.0
Input	2.4	1.5	3.1	0.7	2.6	1.6	0.1	3.4	1.6
2001-12									
TFP	4.9	2.7	1.5	3.1	3.8	2.0	2.1	2.6	2.5
Output	8.0	4.4	5.1	3.4	5.9	2.7	2.6	4.3	4.0
Input	3.1	1.7	3.6	0.3	2.1	0.8	0.5	1.7	1.5
1961-2012									
TFP	0.5	1.3	0.9	2.6	1.6	1.2	1.4	1.7	1.4
Output	2.7	3.7	3.8	4.1	3.7	2.9	3.2	4.2	3.6
Input	2.2	2.4	2.9	1.5	2.0	1.7	1.8	2.5	2.2

Notes: Southeast Asia, as defined here, includes Brunei Darussalam and Timor Leste, but excludes Singapore. TFP: Total Factor Productivity.
Source: USDA (2016a), *International Agricultural Productivity*, www.ers.usda.gov/data-products/international-agricultural-productivity.aspx.

Chapter 2

Agricultural and food security outlook for Southeast Asia

This chapter presents medium-term projections for markets and production of agricultural and fisheries commodities important for the ten Association of Southeast Asian Nations (ASEAN) members, including the implications of these projections for future food security. It also considers longer-term issues for agriculture and food security in the region, related to the possible impacts of climate change. To this end, it first explores projections of prices, production and trade of key regional agricultural and fisheries commodities. Second, it analyses the impacts of these medium-term projections on regional food security within ASEAN. Third, it considers some of the longer term implications of climate change on agriculture and food security.

Key points

- The medium-term outlook for development and growth in the agricultural and fisheries sectors in Southeast Asia is positive. However, the projected improvements will not be sufficient to overcome food insecurity.
- Productivity growth is projected to be the key driver for agricultural sector growth over the medium term, but achieving this requires continued reform efforts to address remaining constraints to productivity growth.
- Over the medium term, if through additional reform efforts ASEAN member states are able to achieve greater than projected growth in incomes and agricultural productivity, undernourishment will fall faster than projected, but implementing policies that improve *access* to food, such as payments to poor households, have the potential to lead to very significant falls in food insecurity.
- Over the longer term, however, climate change is projected to lead to a number of negative effects that will diminish agricultural productivity growth and potentially compromise improvements in food security.
- The projections indicate significant potential for improvements in agricultural infrastructure, such as irrigation; and innovation, such as the development of a new rice variety; to help to offset – or at least moderate – the negative effects of climate change. However, such developments require even greater levels of investment by governments beyond those seen in the past.

2.1 Introduction

Over the past 20 years, the economies and agricultural sectors of countries that comprise the Association of Southeast Asian Nations (ASEAN) have undergone significant change.[1] With growing levels of development, improved agricultural productivity and output, incomes and food security in the region have rapidly improved. The region has also become more connected with international agricultural markets, which are playing an increasing role in the incomes of producers.

In the medium term (2015-24), changes within ASEAN economies, together with changes in other regions, will continue to shape international markets. The medium-term trends will, under a business as usual setting, therefore have implications for the ongoing improvements in food security in the region through changes in the production and prices of agricultural products, as well as in incomes of populations.

However, over the *longer* term (beyond 2025), in addition to rising incomes and increasing and changing demand, other pressures will arise that are likely to further shape markets, economic activity and food security. In particular, the impacts of changes in the climatic conditions in the region, due to global climate change, and the effects of rising incomes and demand will begin to play a greater role in shaping markets.

This chapter explores the medium and long-term outlook for ASEAN agricultural markets. The chapter draws on the 2015 *OECD-FAO Outlook* (OECD-FAO, 2015) publication for the medium-term projections (2015-24).[2] The commodities on which the medium-term analysis is focused are taken from the set of modelled commodities in the Outlook, and, as a consequence, some additional products that are likely to experience changes due to rising incomes, such as processed foods, or which are important for food security, such as fruits, vegetables and nuts, are not included. The possible long-term implications of climate change are assessed with the use of the International Food Policy Research Institute (IFPRI) IMPACT (Robinson et al., 2015) and OECD ENV-Linkages models, which enable the effects of climate change on agricultural production, and therefore economic activity, to be explored towards 2050 and beyond. Climate change adaptation strategies are also considered – however, mitigation is not analysed in detail in this study.

Five ASEAN members (Indonesia, Malaysia, the Philippines, Thailand and Viet Nam) are explicitly modelled in the Outlook. The remaining countries (Brunei Darussalam, Cambodia, Lao PDR, Myanmar, and Singapore) form part of a composite region which includes non-Southeast Asian countries – although the ASEAN countries dominate this grouping with respect to population, gross domestic product (GDP) and agricultural production. As such, individual country results are presented for the five explicitly-modelled countries, but Southeast Asian aggregates contain the composite region.

2.2 Medium-term outlook for agricultural markets

This section explores some key features of the medium-term outlook for agricultural markets. First, projections for the global market are explored, after which projections for macroeconomic changes, production, trade and prices of key commodities for ASEAN are presented.

Global agricultural outlook

Over the medium-term, global agricultural markets are projected to broadly continue along recently observed trends. In aggregate, production is expected to continue to increase across a wide range of agricultural commodities. Moreover, continuing productivity improvements across a broad range of commodities are expected to result in falling costs as production increases, leading to a fall in real prices between 2016 and 2024 (Figure 2.1). This projection means that the projected productivity improvements, based on past trends, will be greater than both increasing levels of demand from higher incomes and growing populations. Improvements in productivity will also outpace an expected rise in the real production costs of a range of inputs. However, on trend, the medium-term projections show that while real prices are expected to fall, they will remain above pre-2007 levels.

Moderating prices are projected for all of the main categories of products over the medium term. The largest levels of moderation are seen for meat products. For cereals, much of the moderation in prices compared with 2010-12 levels has already occurred, and prices are expected to be maintained over the projection period. Dairy products are projected to maintain relatively higher price levels than in the past, despite recent falls from 2010-12 levels.

Figure 2.1. Projected real international price movements over the medium term

Historical and medium-term (2015-24) projections, index values

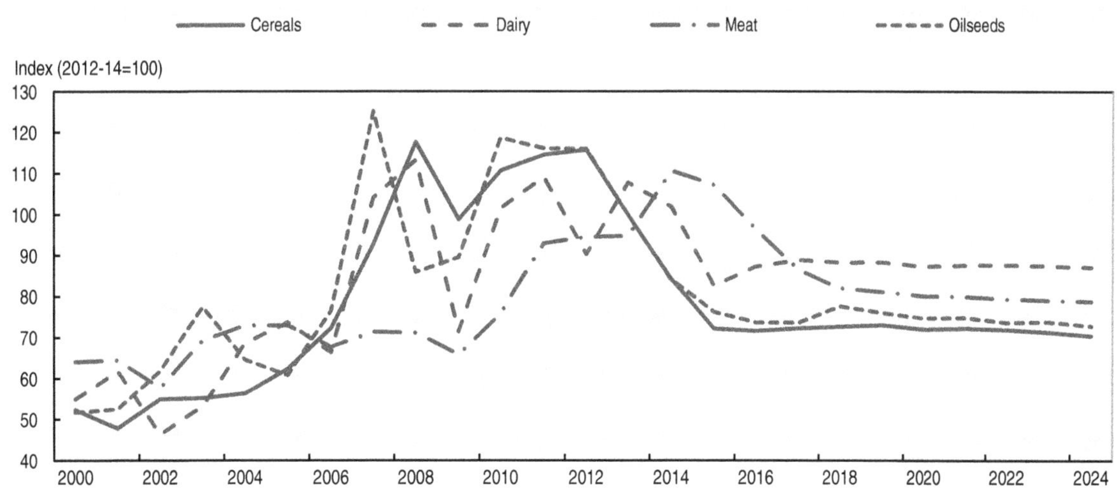

Source: OECD-FAO (2015).

Across broad production types, there is some variability in the projected overall growth in aggregate production (Figure 2.2). Among different crops, coarse grains (including maize) are projected to experience the greatest increases in absolute production – albeit from a much higher base – due to expanding output in both developed and developing countries. Cotton production is expected to increase the least. For the other crops featured in the projections, growth is expected to be lower than for coarse grains but still positive. Moreover, production growth for most crops is dominated by increases within developing and least developed countries. These changes are shifting the balance of international agricultural production in favour of developing and emerging countries.

In terms of relative changes, the largest increases are expected for both oilseeds and protein meals. This is driven by changing patterns of demand and, in particular, rising demand for these products for use in animal feed, in turn driven by increases in animal production.

Figure 2.2. Projected global agricultural production increases over the medium term

Crop and animal products by country groupings, 2015-24 projections

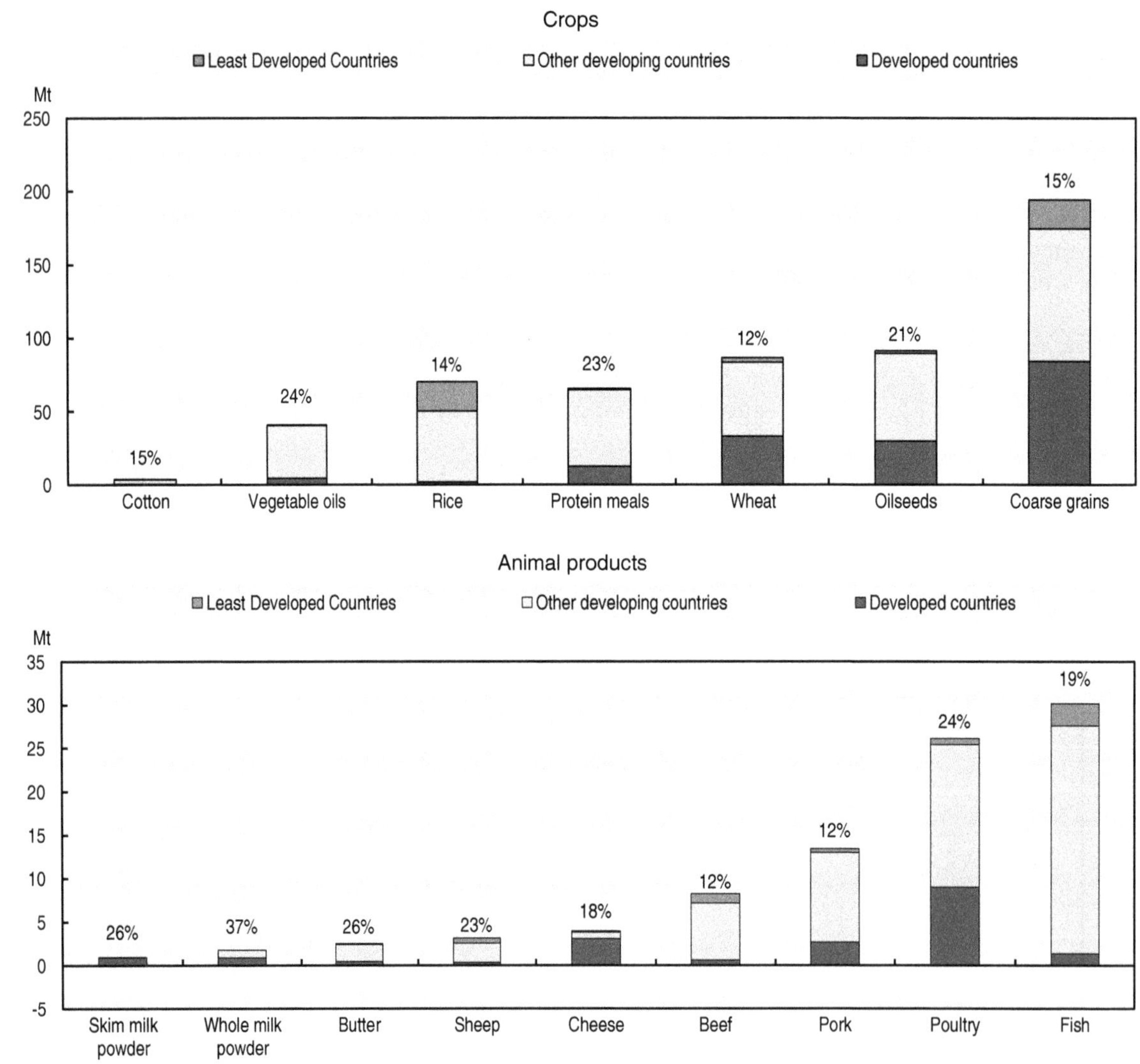

Source: OECD-FAO (2015).

Across animal products, the largest absolute changes in production are observed for fish, poultry and pork. Even more so than for crop production, production increases for most animal products are concentrated in developing regions. Notably, for animal products, there is a significant relative increase in the projected production of dairy-related products. For all of these (skimmed milk powder, whole milk powder, butter and cheese), projected production increases are expected to be around or in excess of 20%, a greater relative amount than for most other products examined.

Coupled with the abovementioned increases in production over the 2015-24 period are projections for increasing relative trade volumes for many products. Increases are projected over a range of commodities, with dairy products, oilseeds, meat and a number of grains all experiencing a rise in the share of production that is traded. For protein meals and vegetable oils, however, production shares traded are expected to *fall*, mostly due to rising domestic demand in the regions in which they are produced.

Macroeconomic projections for ASEAN

While global economic growth is expected to moderately improve in the short term, it is nevertheless projected to remain below past rates for many countries, with a divergence across the major economies.[3] Unemployment is meanwhile set to remain well above pre-crisis levels in many economies.

For countries in ASEAN, however, the outlook over the medium term is more optimistic (Figure 2.3), with GDP growth projected to be equal to or higher than over the past ten years (2005-14). Furthermore, growth rates are expected to exceed many other regions worldwide and, for most ASEAN members, to exceed those of China.

The expected high levels of growth in the Southeast Asian region will have effects on agro-food markets. First, higher growth should reduce poverty levels, which will in turn contribute to the reduction of undernourishment (Section 2.3). Second, higher incomes will also change the *nature* of demand. As incomes grow, there will be a substitution away from some staple crops, such as rice, to other products, particularly animal products. Third, the changes in economic growth are occurring alongside population growth. Higher population levels will lift demand across the board for agro-food products, augmenting the abovementioned income effects for some products.

Figure 2.3. Projected gross domestic product (GDP) growth rates over the medium term

Growth rates, historical (2005-14) and future (2015-24)

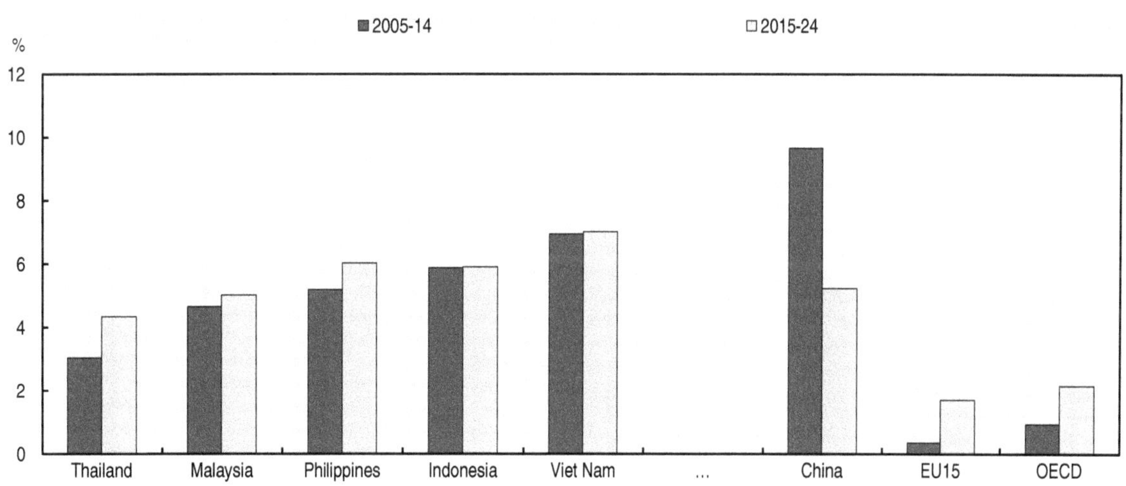

Source: OECD-FAO (2015).

Production and trade in ASEAN

Production

Over the medium term, changes in production can arise from a number of sources. They can relate to changes in quantities of variable inputs used, such as labour, fertilisers, or water; or changes in the use of capital and other fixed inputs, such as land and machinery; but they can also occur due to changes in the productivity of producers. While projected trends include changes in all three for ASEAN countries, some factors are more important than others in driving possible future changes.

Moreover, changes in the international market conditions for a number of products will also feed back into changes in returns and therefore production decisions. ASEAN economies are not immune to such changes, and indeed, for some products, ASEAN countries are themselves among the drivers of the movements in international markets.

For the region as a whole, agricultural production growth is expected to be strong across a wide range of commodities (Figure 2.4). The highest growth areas are sugar, poultry, and vegetable oil production. Importantly, production growth in the region is expected to exceed world averages for almost all products, with the exception of roots and tubers, increasing the importance of Southeast Asia for world supply of agro-food products. It also has the potential to increase the importance of regional and international markets for producers, as the relatively stronger growth in production is likely to be accompanied by greater trade in surpluses.

Across individual countries, however, changes in the production of key commodities are more mixed (Figure 2.5). While the general patterns of increase remain, some changes in the production mix are observed, driven by land constraints and the changing relative returns from other production activities over the medium term.

For the five ASEAN countries examined, production of the key staples of rice, coarse grains (maize), sugar and soybeans is projected to generally increase. The exception is Malaysia where, due to higher returns in the oilseeds (palm oil) and other sectors, a substitution in activities is projected, leading to a fall in coarse grain production. Similarly, the role played by fisheries and aquaculture is expected to increase over the medium term (Box 2.1).

Figure 2.4. Evolution of production in ASEAN

Percentage growth over medium term, and share of world production

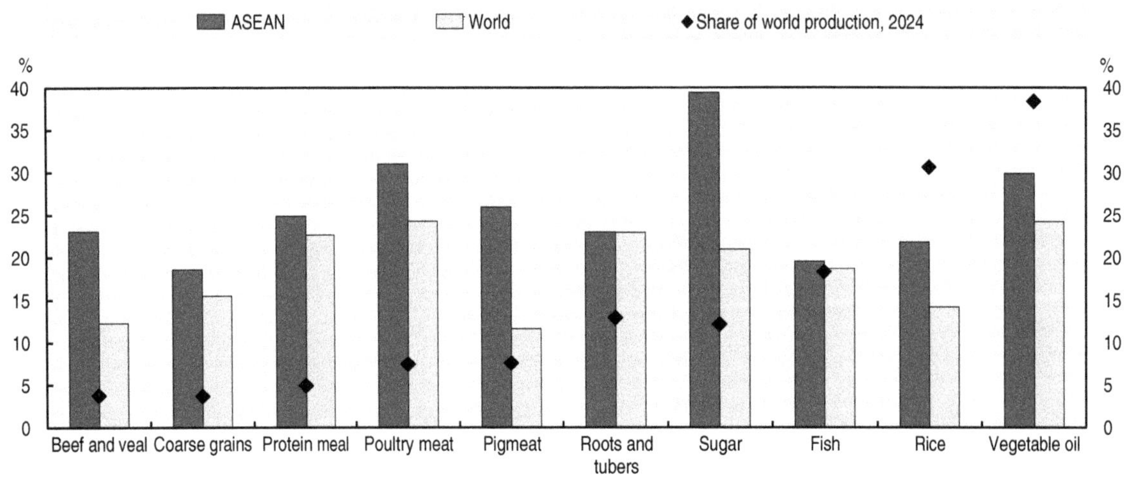

Source: OECD-FAO (2015).

Figure 2.5. Production outlook for selected ASEAN member states

Percentage change 2014-24

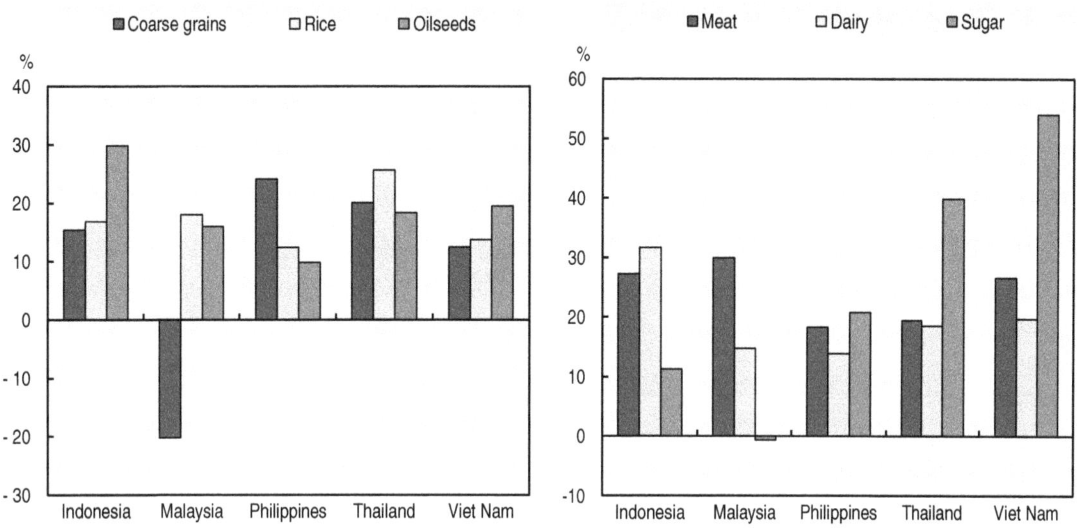

Source: OECD-FAO (2015).

The main driver of production growth in the region is projected productivity improvements, part of which will be driven by improvements in yield (OECD-FAO, 2015). In the ASEAN region, most agricultural land is already in use, and there is limited scope to increase the area of land used in the production of commodities. As such, increases in the harvested area of one crop necessarily reduce the harvested area of others. These land constraints mean that, for a number of major crops, it is yield improvements through increases in *intensification* (greater use of variable inputs, particularly fertilisers or improved seeds) which will be the driver of production growth over the medium term (Figure 2.6).

Figure 2.6. Area and yield factors underpinning production change

Asia and Pacific region: Average annual percentage change 2024 relative to 2012-14

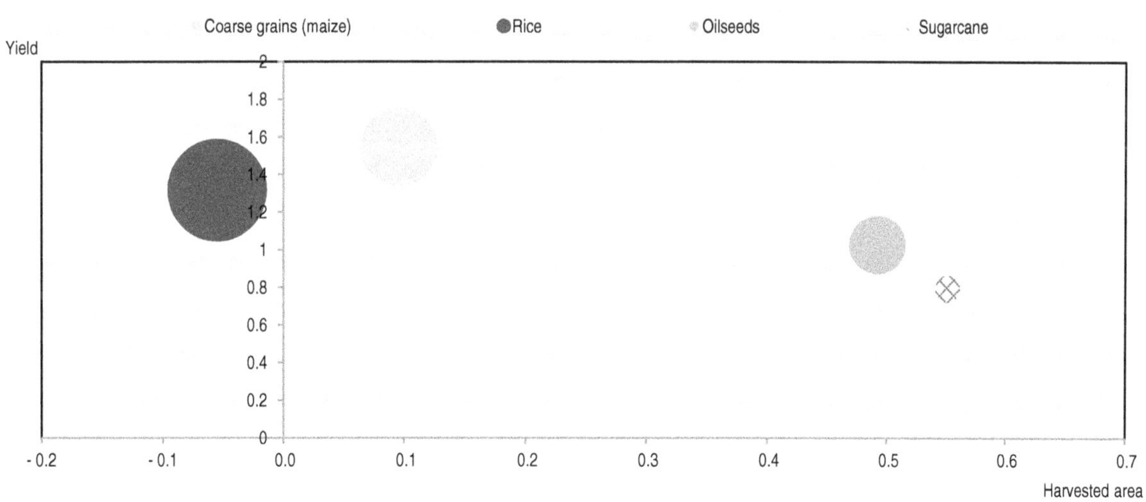

Source: OECD-FAO (2015).

Box 2.1 Fisheries and aquaculture production and trade in ASEAN

Fisheries production – capture fisheries and aquaculture included – will play an important role in food security in the ASEAN region over the medium term. Fish products[1] already form an important part of the regional diet, are an important source of income for many, and are also one of the region's largest exports. Trends in production reveal that for the region overall, increases in output will be driven by increases in production from aquaculture. Production from capture fisheries is projected to stabilise over time due to limits to the potential exploitation of this natural resource. These relative changes are projected to be maintained across ASEAN countries, with the exception of Viet Nam where, over the medium term, some increase in output from capture fisheries is projected –around 30% over the 2015-24 period, although considerable uncertainty exists, as discussed below (Figure 2.7). Indonesia is projected to continue its domination of the regional production of fisheries overall – in the medium term, at least.

Figure 2.7. Fisheries production for selected ASEAN member states over the medium term

Production from capture and aquaculture (tonnes)

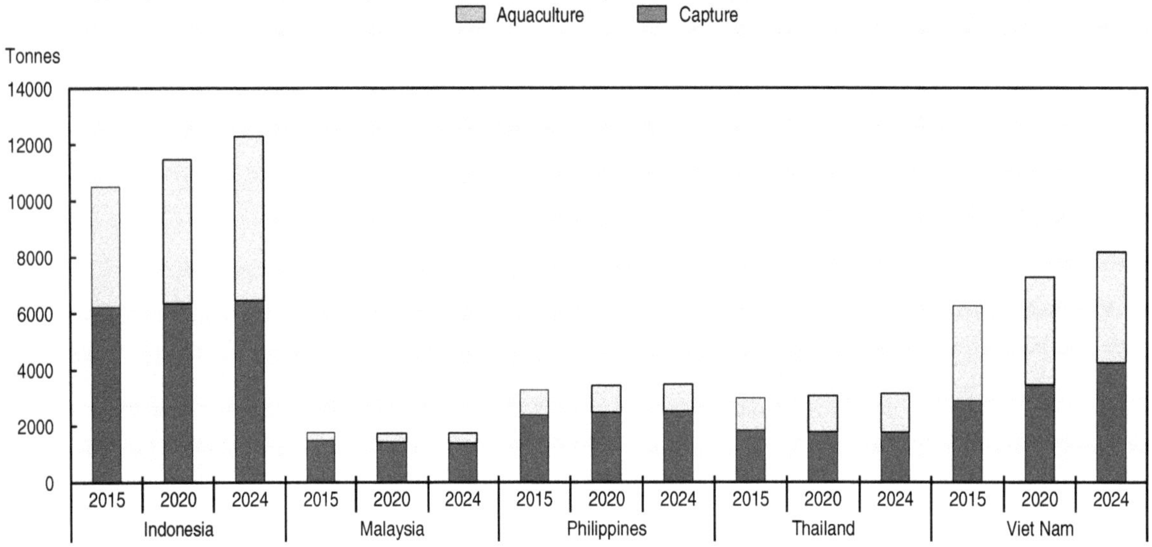

Source: OECD-FAO (2015).

Trade in fish products is projected to increase in the region over the medium term. For the region as a whole, both exports and imports are projected to increase by around 25%, indicating that trade in fish products will become more important for consumers and producers alike – both the share of production exported and the share of consumption imported increase by around 1 percentage point over the period. Across the five ASEAN counties individually modelled in the Outlook, there are contrasting trends (Figure 2.8). For Indonesia, Viet Nam and Thailand, both exports and imports are projected to increase, although for Indonesia, projections may overstate what could be expected without reforms to current limits on imported fish products. For the Philippines and Malaysia, the countries with the lowest trade volumes in fish products in the region, imports are projected to increase but exports to fall.

In terms of consumption, most countries in the region are projected to see demand increase, with the strongest increase in the Philippines. Demand changes are projected to be relatively small in Thailand, with a slight decrease, albeit off a relatively high consumption rate.

That said, there are considerable uncertainties that could undermine these projections. In the case of capture fisheries, a considerable number of stocks in the region are considered to be overfished – that is, fished beyond their biologically-sustainable harvest levels – and sustainable management instruments to control the level of fishing activity are often lacking. As such, even maintaining current levels of production is contingent on policy reforms to address these issues. This is further complicated by high levels of illegal and unreported catch, which would also need to be addressed in order for projections to be realised. For aquaculture, uncertainties exist over the sustainability of current production techniques, and the environmental and human health effects that flow from intensive production. These uncertainties point to a need for robust government policies to ensure that fisheries are able to play their role in contributing to ensuring the region's food security.

Figure 2.8. Changes in trade and consumption of fish products in selected ASEAN member states

Percentage change, 2015-24

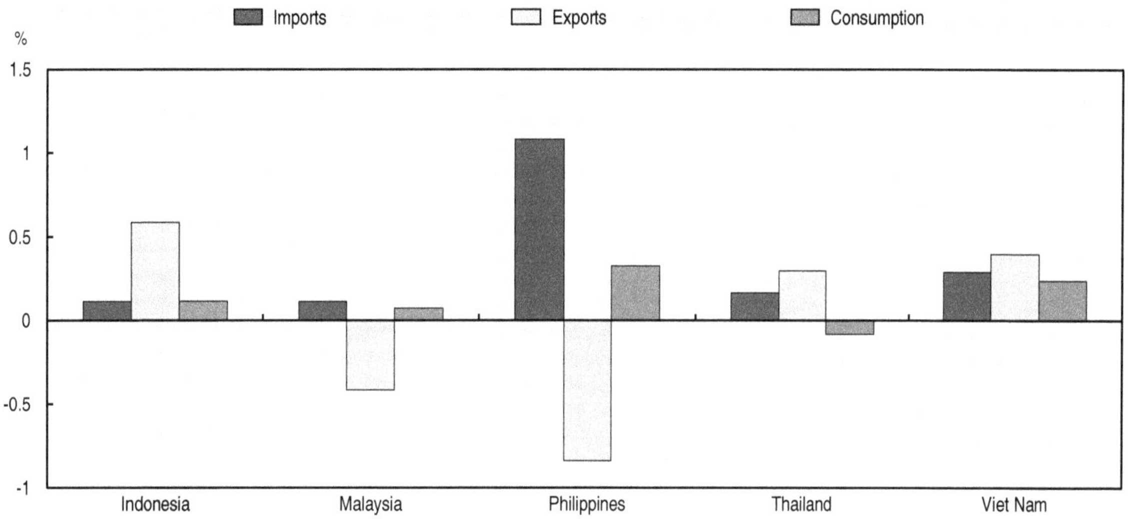

Source: OECD-FAO (2015).

1. Fish products are broadly defined and include fish, crustaceans, molluscs and other aquatic invertebrates but exclude aquatic mammals and aquatic plants.

The importance of yield growth – and productivity growth more broadly – in underpinning ASEAN agricultural production growth has a number of important implications. Changes in yield and productivity require efforts by governments and, importantly, *producers*. Efforts need to be underpinned by a robust enabling environment that facilitates increased knowledge, new production techniques and better incentives for producers to increase their efficiency and productivity. Furthermore, as uncertainties and influences in production are likely to increase with the effects of climate change in future (Section 2.4), the enabling environment will also need to assist adaptation and/or mitigation in order to enable ASEAN to address broader risks to food security and agricultural production in the future (Chapters 5 and 6).

Trade

For individual ASEAN countries, projections of agro-food trade changes over the medium term are mixed (Figure 2.9). Indonesia and Malaysia are projected to increase their levels of oilseeds exports along with Thailand (although off a smaller base), due to the growth of palm oil production in both of these countries (Chapter 1). Meanwhile, rice and sugar exports from Thailand and Viet Nam are also projected to increase. Exports of coarse grains from Indonesia, Malaysia and the Philippines are projected to diminish.

For imports, the picture is more uniform. All five ASEAN countries examined are expected to increase their imports of meat and dairy products, for example. This is driven by the changing nature of demand and rising income levels from higher economic growth. In terms of key grains, there is a projected increase in wheat imports across the region to varying degrees, linked to the same income and demographic changes as for meat and dairy. The picture for rice exports is mixed, particularly for the traditional rice importers of Indonesia, the Philippines and Malaysia. Imports are projected to increase in the Philippines and Malaysia, driven by increasing levels of domestic demand resulting from population growth. In Indonesia, by contrast, rice imports are projected to fall, due to relative demand shifts away from rice and increases in domestic production.

Figure 2.9. Trade outlook for selected ASEAN member states

Percentage change, 2014-24

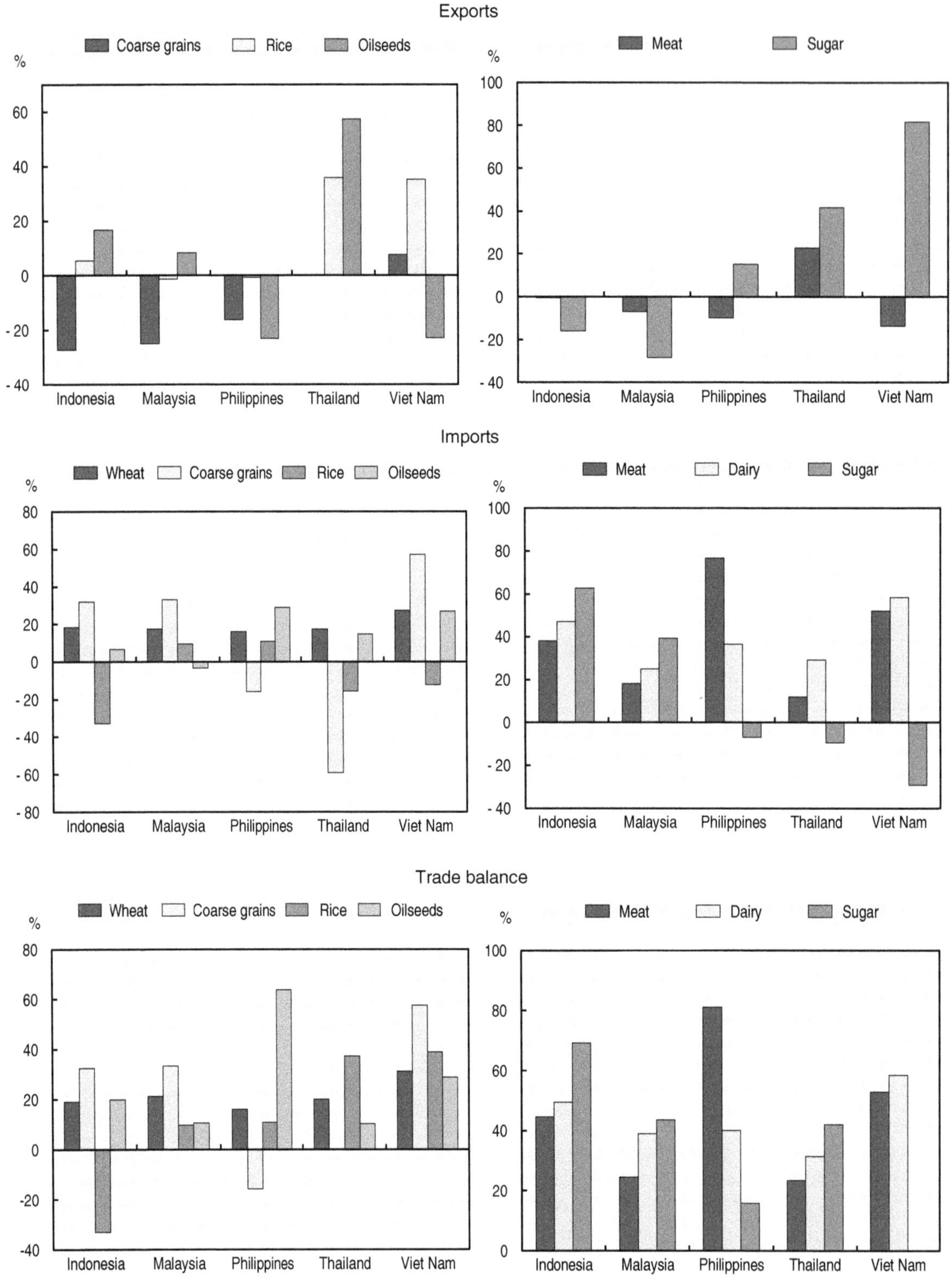

Notes: Trade balance changes show increases compared with current level. As such, if the balance is negative, and becomes more negative, then the percentage change is positive in the figure.

Source: OECD-FAO (2015).

Box 2.2. Risks to regional trade posed by increasing agricultural protectionism

With strong growth in agricultural production and incomes, both producers and consumers in ASEAN countries are becoming more connected to international agro-food markets. The policies in place in other countries, and the rules that underpin international trade in agricultural products, are increasingly important for the region. Recent analysis has shown that increases in agricultural protection can have significant negative effects on agriculture in ASEAN, impacting rural incomes and, ultimately, negatively affecting food security.

Recent analysis has used the OECD's METRO model in order to explore a scenario, termed *policy drift,* in which a number of major agricultural producing countries increase their levels of support to agriculture and reduce access to their markets compared with levels existing around 2011. This was contrasted with a scenario of partial reform of agricultural policies worldwide, termed *trade reform*, which involved the partial removal (up to 50%) of import tariffs, export taxes and domestic support, with developed countries reforming relatively more than developing. For details on the specific scenarios, see OECD (2016). The effects on agricultural production (primary production activities) and food production (the processing of primary materials) were explored.

The results indicate that increasing levels of protection are harmful for the ASEAN region, not only in terms of trade but also in terms of income and gross domestic product (GDP). For some ASEAN countries, which in the scenario increase protection in line with their current practice, total agricultural production was projected to decrease. Falls in total agricultural production led in turn to falls in income. The income effects, together with the effects that interventions in agricultural markets have on domestic prices, suggest that such policies will work against the food security of households, in particular those in rural areas facing higher food prices with fewer income-generating options. For other countries in the region whose policies did not change, changes in international markets drove the negative effects (Figure 2.10).

In contrast, trade reforms that deliver reductions in support and improvements to market access – the *trade reform* scenario – have the potential to more significantly benefit the region. The results indicate that it is in ASEAN members' interests to pursue more open markets, both regionally and internationally, among a wider set of their trading partners. Such policies should provide greater opportunities for their agricultural sectors, ultimately helping to increase incomes in rural communities connected with agriculture, and thus improve food security.

Figure 2.10. Impacts on trade of international trade reform and policy drifts

Percentage change in trade volumes for selected ASEAN members, 2011

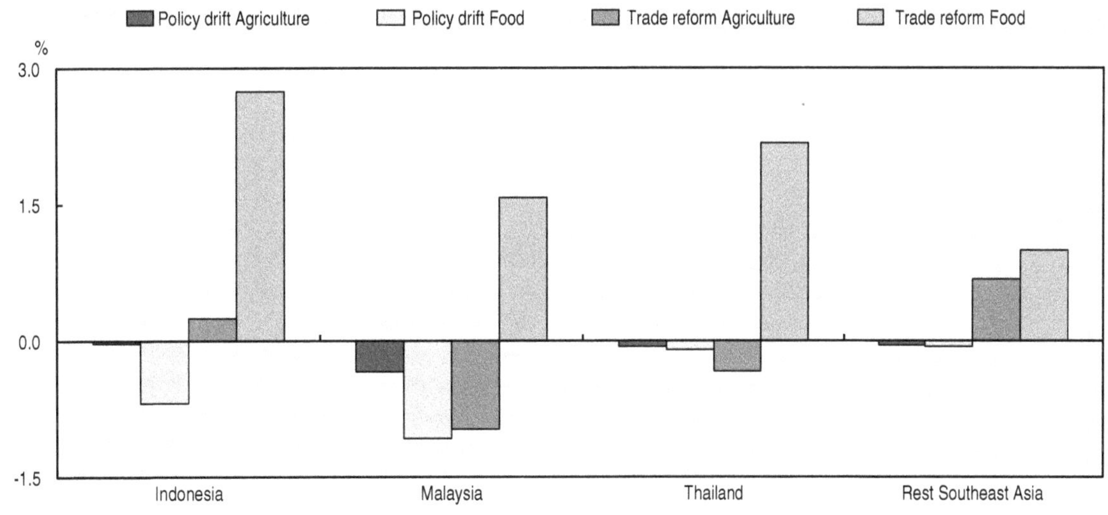

Source: OECD (2016).

These changes in imports and exports point to a general deepening of the existing trade balance (exports less imports for each country with the rest of the world) across most commodities. For net importers, imports increase, while net exporters increase exports relative to imports.

Across a wider set of products, there is a deepening of specialisation in trade in ASEAN countries. Vegetable oil exports will continue to be dominated by Indonesia and Malaysia (palm oil), while oilseeds (soybeans) sugar, wheat and coarse grains will remain their main imports (as for the region in total). Over the outlook period, rice, roots and tubers (sweet potato and yams), along with fish and sugar are increasingly important exports for Thailand and Viet Nam, with oilseeds, wheat and protein meals remaining their main imports. By contrast, for the Philippines, projections over the medium term show only small developments, with the Philippines expected to remain a small net importer of the various commodities.

However, not modelled over the projection period are the possible policy-related changes that influence trade and production. While changes in any direction are uncertain, recent OECD analysis suggests that regional trade and the returns to ASEAN countries could be significantly negatively affected by increasing levels of intervention in agricultural markets, both within the region and in other major markets such as China and India (Box 2.2). As ASEAN members now constitute a larger share of world agro-food trade, there is greater need for less distorted and more transparent international agro-food markets to enable ASEAN producers to exploit their growing competitive advantages in a range of agricultural and food sectors.

Prices of key regional commodities

Across the key food security crops (rice, sugar, soybeans and maize) in the region, world prices are projected to moderate from current levels and decline slightly in real terms compared with 2012-14 levels over the medium-term projection period (Figure 2.11 – soybeans are represented by oilseeds, and maize by coarse grains). That said, prices for each of these crops are projected to remain at levels above those seen in the early 2000s.

A similar picture is seen for other agricultural products important for ASEAN. In particular, world prices of meat products (with the exception of beef) and wheat, which are important imports (and important in production in the case of beef), are projected to decline more significantly from current levels. For these products, world prices are influenced by not only yield and productivity improvements within ASEAN itself, but also from across the world.

Moderating prices and rising incomes, combined with relatively strong economic growth in the region, should lead to a continuation of the past trends of improvements in food security (Section 2.3). However, considerable uncertainty exists over the projection period. In particular, yield and macroeconomic uncertainties may lead to considerable variations in the price paths depicted. As shown in Figure 2.12 for coarse grains, the band in which prices may possibly move is quite large. This band increases over time, as uncertainties in the projection period are compounded.

Figure 2.11. Price outlook for selected commodities important to ASEAN

Real prices, 1994-24

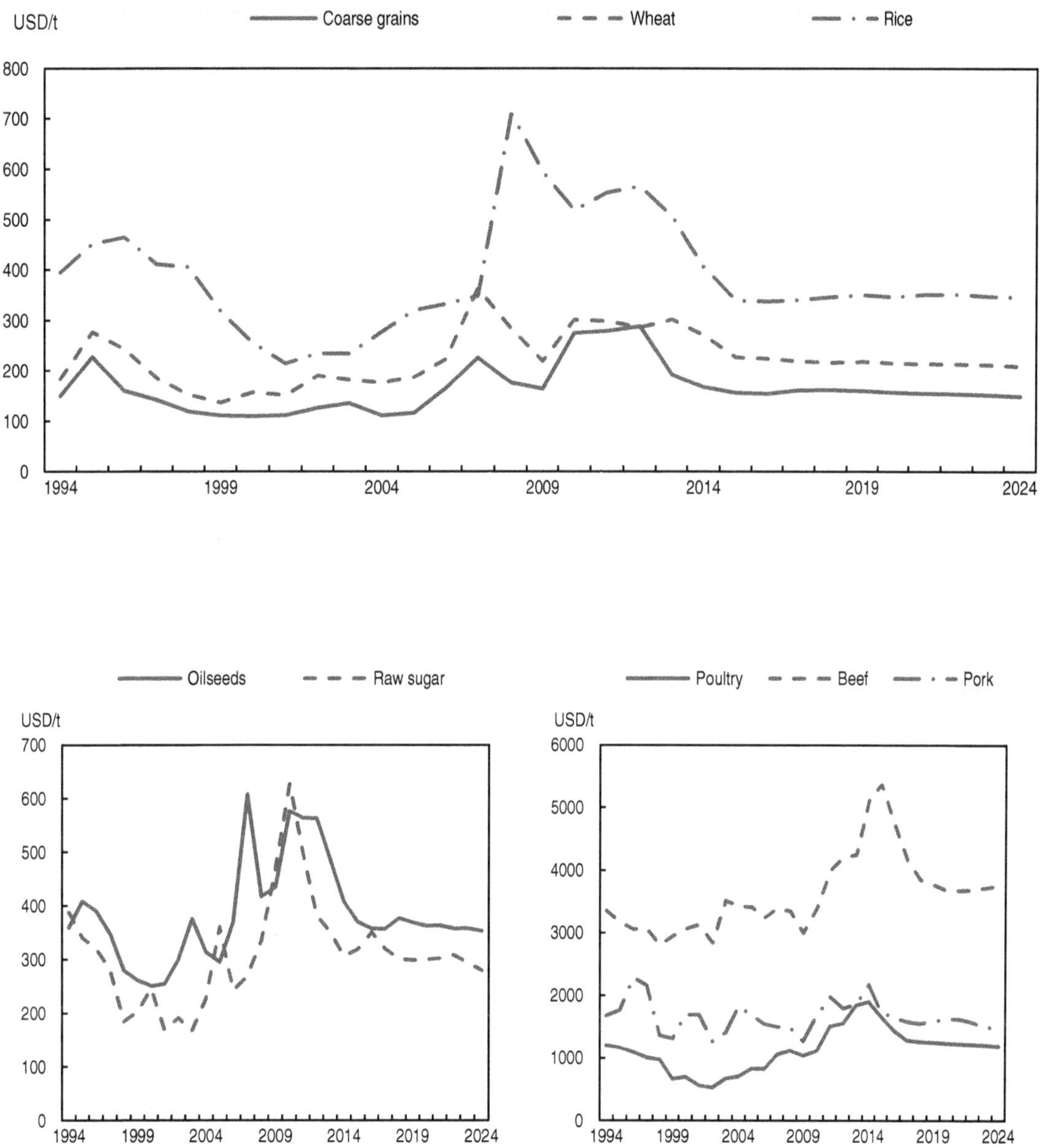

Source: OECD-FAO (2015).

Figure 2.12. International coarse grain price including variation derived from stochastic analysis

Nominal prices, 2004-24

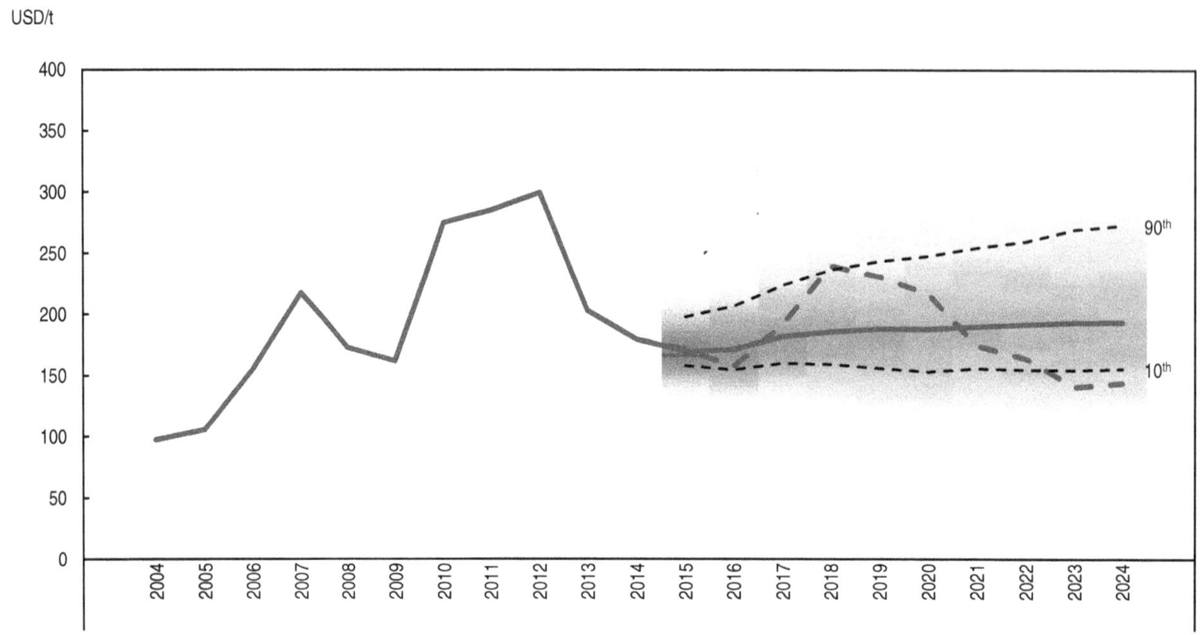

Notes: The smooth blue line represents the historical price evolution and baseline. The dotted blue line represents one arbitrarily-chosen price path out of the 1 000 simulations. Darker shading represents a greater probability that the price will reach a particular level in one specific year. The lower and upper dotted black lines represent the 10th and 90th percentiles, respectively.
Source: OECD-FAO (2015).

2.3 Implications of agricultural market developments for food security

This section analyses the linkages between food security, trade and markets using the OECD-FAO AGLINK-COSIMO model (Box 2.3).[4] In addition to the implications of the medium-term outlook for food security, four alternative scenarios are explored: i) faster income growth relative to the baseline in developing countries, ii) stronger growth in agricultural productivity than the baseline, iii) a combination of a faster income growth and stronger productivity growth, and iv) increased access to available food supplies.

Undernourishment projections from the medium-term outlook

Embodied in the medium-term projections is a significant increase in the supply of calories. Across the world, the agricultural sector will supply an additional 2.8 trillion kcal/day in 2024, with 83% coming from crops. Higher incomes across all income deciles[5] will enable additional consumers to access this food, with the result that by 2024, the number of undernourished individuals is projected to fall by almost 153 million relative to 2015, while the global share of undernourished will drop to 8%.

For ASEAN countries over the medium term, most of the supply of additional calories will come from crops. For example, Indonesia's agricultural sector will supply an additional 128 billion (bln) kcal/day, of which 93% will come from crops. Similarly, Malaysian producers will grow an additional 15 bln kcal/day, with 78% coming from crops, while for the Philippines an additional 42 bln kcal/day will be produced, with 84% coming from crops. In Thailand's case, an additional 17 bln kcal/day are projected with 86% coming from crops, while farmers in Viet Nam will produce an additional 39 bln kcal/day, with 62% coming from crops.

Box 2.3. Using the *OECD-FAO Agricultural Outlook* to calculate the FAO undernourishment indicator

The *OECD-FAO Agricultural Outlook* contains projections for the production, consumption and trade of the major agricultural commodities. It explicitly models the availability of calories at national level for 54 countries and regions covered by the AGLINK-COSIMO model (OECD, 2015a). The FAO undernourishment indicator converts national calorie availability into estimates of undernourishment on the basis of an estimated distribution of people's access to available calories. The indicator measures the probability that an individual from a reference population will consume less than the minimum calorie requirement for an active and healthy life. The Prevalence of Undernourishment (PoU) converts national calorie availability into estimates of undernourishment on the basis of an estimated distribution of people's access to available calories.

Undernourishment, as captured by the FAO indicator, provides a broad cross-country gauge of the number of people not consuming sufficient calories. However, it is only a partial gauge of food security which, according to the FAO definition, exists when "*all people, at all times, have physical, social and economic access to sufficient, safe and nutritious food to meet their dietary needs and food preferences for an active and healthy life*". As such, other aspects important for food security are not captured. That said, it provides a useful means to connect medium-term projections to indicative outcomes on food security (for a more in-depth discussion, see Tallard, Liapis and Pilgrim, 2016).

The calculations of the PoU indicator for this paper follow the same methodology and use the same publicly available data that FAO uses to calculate the PoU as reported in *The State of Food Insecurity in the World* (FAO, IFAD and WFP, 2015). Estimates of future total calories available for human consumption across countries are calculated using calorie conversion tables. Assuming that the distribution of access to total calorie availability remains unchanged over the next ten years, the evolution of the PoU can be determined for the 32 developing countries that are explicitly modelled. This can mean that with population growth, the absolute number of undernourished persons can increase in the presence of no income change. On this basis, projections for undernourishment for most countries that are part of the SEA region are computed (for more information, see Tallard, Liapis and Pilgrim, 2016, Section 2). For the Southeast Asian region, those countries are Indonesia, Malaysia, Philippines Thailand and Viet Nam. Unfortunately, three ASEAN countries – Cambodia, Lao PDR and Myanmar – are not explicitly modelled, but are nevertheless included in a regional aggregate with other countries which precludes calculations of caloric statistics for each of these but allow an estimation of the PoU, number of undernourished and the depth of undernourishment by applying the regional result to the countries. The evolution of the PoU is then projected to 2024 using results from the OECD-FAO AGLINK-COSIMO model (see Tallard, Liapis and Pilgrim, 2016, Annex 2.A1).

Source: Tallard, Liapis and Pilgrim (2016).

Higher income will enable greater access by ASEAN consumers to food, with the result that by 2024, the number and the share of undernourished individuals is projected to fall (Figure 2.13). For the region in total, the number of undernourished individuals is projected to fall by almost 13 million relative to 2015. Increased access to food will not only reduce the number of people facing undernourishment, but will also reduce the *depth* of undernourishment for those who remain. The reduction in the depth of undernourishment will also lower the absolute number of calories needed to further reduce undernourishment, enabling targets such as those set out in the Sustainable Development Goals (SDGs) to be more attainable – for example, the 5% undernourishment target (Annex 2.A1, Table 2.A1.3). Overall, the additional calories generated by agriculture and fisheries are expected to be sufficient to enable more people to consume above the minimum dietary energy requirement, to the extent that the share of people undernourished in 2024 should drop to 6.8% of the population, with Indonesia and Thailand dipping under the 5% threshold (Figure 2.14).[6]

The changes projected over the medium term will see a relatively more significant reduction in the number of undernourished within ASEAN countries compared with other regions worldwide (Figure 2.15). Strong reductions are also seen for other countries in the Asia Pacific region. In contrast, the medium-term outlook predicts that the absolute number of undernourished people in Africa will remain relatively stable, whereas in India, falls in undernourishment are expected to occur, but at a later point.

The fall in the number of undernourished masks differences across countries. While incomes are increasing across the ASEAN region, so too are populations. With a rapidly-growing population, the absolute number of undernourished people in the Philippines, for example, actually rises, even though the undernourished share of the total population falls (Figure 2.14). Such differences mean that food security will remain a concern for a number of countries in ASEAN over the medium term.

Figure 2.13. Evolution of undernourishment in Southeast Asia

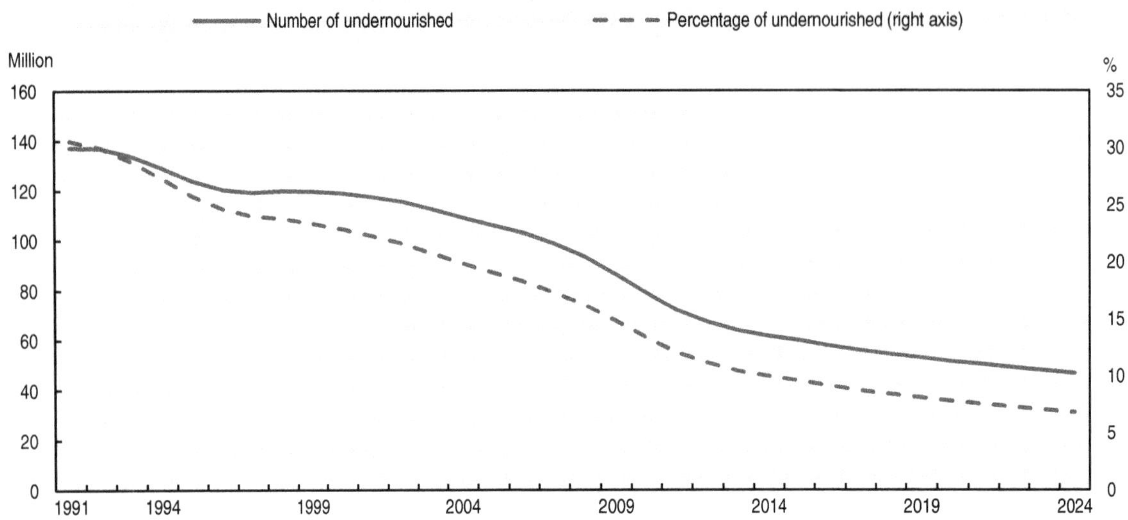

Source: OECD estimates based on AGLINK-COSIMO.

The impacts of alternative medium-term futures on undernourishment

The abovementioned medium-term outlook assumes that current policy settings will remain unchanged, and that past trends – including efforts to raise productivity through improvements in innovation systems and the agricultural enabling environment – will continue with the same intensity. However, it is also worth exploring the effects of *additional* actions that may shift future trends. In essence, these require additional reform and action by governments, producers and consumers beyond the levels observed in the past.

Figure 2.14. Evolution of undernourishment in ASEAN region

Number of undernourished

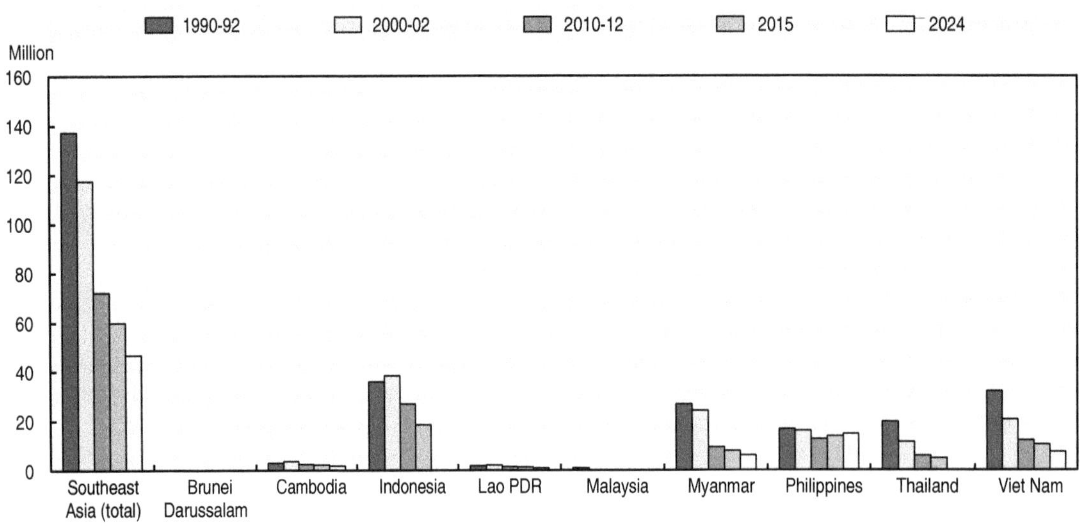

Note: Countries without observations represent those where undernourishment impacts less than 5% of the population, resulting in the model estimates of changes being insignificantly different from each other.
Source: OECD estimates based on AGLINK-COSIMO.

Figure 2.15. Evolution of undernourishment worldwide

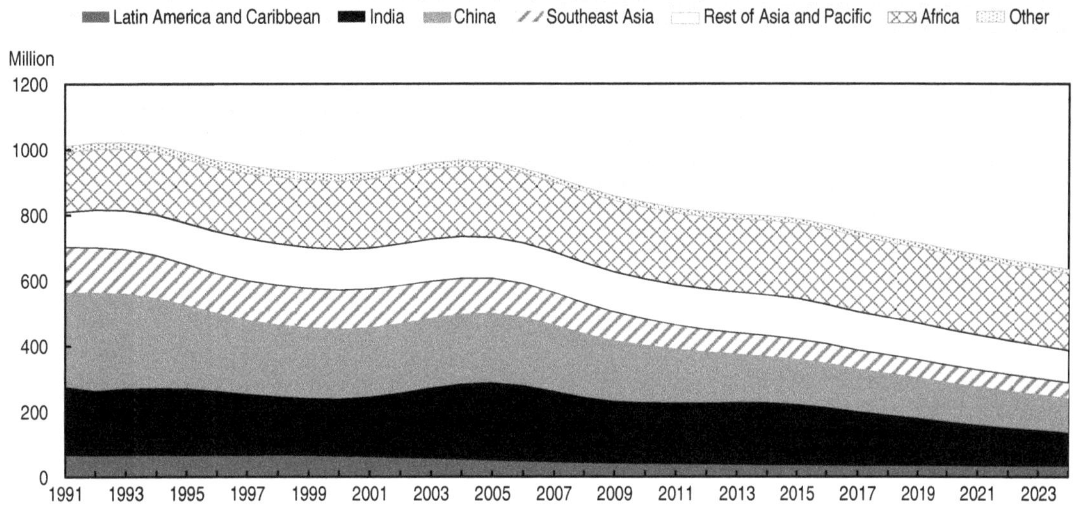

Source: OECD estimates based on AGLINK-COSIMO.

Figure 2.16. Projected number of undernourished in ASEAN under different scenarios

Number of undernourished

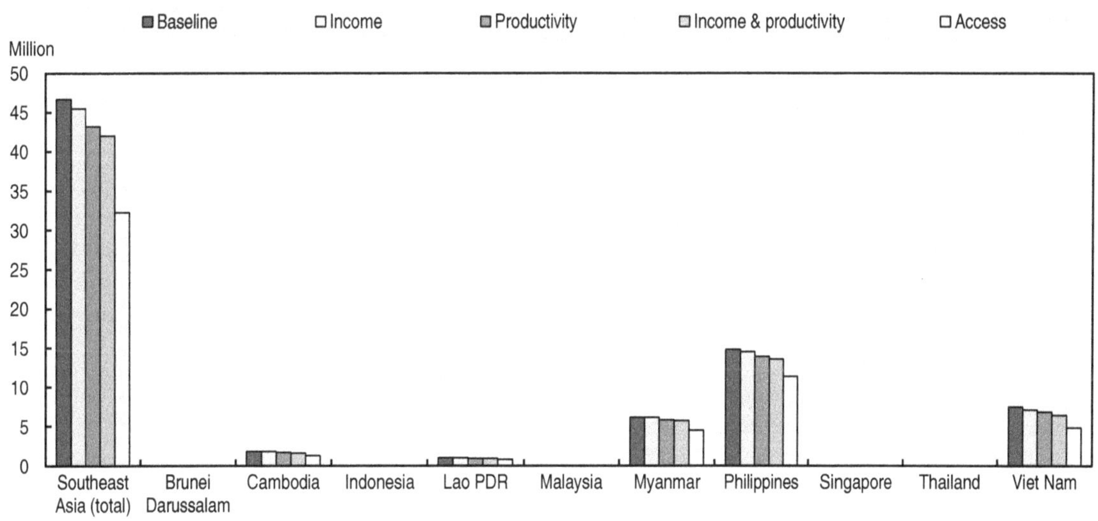

Note: Countries without observations represent those where undernourishment impacts less than 5% of the population, resulting in the model estimates of changes being insignificantly different from each other.
Source: OECD estimates based on AGLINK-COSIMO.

Four different scenarios were explored to examine their impact on food security in the region, each of which impacted the medium term outlook. These scenarios related to:

- *higher income growth*, modelled as a gradual cumulative annual increase in income so that in 2024, the income in each developing country (all developing countries worldwide) is 10% higher than in 2024 relative to the baseline
- *greater agricultural productivity growth*, modelled as a gradual cumulative annual increase in productivity so that in 2024, the productivity in each developing country (again for all developing countries) is 10% higher than in 2024 relative to the baseline
- *combined higher incomes and productivity*
- *increased access to food* by poor households.[7]

The impacts of these differing scenarios on undernourishment in the region is summarised in Figure 2.16.

Higher income growth

In this scenario, higher income growth leads to a more rapid expansion in food demand, resulting in slightly higher prices than under the baseline. Producers respond to these higher food prices and increase production. Meanwhile, consumers with higher incomes worldwide adjust their diets, creating more pronounced shifts in demand towards meat and dairy products. Worldwide consumption in 2024 from plant-based products is close to 141 bln kcal/day (0.7% greater than the baseline), while calories from consuming livestock products is 60 bln kcal/day (1.7% greater). The expansion in consumption is however partially mitigated by the increase in prices.

Where Southeast Asia in particular is concerned, higher incomes increase the number of people with access to food, reducing the number of undernourished people in the region by 1.2 million. In relative terms, this lowers the prevalence of undernourishment by 0.2 percentage points; 6.6% of Southeast Asia's population remains undernourished, down from 6.8%.

Differences in taste and preference and relative prices from rising incomes lead to varying consumption patterns among ASEAN countries. In Indonesia, for example, the population consumes an additional 2.6 bln kcal/day of plant-based calories (0.3% greater than the baseline), while calories from consuming livestock products are 0.7 bln kcal/day (1.3% higher). This shift to livestock products is replicated in Malaysia, the Philippines and Viet Nam. In the case of Thailand, however, there is also more of a complementary increase in demand for plant-based calories.

In this scenario, most of the new demand for plant-based calories in ASEAN countries is fully sourced from local production and, in several cases, there are increased exports of calories, such as those from Indonesia, Malaysia, Philippines and Thailand – the exception being Viet Nam, which imports 71% of the additional plant-based calories compared with the baseline. Of the additional animal-based calories consumed, the share of locally-produced livestock is 93% for Indonesia, 45% for Malaysia, 66% for the Philippines, 87% for Thailand and 72% for Viet Nam, underscoring the importance of trade in these products for meeting future demand and calorie requirements.

Higher agricultural productivity

Higher productivity in developing countries across all commodities and years is projected to increase output, reduce prices and stimulate consumption.[8] Improvements on the supply side (as opposed to demand as with the income scenario) have a slightly higher impact on improving food security – both worldwide and within the ASEAN countries. Across Southeast Asia, the number of undernourished falls by 3.5 million people compared to 1.2 million in the income scenario. For developing countries in general, and for ASEAN countries, the higher output of key staples and ensuing lower prices combine to have a larger impact on improving food security than that of rising incomes, assuming the population growth levels of the baseline.

For ASEAN, higher productivity levels have a lesser effect than increased incomes on changes in the composition of diets. Instead, the resulting lower prices lead to relatively higher reliance on calories from *crops* compared with calories from livestock. The increase in productivity also increases the export volumes of most agricultural products in the region.

Combined effects of rising incomes and productivity

The combined effects of a 10% increase in income and productivity accentuate the impacts of the abovementioned scenarios in an additive way. While higher incomes spur higher prices, the increase in productivity works against this, allowing higher demand to be met *without* price rises.

In the third scenario, compared with the baseline, 82.6 million individuals worldwide would no longer face undernourishment, dropping the global prevalence of undernourishment by 1 percentage point to 6.9% by 2024. A similar impact is seen in Southeast Asia, with the major beneficiaries of the decline in undernourished being Indonesia (1.5 million), Philippines (1.2 million) and Viet Nam (1.1 million). The overall reduction in undernourished people in Southeast Asia would reach 4.7 million – a 0.7% reduction in the prevalence of undernourishment.

Better access to food

This scenario explores the ability of households to better access available food at prices and supply levels seen in the medium-term outlook. Essentially, it is assumed that for a given level of income, the access of poor households to food is improved. This can be achieved through social security or other redistributive efforts, such as food vouchers. The scenario assumes that, without the need for production changes, those who previously consumed too little have greater access to food and consume more with the same level of income as before, shifting consumption more equally across the income distribution.

The results suggest that more equitable access to food has greater impact on food security worldwide than either income or productivity increases, or even the combination thereof. Compared with the baseline, 139 million individuals worldwide no longer face undernourishment, dropping the global prevalence of undernourishment by 1.7 percentage points to 6.2% by 2024.

The worldwide results are replicated in Southeast Asia, where the rate of undernourishment for the region falls below 5%, due to the improved access to food by poor households. Overall, the reduction in the number of undernourished people is 14.4 million, a reduction in the prevalence of undernourishment of 2.1 percentage points. Indonesia (4.7 million), the Philippines (3.4 million) and Viet Nam (2.7 million) experience the largest falls in undernourishment. For Viet Nam, whose prevalence of undernourishment remained above 5% in the other scenarios, the rate of undernourishment falls below 5%. These results suggest that the promotion of more equitable access is the most effective way to reduce undernourishment.

Some implications

Higher incomes and improved productivity – through the effects on output prices – and a combination of both, all have a beneficial effect on undernourishment. What the analysis indicates is that a given percentage increase in productivity growth has a *larger* effect on undernourishment than the same percentage increase in income. This is due to the effect that income has on prices (in other words, increasing prices) compared with productivity (which decreases prices). Overall, improving people's access to food through non-market-distortionary means enables more people in more countries to become food secure compared with any of the other scenarios explored. For policy makers, actions taken will focus on all three aspects, meaning that such changes are not alternatives but rather complementary steps taken to address food insecurity.

The scenarios confirm that it is not a lack of available food that is the fundamental problem, but rather effective *access* to that food. Nevertheless, while improved access to food helps to reduce the prevalence of undernourishment for the region to below 5%, it does not do so for all countries. For some countries such as Cambodia, Lao PDR, Myanmar and the Philippines, the depth of undernourishment is initially so severe that more drastic action to raise access to food among the poorest will be needed.

The scenarios explored in this chapter also confirm that trade contributes to the food security of countries by moving production from surplus to deficit countries. The role of trade depends on the circumstances of each country, although in many cases, much of the additional consumption is sourced locally.

2.4 Implications of climate change for agriculture and agricultural markets[9]

The medium-term outlook for markets and the resulting implications for food security are, on balance, relatively positive. Many of the drivers of past ASEAN performance, from yield improvements to continued regional and international market opening, will deliver a number of benefits to agricultural producers and help to promote regional food security.

However, over the longer term, in the lead-up to 2050, climate change will play a larger role in determining outcomes for the agricultural sector.[10] Some of the negative effects of climate change and climate variability are already visible in the region. Extreme weather events have not only been occurring more frequently, but are also occurring with greater intensity (Yusuf and Francisco, 2009; Hijioka et al., 2014). In the last few decades, sea levels in the region have risen by 1-3 mm/year, marginally higher than the global average (ADB, 2009). The number of floods, cyclones and periods of drought has also increased, leading to a decline in water, soil and land resources. Over the longer term, these shocks to production are expected to not only lead to temporary disruption but also to have effects on the observed *trends*. This impact on trends over the longer term will have a compounding effect on the incomes of producers and on the prices of agricultural and food products. This impact is explored in this section.

The climatic changes already observed are likely to intensify in future. Climate change-induced changes in rainfall are expected to continue to grow in the lead-up to 2040. Tropical cyclone intensities are likely to increase between 10% and 20% from current levels by 2050, and global temperatures are projected to be around 0.7–0.9°C higher than current levels by 2050 (Cruz et al., 2007). By 2100, in Indonesia, the Philippines, Thailand and Viet Nam, the annual mean temperature is projected to rise by 4.8°C. At that point, the global mean sea level could increase by 70 cm (ADB, 2009).

In this context, the likelihood of short-run crop failures and long-run production declines increases. Droughts, currently the climate feature with the greatest negative effect on annual production in the region, are expected to increase in intensity and frequency. Rice varieties grown in Southeast Asia are highly sensitive to drought stress and, as such, drought alone could have a significant negative effect on regional production and farm household incomes. In Cambodia, for example, droughts over the 1998-2002 period resulted in a 20% drop in rice production (ADB, 2014). Similarly, in 2010, Thailand experienced USD 450 million in crop damages, due to a severe drought. At the other end of the spectrum, flooding events are also likely to increase. In Thailand, following the 2010 drought, flash floods decimated rice crops, causing USD 40 billion in damage.

Gaining a better picture of the impact of climate change on regional agriculture is critical to enabling ASEAN countries to better design and implement strategies to ensure food security over the long term through the better management of climate-related risks.

A number of uncertainties exist concerning the potential agricultural and socio-economic impacts of climate change. In particular, uncertainty exists over the extent of change in future emissions and the impact that these will have on the climate – uncertainties exist over the extent of global action, the existence of tipping points in natural systems, and the potential role of contributing and countervailing natural influences. In this chapter, two separate models are used (the IMPACT and ENV-Linkages models) to explore the possible impacts on the Southeast Asian region. IMPACT is a global agricultural partial equilibrium model developed by IFPRI, and ENV-Linkages is a recursive dynamic neo-classical computable general equilibrium model, developed at the OECD. Both models can be used to assess both the possible long-term outlook for the region and possible policy responses. The current ENV-Linkages model is well suited for the analysis of policies, productivity and trade analysis; however, it only has limited coverage of some climate adaptation policies (Château, Dellink and Lanzi, 2014; Ignaciuk and Mason-D'Croz, 2014). The IMPACT model, on the other hand, is designed for productivity and climate change adaptation analysis. It can integrate more detailed information on agricultural products, markets, production technologies, environment, land use, and policy shocks or interventions, but cannot fully assess the implications of climate change and adaptation policies for a wider set of economic activities. Linking both models provides quantitative results on the outlook for Southeast Asia in the long term.

The potential costs of climate change for agriculture in Southeast Asia

On production

Looking ahead to 2050, climate change has the potential to significantly impact agriculture in the absence of policy responses to support either adaptation or mitigation. That said, global action has begun, and governments have taken steps, with the agreements reached at the United Nations Framework Convention on Climate Change (UNFCCC) 21st Conference of the Parties (COP21) in 2015, to address some of the potential negative consequences of climate change. The projections presented here should be seen as indicative of the possible consequences of inaction or delay.

Climate change is expected to negatively affect agriculture in ASEAN countries as a whole. Both rain-fed and irrigated crops will be affected through the expected influence on yields, although generally, the output and yield of rain-fed production is likely to be more negatively impacted due to the limited ability of producers to smooth water supply. With a large share of staple crop production dominated by rain-fed fields – approximately 55% of total rice production in Southeast Asia is rain-fed – the overall effect on production could be significant.

The projected effects on yield are calculated relative to a baseline with no climate change impact. That is, between now and 2050, yields will still increase, but to a *lesser* extent than may have otherwise occurred. The projected effects of climate change on irrigated and rain-fed yields in 2050 in the Southeast Asia region is shown in Figures 2.17 and 2.18. For most crops, including the key staples of rice, maize and cassava, projections highlight a decline in yield growth. The cumulative effect of this decline by 2050, if the average result of two climate models is taken, could be yields of rain-fed (irrigated) rice that are as much as 17% (16%) lower than they would have been in the absence of climate change, maize yields that are 16% lower (29%), and cassava yields that are 21% lower (14%). Vegetables, bananas, coffee and tea will also be affected, but to a lesser extent, with a lower yield of between 0% and 13% for irrigated and rain-fed areas. Beans, cocoa and palm oil are the only crops that would benefit from the shifts resulting from climate change, with a potentially higher yield of 3% for both irrigated and rain-fed cocoa and beans, and 2% for both irrigated and rain-fed palm oil compared with the baseline.

The shifts in relative returns resulting from the changes in yields – combined with the effects of price changes, discussed below, and demand – are projected to influence relative land use within the region. Producers are expected to respond to the changes in production costs and prices and – in the longer term – to be able to shift their relative production mix. With strong and inelastic demand for agricultural products, and in particular for staples, lower yield increases in the major staple crops of rice, maize, and cassava will drive *increased* allocation of land to these activities. This reallocation then reduces the land available for other crop or livestock activities, with changes projected to be particularly pronounced for bananas, beans and coffee production. The effects (again, the average result of two climate models) range from area increases of 2% to 6% for rain-fed rice, maize and cassava, to declines of 3% to 7% for rain-fed vegetables, bananas, beans and teas in particular (Figures 2.17 and 2.18). The area variation is strongest for rain-fed areas. As most crop cultivations in Thailand, Myanmar, Lao PDR and Cambodia are rain-fed and not irrigated, these countries experience the greatest changes in area allocations.

Not all ASEAN countries will be equally affected.[11] Countries within the region face different climate conditions and, as such, the impacts of climate change are not uniform.[12] Climate change is projected to have greater impacts on some crops in some countries, thereby creating differences in land use change across the region, which are further influenced by the abovementioned differences in the relative mix of irrigated and rain-fed production. The diverse impacts of climate change suggest that there will be greater benefits from open regional markets for agro-food products, which can enable individual ASEAN countries to offset some of the negative effects on production and domestic prices through a regional spreading of climate risk.

Figure 2.17. Impacts of climate change on rain-fed crops in Southeast Asia

Percentage change in yields, area and production in 2050, compared with situation of no climate change

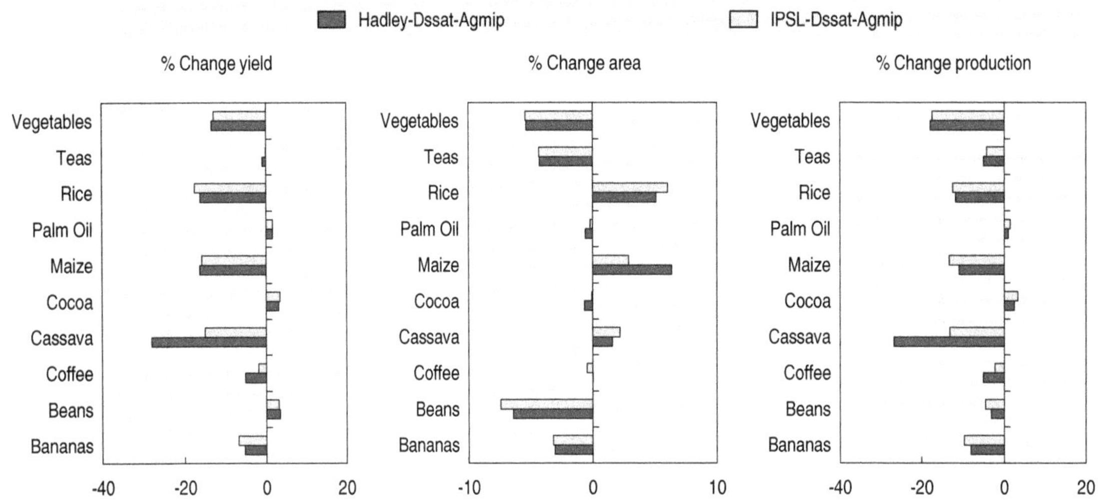

Source: OECD estimates based on IFPRI IMPACT model.

Figure 2.18. Impacts of climate change on irrigated crops in Southeast Asia

Percentage change in yields, area and production in 2050 compared with situation of no climate change

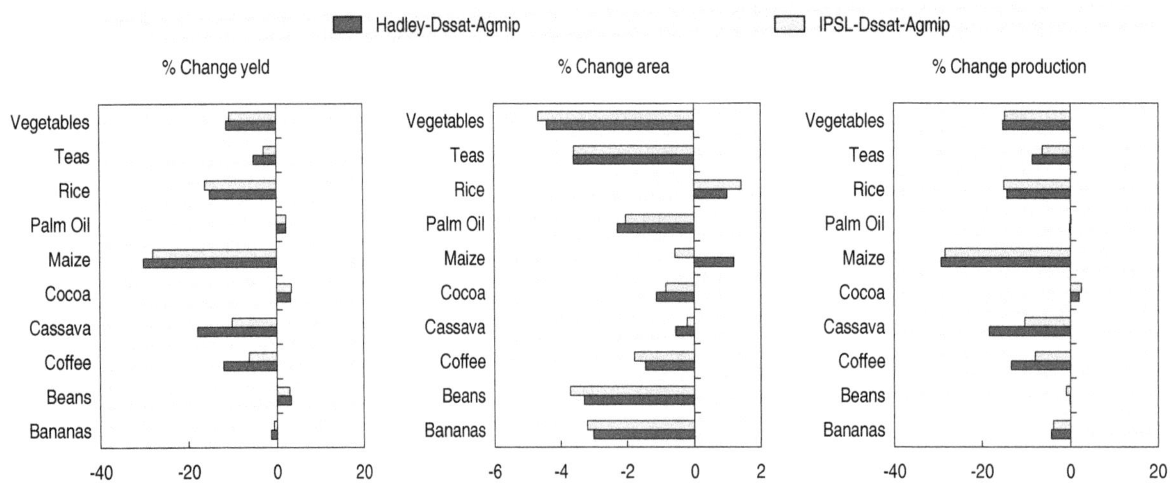

Source: OECD estimates based on IFPRI IMPACT model.

The non-uniform effect of climate change is highlighted in the case of two of the key staple crops for the region: rice and cassava. By 2050, the negative effects on rain-fed rice yield would be the largest in Thailand, while rain-fed rice yield in Malaysia would be the *least* affected (Figure 2.19). Currently, much of the rice production that occurs in Thailand is rain-fed and, as such, without changes in the level of irrigation infrastructure for rice, the production effect is projected to be significant. For cassava, relative yield decreases are also expected to be greatest across the Mekong delta (in Thailand in particular and extending to its Malay Peninsula – Figure 2.20). Myanmar and parts of Malaysia and northern Viet Nam, on the other hand, will be least affected, although still negatively. It should be noted, however, that while these projections indicate relative yield declines compared with baseline levels, when compared with 2015 levels, yields are still expected to increase through productivity improvements over time.

On prices

The effects of climate change on production will flow through to markets. Across the world and in the Southeast Asian region, climate change is expected to increase real prices for agricultural commodities compared with a situation of no change. These patterns emerge as the cumulative effects of shocks and weather changes translate into changes in real price trends.

The increases in prices are predominately driven by the changes that are seen in yields. Significantly lower yield growth, particularly for staples, combined with inelastic demand, leads to higher prices when compared with a situation of no climate change. Prices worldwide are projected to rise for all agricultural products examined. However, in Southeast Asia, prices for staple crops such as rice, maize and cassava are expected to rise by more than other regions worldwide (Figure 2.21). The prices of rice, maize and cassava are, on average, between 45% and over 55% higher when climate effects are included in the projections, compared with a baseline of no climate change.

Figure 2.19. Country-level impacts of climate change on rain-fed rice yield

Percentage change in yields in 2050 compared with situation of no climate change

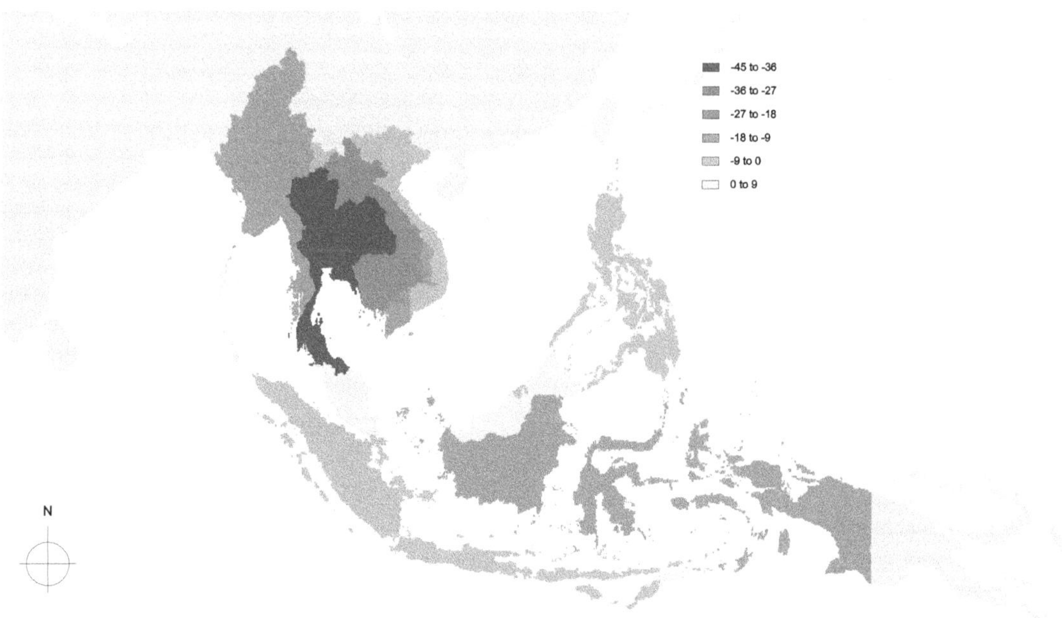

Note: Estimates are final yield estimates derived from the model.
Source: OECD estimates based on IFPRI IMPACT model (Hadley-Dssat-Agmip climate change scenario).

Figure 2.20. Country-level impacts of climate change on rain-fed cassava yield

Percentage change in yields in 2050 compared with situation of no climate change

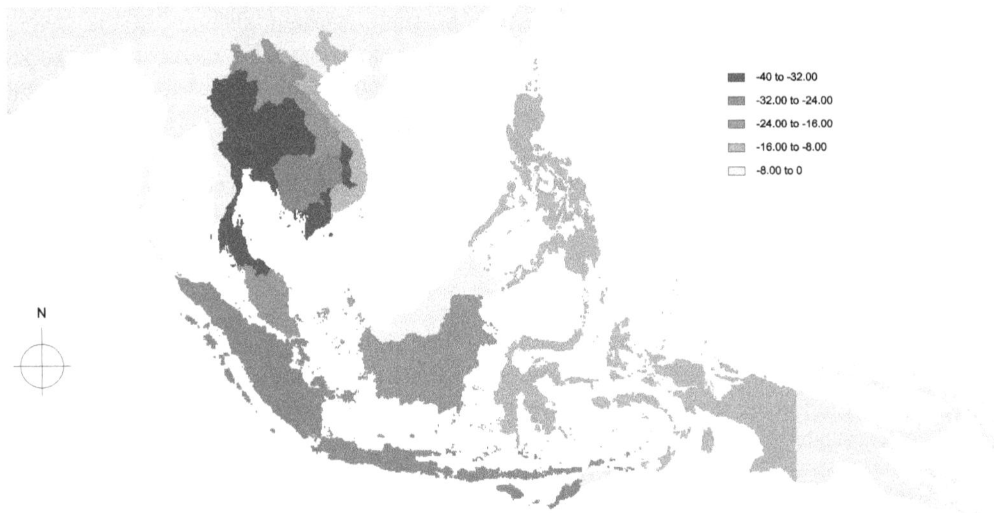

Notes: The climate change shock for cassava in the IMPACT model is based on the average shock for C3 crops – those of wheat, rice, potato, groundnut and soybean. Estimates are final yield estimates derived from the model.
Source: OECD estimates based on IFPRI IMPACT model (Hadley-Dssat-Agmip climate change scenario).

Figure 2.21. Projected price effects across the world and in ASEAN

Percentage change in prices in 2050, compared with situation of no climate change

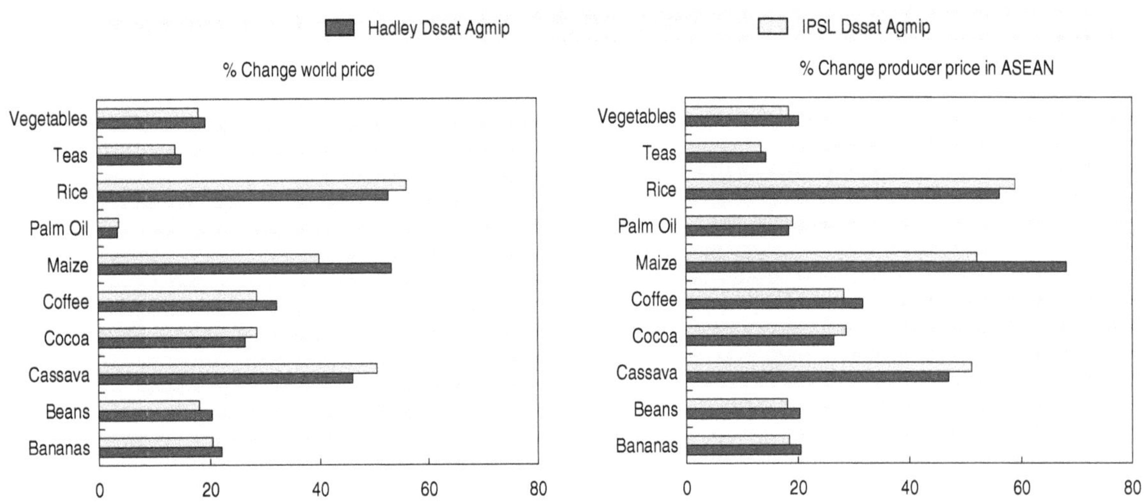

Source: OECD estimates based on IFPRI IMPACT model.

The disproportionate effect in Southeast Asia on prices for staples reflects the impact of regional policy settings. The projections include no changes in the current policy settings of ASEAN members. Some of the projected price increases due to climate change are compounded by current policies and market rigidities in certain countries, such as those related to trade restrictions, suggesting that reform options should be considered.

On trade

Following the production and price effects of climate change, the projections suggest that agricultural trade between ASEAN and the rest of the world will also diminish (Figure 2.22). The effect on trade for individual products is both direct and indirect. The *direct* effects result from the changes in crop yield and production, generating fewer surpluses for trade on world markets. The *indirect* effects relate to the use of higher-priced inputs in the production of other agricultural products, such as higher feed crop prices for livestock and meat production, thereby reducing the relative competitiveness of these industries. Similarly, for processed food products, the increasing costs of inputs drive prices up and results in a relative reduction in the region's competitiveness. A further contributing factor in this effect is the maintenance of current restrictive trade policies in some countries which amplify some of the price movements as they limit the ability of trade to moderate rising prices (see Chapter 3 for a review of current trade policies).

Rice trade is significantly negatively affected by climate change. This is due to the abovementioned climatic effects – decreasing the surplus of rice that is available for trade with the rest of the world – which are in turn compounded by the influence of trade and domestic policy settings. The rice market and rice trade are also negatively affected to a large degree by a number of different government policy interventions, both across the region and the world more generally (Chapter 3). This fall in trade volumes, in already thin markets, can have implications for regional price-setting and the ability of individual countries to deal with shocks that can disrupt domestic production, thereby negatively influencing food security.

Figure 2.22. Projected ASEAN trade effects resulting from climate change

Percentage change in exports and imports in 2050, compared with situation of no climate change

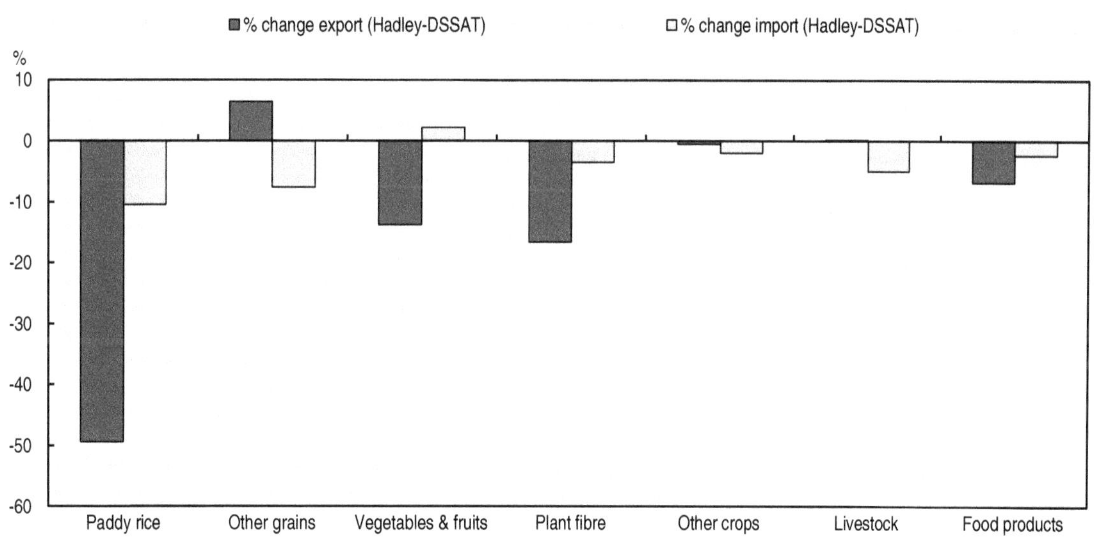

Source: OECD estimates based on ENV-Linkages Model.

Beyond agriculture: The potential economy-wide impacts of climate change in ASEAN

Beyond agriculture, climate change will have more pervasive impacts across ASEAN economies. For example, across Southeast Asia, compared with a situation without climate change, worker productivity would be expected to be lower – from occupational heat stress in outdoor activities – and healthcare expenditures higher. Climate change has the potential to alter the composition of the economy and therefore can influence growth rates over the longer term.

The spread of effects also varies across industries and countries alike, and indeed populations and areas within countries. ASEAN countries are projected to suffer significantly from coastal zone damage, mostly from sea-level rises. Rising sea levels would not only reduce land surface but also destroy some production capacity and physical capital located within coastal areas. The negative impacts of coastal zone damage on GDP are projected to increase from current levels by 16% in 2035, and by up to 19% in 2060. However, these estimates do not include the potential costs from increased and more severe flooding. As such, they are likely to be underestimates (OECD, 2015b).

Overall, the negative effects on GDP are projected to be greater in ASEAN than either worldwide or in developed OECD countries (Figure 2.23). Within ASEAN, those countries that are highly-populated and with a high concentration of the population and economic activities located in coastal areas are likely to be relatively more affected by climate change than others. Given this, the effects for Indonesia are particularly pronounced. The losses in terms of lower GDP for Indonesia are projected to be 1.7% by 2050 and 2.1% by 2060, relative to a situation with no climate change. For other ASEAN countries, losses are projected to be marginally lower in 2050, at 1.6% of GDP, but the same in 2060, at 2.1% of GDP.

Figure 2.23. Projected effects on regional gross domestic product (GDP)

Percentage change in GDP compared with situation of no climate change

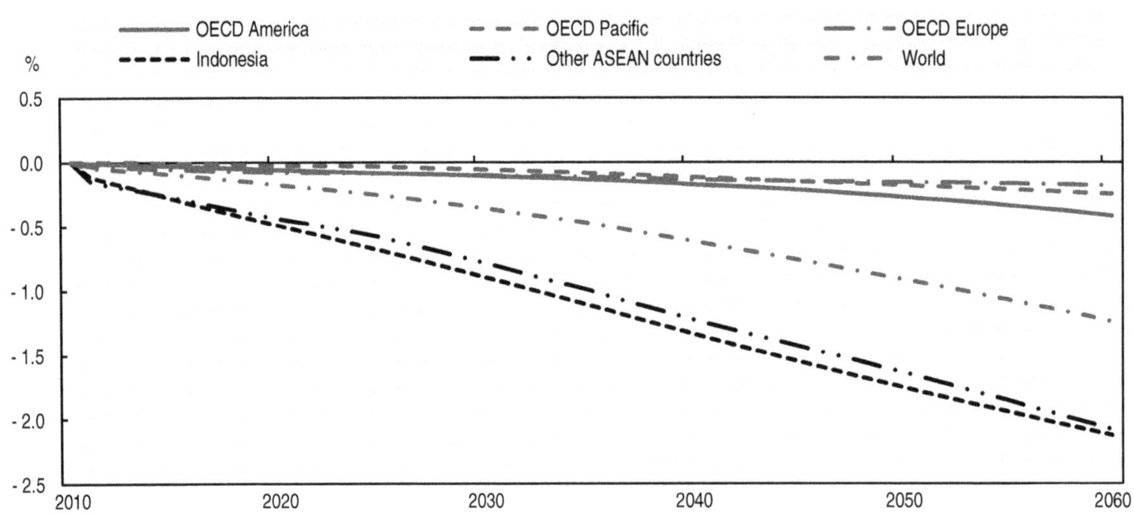

Source: OECD estimates based on ENV-Linkages.

The role of adaptation

Given these potentially significant negative impacts of climate change on the region, it is worth exploring some of the potential benefits of government action. Specifically, climate change adaptation strategies have the potential to avoid, or at least lessen, some of the projected impacts on countries and markets. Beyond adaptation, mitigation efforts related to soil carbon sequestration, although not explored in detail in this section, also have potential to positively impact food security (Box 2.4).

Key to adaptation actions is the development of new technologies and application of new (or greater use of existing) techniques which either reduce the costs of adaptation or provide new ways in which adaptation can occur. Such technologies could potentially play a pivotal role in sectors like agriculture, which is vulnerable to the effects of climate change (Francisco, 2008). This role is being recognised – climate change adaptation represents one of the main objectives of research and development (R&D) policies around the world, including within ASEAN economies.

Several types of adaptation practices and technologies are needed to facilitate the adaptation of the agricultural sector (Table 2.1). These vary from the development of new crop varieties to the improved use of available natural resources, via better irrigation techniques, for example. The scenario explored below (Figure 2.24) examines one of these adaptation technologies – the development of new resilient crop varieties able to maintain higher yields in the face of changing climatic conditions. Several research organisations are currently working to develop new or improved crop varieties with greater resilience to the effects of climate change, and which are also capable of satisfying future demand for food. Privately-funded research centres are also becoming more active in this field. An example of these new varieties is C4 rice, which is being developed by the International Rice Research Institute (IRRI) in the Philippines. It is expected that this technology could increase rice yields by between 30% and 50% while improving the resistance of rice to weather variations (GRiSP, 2013; Sheehy and Mitchell, 2015).

As an example of the role that adaptation technologies can play, a scenario that simulates the effects of the introduction of a new rice variety similar to the C4 rice variety developed by IRRI was explored. The scenario considers the impacts on ASEAN overall. The new variety is assumed to increase the yields of rice by 15% by 2050, based on work by Sheehy and Mitchell (2015). To reflect the natural process of technology diffusion, an 80% maximum adoption rate is expected to be reached by 2045. The IMPACT model assumes that local producers will not face higher production or input costs following the implementation of this new rice variety. The scenario also assumes zero development, implementation, or opportunity costs.

In this scenario, the adoption of a new and improved rice variety increases production by 14%, but, importantly, also has significant effects on land use. Despite the effects of climate change, the model projections indicate an overall increase in rice supply in the next few decades as a result of the new rice variety. The substitution of land away from other crops to rice is also significantly lessened. Indeed, compared with the no climate change baseline, the irrigated area dedicated to rice actually *falls*, by almost 1%, with even larger falls observed for land allocated to rain-fed rice production – almost 2%. These changes increase the land planted with other crops, and therefore also increases the production of these crops, limiting the pressure on commodity prices overall.

Box 2.4. Climate change mitigation and food security

Alongside adaptation activities, activities that seek to mitigate the effects of climate change can also have a positive influence on food security. In the first instance, they have the potential to lessen the change likely to occur from the increase in atmospheric greenhouse gases. In the second, the activities themselves may also influence agricultural and food production.

A number of studies have examined the role that soil carbon sequestration can play in both mitigating the effects of climate change and promoting agricultural and food production – a feature recognised in the agreements reached at the United Nations climate change conference, COP21, in 2015. Lal (2004; et al., 2007), for example, suggests that addressing degraded agricultural soils can both help to mitigate the effects of climate change and improve agricultural production. To do so, a range of soil management and agricultural practices would need to be adopted, including no-till farming, cover crops, nutrient management, manure and sludge application, improved grazing, water conservation and harvesting, efficient irrigation, agroforestry practices, and the cultivation of energy crops on spare lands rather than displacing food production elsewhere. Lal (2004) suggests that, notwithstanding variation across soil types, increases in soil carbon levels would also increase yields for crops such as wheat, maize and cowpeas. Furthermore, the sequestration potential is large, at 5 to 15% of global fossil fuel emissions (in 2004).

Sources: Lal (2004), Lal et al. (2007), Lal (2010).

Table 2.1. Agricultural technologies and techniques for climate change adaptation

Projected impact of climate change	Adaptation technology or technique
Reduced crop yields resulting from higher temperatures	New crop varieties with greater heat tolerance
Reduced crop yields in rain-fed agriculture, due to less precipitation	New crop varieties with lower water requirements
	Improved water collection, storage, and distribution techniques
	Improved irrigation techniques
Reduced crop yields in irrigated agriculture, due to reduced availability of irrigation water	Improved irrigation efficiency
	New crop varieties with lower water requirements
	Real-time and remote-sensing capabilities to improve water management and efficiency of use (e.g., soil moisture, evapotranspiration)
Reduced irrigation water availability, due to saltwater intrusion	Barriers to saltwater intrusion
	Increased sustainable aquifer recharge
	New crop varieties with greater salinity tolerance
	Improved water collection, storage, and distribution techniques
Reduced crop yields from increased flooding or waterlogging	New crop varieties with higher moisture tolerance
	Improved drainage or flood control techniques
Increased incidence of crop pests and diseases	New crop varieties with improved pest and disease resistance
	Improved pest and disease management techniques
Loss of crops due to extreme weather events	Improved extreme weather event prediction and early warning systems
	Improved techniques to increase resilience of crops to extreme weather events

Source: ADB (2014).

These production effects in turn impact prices. The adoption of the new rice variety leads to falls in world market prices to the extent that price levels are almost equivalent to those observed in a world without climate change (Figure 2.25). It is projected that world rice prices would fall by 7%, compared with a situation of no adaptation. Moreover, due to greater land availability, the prices of other crops would also fall, by slightly less than 1%.

For the region, the price and production effects are also projected to increase trade. However, the relative impacts on trade, and therefore some of the returns to the new variety, are of course dependent on what happens elsewhere in the world – for example, whether or not other new varieties are developed in other regions.

Figure 2.24. Land use changes in ASEAN arising from improved rice variety adoption

Percentage change compared with situation of no climate change

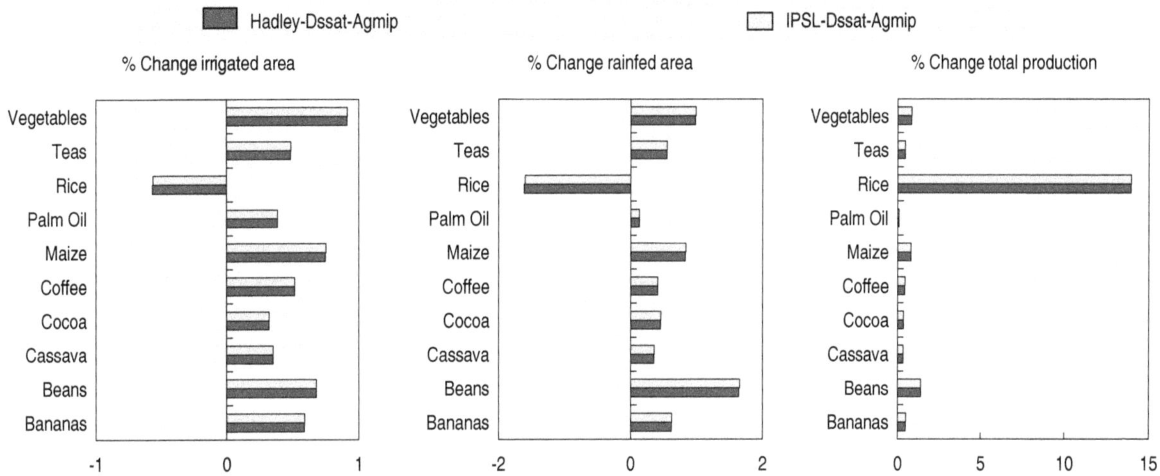

Source: OECD estimates based on IFPRI IMPACT model.

Figure 2.25. World market price changes resulting from improved rice variety adoption

Percentage change compared with no adaptation scenario

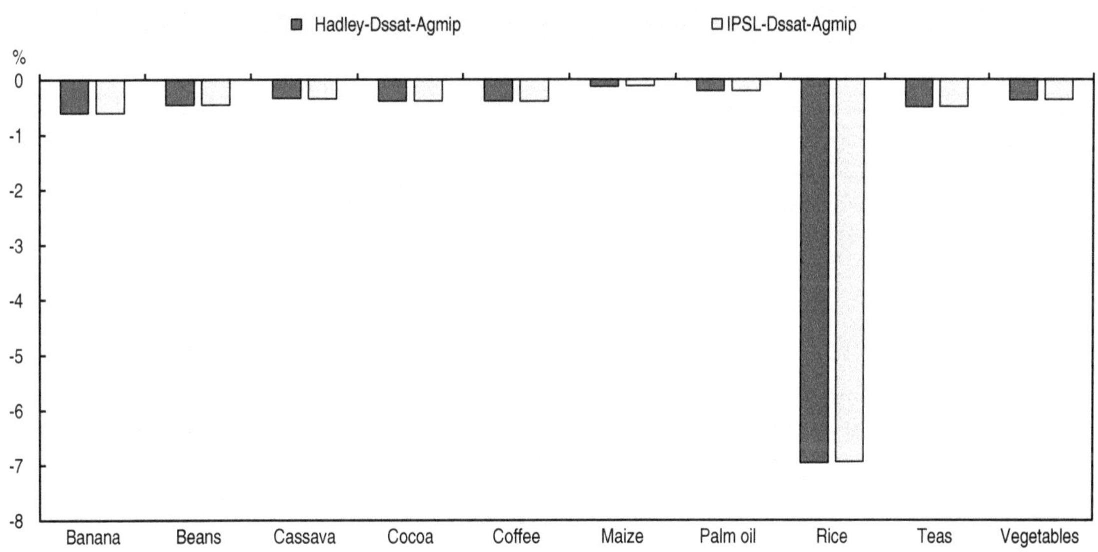

Source: OECD estimates based on IFPRI IMPACT model.

2.5 Concluding comments

In formulating policy recommendations, and in taking policy action, governments must be aware that the development of any new technology, such as a new rice variety, requires investments. These investments need to occur not only directly in the research and development (R&D) system, but also in the other areas of the wider economy and society that support greater levels of innovation – general education, for example. Furthermore, producers must also make investments of their own. These include those related to learning about the new technology and acquiring the necessary skills or products for its adoption by, for example, purchasing the new seed variety that is likely to be at a price premium compared with current varieties. If the effects are to be widespread, these investments by producers also need to be widespread to ensure significant uptake.

It should also be noted that if either the costs to governments or producers are high, or the decisions to make these investments are limited by policy in some way, the potential for a new technology to moderate price increases will be lessened. In the example above, this means that the potential benefits will be less than those depicted.

In the scenario explored, while the costs of investments required to create a new variety of rice that is able to offset the impacts of climate change are unknown, the scenario highlights that the gains would be significant. In particular, it highlights the positive effects of developing adaptation technologies in significant industries, in addition to the indirect effects that this can have on other agricultural activities, and, ultimately, on prices paid by consumers.

More broadly, the role that R&D has played in promoting productivity growth in agriculture over time has been well documented. Nevertheless, Smeets Kristkova, Van Dijk and Van Meijl (2016) demonstrate that if past relationships between R&D investments and productivity growth were to hold into the future, current declines in R&D investments will mean that the assumption of yield growth that underpins many climate change models, including the ones used in this study, are overly optimistic. This highlights the need for additional action *beyond* the current trend of R&D investment in the sector. Importantly, while the work by Smeets Kristkova et al. (2016) did not assume a costless R&D investment path, it nevertheless continued to show that significant benefits for national and global economies flow from these investments.

Finally, in addition to innovation and R&D, the projections also highlight the effects of other policies. For ASEAN members, market-distorting rice and other agricultural policies, if left unchanged, have the potential to amplify the negative effects of climate change – particularly with respect to price. This amplification would have consequences, both for food security in the region as a whole, and for the ability of individual households to manage any transitory risks to food security.

Notes

1. The ten ASEAN member states are Brunei Darussalam, Cambodia, Indonesia, Lao People's Democratic Republic (PDR), Malaysia, Myanmar, Philippines, Singapore, Thailand and Viet Nam.

2. In this chapter, the *2015-2024 OECD-FAO Outlook* is used to present the medium-term projections for Southeast Asia. This was done to ensure consistency with the modelling of improvements in food security presented in Section 2.3. The *Outlook* publication is updated annually based on changes in world markets and the policy positions of agricultural producers. Given this, recent projections are likely to differ from past – however, for most commodities, global trends are maintained. The latest projections can be accessed via: www.agri-outlook.org/.

3. The projections of economic growth for Southeast Asia and the rest of the world are derived from the *OECD Economic Outlook* (OECD, 2014) and the International Monetary Fund's *World Economic Outlook* (IMF, 2014). It should be noted that these projections are subject to uncertainty.

4. The AGLINK-COSIMO model explores the projected evolution of undernourishment in Southeast Asia under the baseline of the 2015 *OECD-FAO Agricultural Outlook*, using FAO methodology for the calculation of undernourishment. This section is based on an earlier OECD publication (Tallard, Liapis and Pilgrim, 2016).

5. Higher incomes per se may not translate into greater food consumption or purchasing power if income growth is skewed towards higher income brackets. For the projections provided here, unless otherwise stated, income growth is associated with growth across *all* income groups, thereby maintaining both the current income distribution and the relevant inequalities.

6. It should be noted that it is not just energy supply from food that is an issue for food security. While this metric is the focus of this study, other aspects of food supply related to the nutrient content of the food that is consumed and the mix between different components, such as proteins and carbohydrates, are important in addressing both food security and issues associated with malnutrition.

7. This was modelled as a gradual 1% cumulative annual reduction in the Coefficient of Variation (CV), so that in 2024, the CV in each developing country is 10% lower than in 2015. This means that poorer households consume *more* for a given level of income, with those at the other extreme consuming less.

8. This scenario assumes that higher agricultural sector productivity – in the form of higher yields in crops and livestock – translates into *increased* output as opposed to producing the same output more efficiently.

9. This section benefited from contributions by Laetitia Leroy, Jean Chateau, and Gerardo Aragon, in addition to Daniel Mason-D'Croz from the International Food Policy Research Institute (IFPRI).

10. In this chapter, the effects of climate change are explored based on projections from two different climate models – the Hadley and the IPSL. These are models developed by the Hadley Centre for Climate Prediction and Research at the United Kingdom Met Office (Hadley) and from the Institut Pierre Simon Laplace (IPSL) in France. The effects of the different climate projections, in terms of changes in the monthly averages of regional temperatures and precipitation, are translated into possible effects on yields by using the DSSAT crop model. Information is also drawn from the international Agricultural Model Inter-comparison and Improvement Project (AgMIP), which has developed a harmonised set of climate and policy scenarios. These elements give rise to the Hadley-DSSAT-Agmip and IPSL-DSSAT-Agmip projections in each of the scenario results presented. Both models are based on the "Representative Concentration Pathway" (RCP) 8.5. The RCPs, of which there are four, represent different carbon dioxide concentration trajectories developed by the Intergovernmental Panel on Climate Change in its fifth assessment report released in 2014 (IPCC, 2014). The RCP 8.5 scenario represents a 2 degree Celsius warming by 2046-65.

11. Different climate models produce different estimates of yield changes across the region, with the result that there is a degree of uncertainty in the individual country effects. Despite this, at regional level, the overall expected changes are relatively consistent.

12. Projected changes in crop yields by the two climate models (Hadley and IPSL) are very similar.

References

ADB (2009), *The Economics of Climate Change in Southeast Asia: A Regional Review*, Asian Development Bank, Manila.

ADB (2014), *Technologies to Support Climate Change Adaptation in Developing Asia*, Mandaluyong City, Philippines.

Château, J., R. Dellink and E. Lanzi (2014), "An overview of the OECD ENV-Linkages model: Version 3", *OECD Environment Working Papers*, No. 65, OECD Publishing, Paris, http://dx.doi.org/10.1787/5jz2qck2b2vd-en.

Cruz, R.V., H. Harasawa, M. Lal, S. Wu, Y. Anokhin, B. Punsalmaa, Y. Honda, M. Jafari, C. Li and N. Huu Ninh (2007), "Asia" in *Climate Change 2007: Impacts, Adaptation and Vulnerability. Contribution of Working Group II to the Fourth Assessment Report of the Intergovernmental Panel on Climate Change,* M.L. Parry, O.F. Canziani, J.P. Palutikof, P.J. van der Linden and C.E. Hanson (eds.), Cambridge University Press, Cambridge, United Kingdom, pp. 469-506.

FAO, IFAD and WFP (2015), *The State of Food Insecurity in the World. Meeting the 2015 International Hunger Targets: Taking Stock of Uneven Progress*, Food and Agriculture Organization, International Fund for Agricultural Development, and World Food Programme, Rome.

Francisco, H.A. (2008), "Adaptation to Climate Change: Needs and Opportunities in Southeast Asia", *ASEAN Economic Bulletin*, Vol. 25/1, pp. 7-19.

GRiSP (2013), *Rice Almanac*, 4th edition, Global Rice Science Partnership, International Rice Research Institute, Los Baños, Philippines, 283 pp.

Hijioka, Y., E. Lin, J.J. Pereira, R.T. Corlett, X. Cui, G.E. Insarov, R.D. Lasco, E. Lindgren, and A. Surjan (2014), "Asia", in *Climate Change 2014: Impacts, Adaptation, and Vulnerability. Part B: Regional Aspects, Contribution of Working Group II to the Fifth Assessment Report of the Intergovernmental Panel on Climate Change*, Barros, V.R., C.B. Field, D.J. Dokken, M.D. Mastrandrea, K.J. Mach, T.E. Bilir, M. Chatterjee, K.L. Ebi, Y.O. Estrada, R.C. Genova, B. Girma, E.S. Kissel, A.N. Levy, S. MacCracken, P.R. Mastrandrea, and L.L. White (eds.), Cambridge University Press, Cambridge, United Kingdom and New York, United States, pp. 1327-1370.

IGES, SEARCA (2012), *A Review of Issues and Challenges in Climate Change and Agriculture in Southeast Asia*, Institute for Global Environmental Strategies, Hayama, Japan, and Southeast Asian Regional Center for Graduate Study and Research in Agriculture, Los Baños, Philippines.

Ignaciuk, A. and D. Mason-D'Croz (2014), "Modelling adaptation to climate change in agriculture", *OECD Food, Agriculture and Fisheries Papers*, No. 70, OECD Publishing, Paris, http://dx.doi.org/10.1787/5jxrclljnbxq-en.

IMF (2014), *World Economic Outlook: Legacies, Clouds, Uncertainties,* International Monetary Fund, Washington, DC.

IPCC (2014), *Climate Change 2014: Impacts, Adaptation, and Vulnerability, Part B: Regional Aspects. Contribution of Working Group II to the Fifth Assessment Report of the Intergovernmental Panel on Climate Change,* Cambridge University Press, New York.

Lal, R. (2010), "Beyond Copenhagen: Mitigating climate change and achieving food security through soil carbon sequestration", *Food Security*, Vol. 2/2, pp. 169-177.

Lal, R. (2004), "Soil carbon sequestration impacts on global climate change and food security", *Science*, Vol. 304/5677, American Association for the Advancement of Science, Washington, DC, pp. 1623-27.

Lal, R., R.F. Follett, B.A. Stewart, and J.M. Kimble (2007), "Soil carbon sequestration to mitigate climate change and advance food security", *Soil Science*, Vol. 172/12, pp. 943-956.

OECD (2016), *Evolving Agricultural Policies and Markets: Implications for Multilateral Trade Reform*, OECD Publishing, Paris, http://dx.doi.org/10.1787/9789264264991-en.

OECD (2015a), *Aglink-Cosimo Model Documentation*, www.agri-outlook.org/abouttheoutlook/Aglink-Cosimo-model-documentation-2015.pdf.

OECD (2015b), *The Economic Consequences of Climate Change*, OECD Publishing, Paris, http://dx.doi.org/10.1787/9789264235410-en.

OECD (2014), *OECD Economic Outlook*, Vol. 2014/2, OECD Publishing, Paris, http://dx.doi.org/10.1787/eco_outlook-v2014-2-en.

OECD-FAO (2015), *OECD-FAO Agricultural Outlook 2015*, OECD Publishing, Paris, http://dx.doi.org/10.1787/agr_outlook-2015-en.

Robinson, S., D. Mason-D'Croz, S. Islam, T.B. Sulser, R. Robertson, T. Zhu, A. Gueneau, G. Pitois, and M. Rosegrant (2015), "The International Model for Policy Analysis of Agricultural Commodities and Trade (IMPACT); Model description for version 3", *IFPRI Discussion Paper*, International Food Policy Research Institute, Washington, DC.

Sheehy, J. and P.L. Mitchell (2015), "Calculating maximum theoretical yield in rice", *Field Crops Research*, Vol. 182, Elsevier, Amsterdam, pp. 68-75.

Smeets Kristkova, Z., M. Van Dijk and H. Van Meijl (2016), "Projections of long-term food security with R&D driven technical change – A CGE analysis", *NJAS – Wageningen Journal of Life Sciences*, Vol. 77, Elsevier, Amsterdam, pp. 39-51.

Tallard, G., P. Liapis and G. Pilgrim (2016), "The implications of agricultural trade and market developments for food security", *OECD Food, Agriculture and Fisheries Papers*, No. 95, OECD Publishing, Paris, http://dx.doi.org/10.1787/5jlr579rkqwk-en.

Yusuf, A.A. and H.A. Francisco (2009), *Climate Change Vulnerability Mapping for Southeast Asia*, Economy and Environment Program for Southeast Asia, Singapore.

Annex 2.A1

Detailed undernourishment projections

Table 2.A1.1. Number of undernourished by region (millions)

	Baseline					Income		Productivity		Inc+Prod		Access	
	1990-92	2000-02	2010-12	2015	2024	2024	Diff	2024	Diff	2024	Diff.	2024	Diff.
World	1 010.6	929.6	820.7	788.3	635.7	603.4	32.3	582.1	53.6	553.0	82.7	497.0	138.7
Developing	990.7	908.4	805.0	773.7	621.1	588.8	32.3	567.5	53.6	538.4	82.7	482.4	138.7
Africa	181.7	210.2	218.5	228.9	234.9	225.7	9.2	230.3	4.7	221.3	13.6	207.0	27.9
Latin America and Caribbean	66.1	60.4	38.3	35.0	ns	ns	1.5	ns	2.3	ns	3.6	ns	8.5
Asia and the Pacific	742.9	637.8	548.2	509.8	354.0	332.5	21.5	307.4	46.6	288.5	65.5	251.8	102.2
China	289.0	211.2	163.2	137.8	105.4	100.3	5.0	ns	35.5	ns	39.3	ns	33.7
India	210.1	185.5	189.9	187.5	104.4	94.5	9.9	101.4	3.1	92.6	11.8	ns	34.9
Southeast Asia	137.2	117.4	72.1	60.0	46.7	45.5	1.2	43.2	3.5	42.0	4.7	ns	14.4
Brunei Darussalam	ns	ns	ns	ns	ns	ns	0.0	ns	0.0	ns	0.0	ns	0.0
Cambodia	3.0	3.6	2.5	2.3	1.8	1.8	0.0	1.7	0.1	1.6	0.1	1.3	0.5
Indonesia	35.9	38.3	26.9	18.4	ns	ns	0.3	ns	1.2	ns	1.5	ns	4.7
Lao PDR	1.9	2.1	1.4	1.3	1.0	1.0	0.0	0.9	0.1	0.9	0.1	0.8	0.2
Malaysia	1.0	ns	ns	ns	ns	ns	0.0	ns	0.1	ns	0.1	ns	0.2
Myanmar	26.8	24.3	9.4	8.0	6.1	6.1	0.0	5.8	0.4	5.7	0.4	4.5	1.7
Philippines	16.7	16.1	12.7	13.9	14.8	14.5	0.3	13.9	0.9	13.6	1.2	11.4	3.4
Thailand	19.8	11.6	6.0	4.9	ns	ns	0.1	ns	0.2	ns	0.3	ns	1.0
Viet Nam	32.1	20.7	12.2	10.5	7.5	7.1	0.4	6.8	0.7	6.4	1.1	ns	2.7

Note: 'ns' refers to countries with levels of undernourishment below 5%.
Source: OECD estimates based on AGLINK-COSIMO.

Table 2.A1.2. Percentage of undernourished by region (%)

	Baseline					Income		Productivity		Inc+Prod		Access	
	1990-92	2000-02	2010-12	2015	2024	2024	Diff	2024	Diff	2024	Diff.	2024	Diff.
World	18.7	15.0	11.7	10.8	7.9	7.5	-0.4	7.3	-0.7	6.9	-1.0	6.2	-1.7
Developing	23.3	18.2	14.0	12.8	9.3	8.8	-0.5	8.5	-0.8	8.1	-1.2	7.2	-2.1
Africa	28.1	25.4	20.7	19.6	16.4	15.8	-0.6	16.1	-0.3	15.5	-1.0	14.5	-2.0
Latin America and Caribbean	14.6	11.3	6.4	5.6	<5.0	<5.0	-0.2	<5.0	-0.3	<5.0	-0.5	<5.0	-1.2
Asia and the Pacific	23.6	17.5	13.4	12.0	7.7	7.2	-0.5	6.7	-1.0	6.3	-1.4	5.5	-2.2
China	23.9	16.0	11.7	9.6	7.1	6.8	-0.3	<5.0	-2.4	<5.0	-2.7	<5.0	-2.3
India	23.7	17.5	15.6	14.6	7.4	6.7	-0.7	7.2	-0.2	6.6	-0.8	<5.0	-2.5
Southeast Asia	30.6	22.3	12.1	9.6	6.8	6.6	-0.2	6.3	-0.5	6.1	-0.7	<5.0	-2.1
Brunei Darussalam	<5.0	<5.0	<5.0	<5.0	<5.0	<5.0	0.0	<5.0	-0.1	<5.0	-0.2	<5.0	-0.6
Cambodia	31.9	28.8	17.1	14.6	9.9	9.9	0.0	9.3	-0.6	9.3	-0.6	7.3	-2.6
Indonesia	19.7	18.1	11.0	7.2	<5.0	<5.0	-0.1	<5.0	-0.4	<5.0	-0.5	<5.0	-1.7
Lao PDR	43.2	38.2	21.5	18.6	12.5	12.4	0.0	11.7	-0.8	11.7	-0.8	9.6	-2.9
Malaysia	5.3	<5.0	<5.0	<5.0	<5.0	<5.0	0.0	<5.0	-0.1	<5.0	-0.2	<5.0	-0.6
Myanmar	62.6	49.7	17.9	14.7	9.9	9.9	0.0	9.4	-0.6	9.3	-0.6	7.3	-2.7
Philippines	26.3	20.3	13.4	13.6	12.6	12.4	-0.2	11.8	-0.8	11.6	-1.0	9.7	-2.9
Thailand	34.7	18.4	9.0	7.3	<5.0	<5.0	-0.1	<5.0	-0.3	<5.0	-0.4	<5.0	-1.5
Viet Nam	45.6	25.3	13.6	11.3	7.5	7.1	-0.4	6.8	-0.7	6.4	-1.1	<5.0	-2.7

Note: "<5.0%" refers to countries with levels of undernourishment below 5%.
Source: OECD estimates based on AGLINK-COSIMO.

Table 1. Table 2.A1.3. Depth of undernourishment by region (kcal/person/day)

	Baseline					Income		Productivity		Inc+Prod		Access	
	1990-92	2000-02	2010-12	2015	2024	2024	Diff	2024	Diff	2024	Diff.	2024	Diff.
Asia and the Pacific													
China	473	341	227	172	83	69	-14	0	-83	0	-83	0	-83
India	387	345	288	270	95	70	-25	88	-8	65	-30	0	-95
Southeast Asia													
Brunei Darussalam	0	0	0	0	0	0	0	0	0	0	0	0	0
Cambodia	450	379	290	246	151	150	-1	136	-15	135	-16	70	-80
Indonesia	343	338	147	77	0	0	0	0	0	0	0	0	0
Lao PDR	631	545	317	294	198	197	-1	183	-15	182	-16	125	-73
Malaysia	0	0	0	0	0	0	0	0	0	0	0	0	0
Myanmar	727	617	286	249	150	149	-1	135	-15	134	-16	66	-84
Philippines	447	356	246	259	238	232	-5	221	-17	215	-22	151	-86
Thailand	531	339	101	87	0	0	0	0	0	0	0	0	0
Viet Nam	730	387	262	201	86	69	-16	57	-28	41	-45	0	-86

Source: OECD estimates based on AGLINK-COSIMO.

Table 2.A1.4. Market impacts for crop products

	Food Consumption					Net Trade				
	Baseline (bln kcal/day)		Inc.	Prod.	Inc+ Prod	Baseline (bln kcal/day)		Inc.	Prod.	Inc+ Prod
	2015	2024	Difference from 2024 baseline (%)			2015	2024	Difference from 2024 baseline (%)		
World	16 880.0	19 202.0	0.7	2.3	3.0					
Developing	13 732.6	15 971.0	0.9	2.3	3.2	-1 805.3	-2 233.7	1.2	-15.2	-14.7
Africa	2 444.4	3 101.7	1.1	0.8	1.9	-878.3	-1 140.3	0.1	-1.8	-1.9
Latin America and Caribbean	1 418.5	1 586.3	1.1	2.4	3.4	682.8	982.7	2.3	-2.9	0.0
Asia and the Pacific	9 851.6	11 262.6	0.9	2.7	3.6	-1 598.4	-2 063.4	2.4	-16.9	-14.9
China	3 466.6	3 688.3	0.4	7.0	7.5	-1 298.3	-1 561.6	2.8	13.0	16.0
India	2 698.7	3 283.3	1.6	0.1	1.6	-207.2	-365.8	11.6	-58.3	-48.0
Indonesia	753.7	872.6	0.3	1.2	1.6	457.5	560.0	5.2	37.6	43.4
Malaysia	67.4	79.1	0.2	1.1	1.2	345.4	371.0	4.0	27.7	32.0
Philippines	218.5	253.6	0.2	1.3	1.5	-37.9	-57.6	-0.5	1.8	1.4
Thailand	176.7	191.6	0.3	1.0	1.3	252.6	322.7	0.4	14.4	15.0
Viet Nam	233.1	257.3	0.5	1.9	2.5	22.8	21.2	-20.9	63.2	42.2

Note: "Crop products" refers to wheat, coarse grains, rice, vegetable oil, roots and tubers, sugar, sweeteners.
Source: OECD estimates based on AGLINK-COSIMO.

Table 2.A1.5. Market impacts for livestock products

	Food Consumption					Net Trade				
	Baseline (bln kcal/day)		Inc.	Prod.	Inc+ Prod	Baseline (bln kcal/day)		Inc.	Prod.	Inc+ Prod
	2015	2024	Difference from 2024 baseline (%)			2015	2024	Difference from 2024 baseline (%)		
World	3 096.2	3 569.6	1.7	4.0	5.7					
Developing	2 163.0	2 578.8	2.3	5.4	7.8	-33.7	-65.4	14.9	-57.7	-46.2
Africa	159.9	210.3	2.9	0.4	3.3	-20.5	-35.7	4.1	-10.7	-6.6
Latin America and Caribbean	329.0	380.5	2.0	2.7	4.7	29.7	36.6	-1.5	25.5	24.0
Asia and the Pacific	1 672.8	1 986.7	2.3	6.5	8.9	-42.4	-65.8	11.8	-37.4	-29.1
China	942.8	1 055.7	2.6	10.0	12.8	-6.3	-10.8	45.8	-106.2	-77.9
India	262.6	358.2	2.3	3.8	5.8	10.3	10.7	-2.4	10.2	8.8
Indonesia	44.1	53.3	1.3	1.1	2.4	0.5	1.1	-4.3	13.6	10.3
Malaysia	15.2	18.5	0.9	1.8	2.6	-1.4	-2.2	3.9	-21.1	-17.4
Philippines	33.6	40.1	1.9	0.8	2.7	-1.9	-4.1	6.3	-37.9	-32.9
Thailand	21.8	24.2	1.6	0.7	2.3	5.1	6.4	-0.8	24.0	22.9
Viet Nam	53.3	68.0	3.0	1.4	4.4	-2.7	-5.8	9.8	-16.9	-8.4

Note: "Livestock products" refers to bovine, pork, poultry, sheep, eggs, fresh dairy products, butter, cheese, casein, milk powders, fish.

Source: OECD estimates based on AGLINK-COSIMO.

Chapter 3

Stocktake of food security polices in ASEAN

This chapter presents a stocktake of the range of agriculture-related food security policies used across the ten Association of Southeast Asian Nations (ASEAN) members. It first presents the policy landscape, describing the range of interventions related to self-sufficiency targets, trade policies, public stockholding programmes and various interventions in domestic markets. Second, it provides a review of findings from existing research on the effectiveness and efficiency of these policies. Third and finally, a brief review of regional interventions for food security is presented.

Key points

- Association of Southeast Asian Nations (ASEAN) food security policy can be characterised as "rice-centric", meaning that rice production by both large and small rice producers is emphasised.
- The policies used are generally related to attempts to spur domestic production in importing countries through the use of price support, trade barriers and input subsidies; and, for exporting countries, interventions in export markets – via taxes, bans and licencing arrangements, for example – along with attempts to "lock-in" a certain quantity of rice production.
- Reviews have suggested that many current policies create inefficiencies in resource allocation within the economies, discourage private investment by creating greater uncertainties, and impose significant budgetary costs on governments – for which there are significant opportunity costs.
- The price effects of a number of policies also suggest that they are unlikely to be effective in helping to address food security for vulnerable households.
- Moreover, the ineffective nature of the support on farm incomes of the poorest, and in a number of cases the incidence of support accruing to otherwise food-*secure* households, suggests that even for poor rural producers, the long-run impacts on food security are questionable.

3.1 Introduction

This chapter provides a stocktake of agricultural, agro-food trade and other agricultural commodity-based policies used by the Association of Southeast Asian Nations (ASEAN) countries to pursue food security objectives.[1] Policies within the scope of the chapter suggest that governments in the region most often view food security from an availability perspective and, within that, in terms of the availability of domestically-produced rice (Alavi et al., 2012). This perspective has spurred the use of policies targeted towards achieving self-sufficiency in at least rice or across a number of staple products in some countries. The use of such policies has increased since the 2007/08 food price crisis, with governments often framing the push for self-sufficiency around a desire to reduce their vulnerability to world price movements of the like observed during this period.

It should be noted, however, that it is often the case that more efficient and effective responses to food security are to be found *outside* agriculture-related policies. As such, policy makers exploring alternatives to those policies that are reviewed here and found to be ineffective or inefficient, should cast the "policy net" wider if better outcomes are to be achieved.

This chapter is structured as follows: Section 3.2 provides a review of existing agricultural, agro-food trade and other agricultural commodity-based food security policies. The effects of these policies as detailed in existing research is explored in Section 3.3. Section 3.4 describes some regional responses to food security, such as the ASEAN Plus Three Rice Reserve scheme, along with other policy measures of note. Finally, concluding comments are made in Section 3.5.

3.2 The policy landscape

Across the ASEAN region, policies targeted at food security tend to focus on the availability of domestically-produced rice (Alavi et al., 2012). The focus on rice has spurred the use of policies that are targeted towards achieving self-sufficiency in at least rice, or across a number of staple products. The interest in rice by governments in the region is driven by its importance to both consumers and producers. High rates of rice consumption and the dominance of the sector in overall production (Chapter 1) mean that rice and its price are critical for incomes (for producers) and consumption (all households).

Food security policies in the region also tend to have multiple objectives. On one hand, they attempt to spur production towards self-sufficiency through price and input incentives for producers. On the other, they are also often directed at ensuring food availability and access for vulnerable consumers. These goals can compete, creating a tension between the interests of producers in moving towards self-sufficiency, and the interests of vulnerable consumers and their ability to access affordable food (Gillson and Fouad, 2015).

Self-sufficiency targets

The use of policies directed at achieving some level of self-sufficiency has increased since the world food price crisis of 2007/08. The push towards self-sufficiency has often been framed around a desire to no longer be vulnerable to world price movements similar to those that were seen during this period – especially for rice – despite the fact that it was largely *policy* factors, and not global imbalances in supply and demand, that explained the food price spike (Box 3.1).

Policies that look to domestic solutions for food security have continued to be the major policy response by ASEAN countries since the crisis. Indonesia, for example, has continued in its push for self-sufficiency, not only in rice, but also soybeans, maize, sugar and beef (OECD, 2012). Other countries, such as Viet Nam, the Philippines and Malaysia, are also committed to either maintaining a certain level of rice production, in the case of Viet Nam, or increasing it towards self-sufficiency levels, such as in the Philippines and Malaysia. While all these countries also use less-distortionary policy interventions, ranging from investments in research, development and extension to investments in infrastructure, the major policy tools employed relate to either input subsidies or market price support or a combination of both (OECD, 2012; OECD, 2017; Alavi et al., 2012).

Self-sufficiency policies are often supported by production targets for a particular commodity or set of commodities. Across the ASEAN group, almost all countries have some form of self-sufficiency related target (Table 3.1). Within this, Indonesia has the most ambitious set of targets, aiming for self-sufficiency across all main staple products. The Philippines is the only country which has coupled a drive for self-sufficiency in its two main staple crops (rice and maize) with attempts to diversify individual diets by encouraging consumption of a wider set of food products (Philippines Government, 2011).

Box 3.1. The 2007/08 world food price spike

Analysis by the OECD and others has concluded that market imbalances were not sufficient to explain the price spikes that occurred worldwide in 2007/08. Instead, government policy actions contributed significantly.

Food price rises were driven by a confluence of mutually re-enforcing longer-term structural changes, short-term market shocks, and policy responses (OECD, 2008; Piesse and Thirtle, 2009; Naylor and Falcon, 2010; Headey, 2011). In addition to underlying structural changes to world agricultural markets resulting from rising levels of food and feed demand, falling stock-to-use ratios and the channelling of increasing shares of production to biofuels production, world markets were hit by a number of short-term shocks that placed further upwards pressure on prices. Droughts in key grain-producing regions and other weather effects, exchange rate movements and hoarding and panic buying by private agents, all helped to spur already rising prices. Added to this, government policy interventions through trade restrictions and import measures, coupled with panic purchases by some governments, helped to create the spike in prices that was observed. Government policies surrounding biofuels mandates and subsidies also contributed. Price rises were particularly evident for wheat, coarse grains, rice and oilseed crops, all of which experienced strong real price growth between 2005 and 2010.

In terms of rice, Alavi et al. (2012) suggest that government policy was the key driver of the rapid price increases experienced in the ASEAN region (and more generally: Anderson, Ivanic and Martin, 2013; Headey, 2011). They point to strong production growth in the lead up to 2007, which continued throughout the crisis. Indeed, production growth continued to grow rapidly and exceed demand growth in the region. Preceding the crisis, stock levels were also reported to be sufficient, with stock-to-use ratios for rice in particular not especially low by past standards (Alavi et al. 2012; Dawe and Slayton, 2010). In fact, by the end of the crisis, stock-to-use levels had actually increased.

Of particular importance were the effects of the initiation of export restrictions by major producers. India was the first major producer to put such restrictions in place in October 2007. Between then and the first quarter of 2008, other exporters, including Viet Nam, Brazil, Cambodia, China, Egypt and Pakistan all put some form of export restriction in place (Alavi et al., 2012). These actions, coupled with short-term buying by the Philippines, were the major drivers of the price spike. Ultimately, the actions that were intended to place downward pressure on rice prices actually caused them to rise *higher* than would likely have occurred in the absence of government intervention (Anderson, Ivanic and Martin, 2013). As such, these policies actually worked *against* food security.

These targets are further underpinned by a wide variety of output, input and trade-related interventions. Beyond the supply side, some countries have also sought to intervene in markets with the expressed aim of stabilising prices for the benefit of both producers and consumers. This intervention has taken the form of public stockholding policies, most notably in the Philippines and Indonesia. The following section briefly reviews, by broad programme classification, the major policies used in the region that are framed by governments as directly targeting food security.

Table 3.1. Self-sufficiency targets of ASEAN members

Country	Self-sufficiency target
Brunei Darussalam	Rice self-sufficiency of 20% by 2015 and 60% over the longer term (2035)
Cambodia	**No specific self-sufficiency targets**
Indonesia	Complete self-sufficiency (100% of domestic production) targets for rice, maize and soybeans by 2017 and beef and sugar by 2019
Lao PDR	Production targets for rice ~ 4.2 million tonnes by 2015 and rate of increase targets for other products. Absolute quantity targets of food production for some commodities
Malaysia	Self-sufficiency targets for rice of 90% of domestic consumption plus other production targets
Myanmar	**No specific self-sufficiency targets**
Philippines	Self-sufficiency in rice previously set for 2013 but later abandoned set year target. Self-sufficiency in maize production by 2013
Singapore	Increase self-sufficiency levels to 30% for eggs, 15% for fish and 10% for leafy vegetables
Thailand	**No specific self-sufficiency targets**
Viet Nam	Maintain a 2.5% rice yield increase per year until 2020, and the set aside of 3.8 m ha of land specifically for rice production

Sources: APEC Policy Support Unit (2012), MOA (2015), various government websites.

Trade-related policies

Some ASEAN countries have adopted, either formally or on an *ad hoc* basis, policies to manage the import or export of rice – and, at times, of other products – either as a means to limit domestic price growth or volatility, or to ensure adequate supplies on domestic markets.

Import barriers

The majority of ASEAN countries, with the exception of Singapore, apply a range of tariffs on agro-food imports (Figure 3.1). Of those with tariffs, rates are highest for Thailand and Lao PDR, and lowest for Malaysia. However, for most products, the application of tariffs is not due to concerns over food security. For some, applied rice tariffs are also high, including in a number of strongly-exporting countries, such as Thailand and Viet Nam, despite little imports.

Some trade barriers, however, *do* relate to food security. In this respect, import barriers in the form of tariffs or restrictions are often used to support self-sufficiency policies so as to create a more favourable environment for domestic producers. In Indonesia, import restrictions, in the form of tariffs, restrictions such as quotas and import bans, and licensing arrangements are used to support the self-sufficiency targets related to rice, maize, soybeans, beef and sugar. For example, beef imports are controlled via quotas, while rice imports are controlled centrally and not based on market demand. Furthermore, Indonesia has also been able to significantly reduce and control rice imports through its licensing schemes.

Policies like those of Indonesia are also in place in Malaysia and the Philippines. Rice imports are controlled by state-owned enterprises, or regulated monopolies, that control the quantity of imports. Imports are managed to avoid domestic shortages and to manage domestic prices. These policies limit the supply of rice to the market, and provide a means for governments to increase producer prices whilst also controlling excessive price increases by supplementing domestic supply with imports.

Not all policy action around imports is directed at limiting trade flows. During the 2007/08 food crisis, a number of countries altered pre-existing tariffs in an attempt to make food more affordable for domestic consumers. For example, both Indonesia and Viet Nam reduced import tariffs on specific commodities in order to limit price escalation (Jones and Kwieciński, 2010). Similar action has been repeated at other times of high prices. Indonesia, for example, suspended its 5% import duty on wheat in 2011, in the face of high domestic prices (APEC Policy Support Unit, 2012).

Figure 3.1. Import tariffs in ASEAN

Most Favoured Nation applied rates, selected ASEAN members, 2013

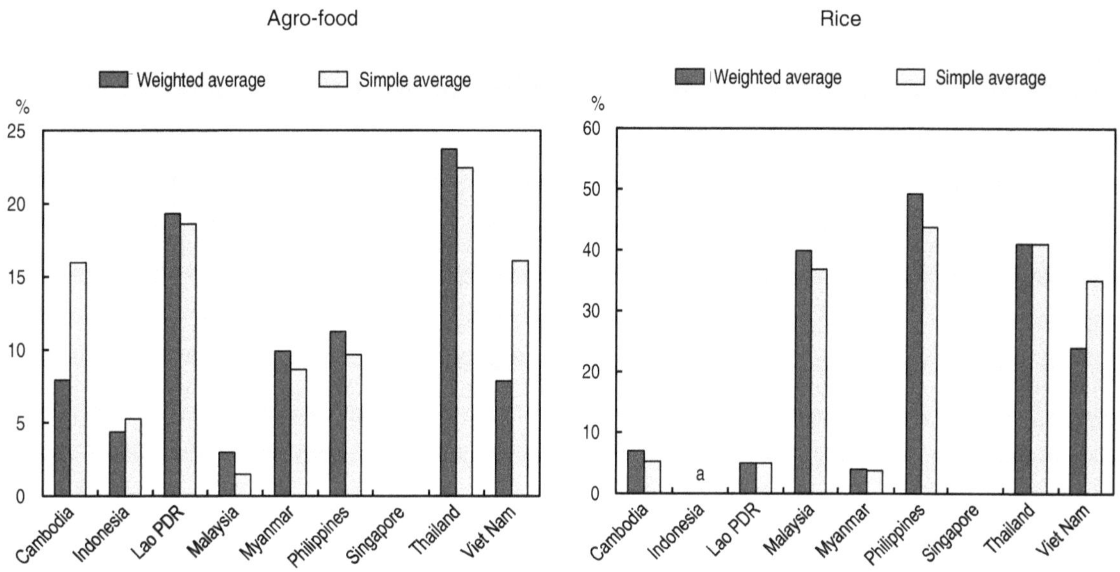

Notes: Data for Lao PDR is from 2008 and 2012 for Cambodia. a: While rice tariffs are not used in Indonesia, import barriers exist, with imports centrally managed. Singapore does not impose any tariffs.
Source: WITS (2016), *World Integrated Trade Solution,* https://wits.worldbank.org/WITS/WITS/Restricted/Login.aspx.

Export restrictions

Some countries with net surpluses have used export restrictions in response to price rises on either domestic or international markets. Myanmar, for example, imposed export restrictions on rice, despite no formal export-related food security policy being in place. Export bans were applied in 2004 and 2008 after a natural disaster, and limits to previously-licensed export amounts were put in place in 2011 and 2013. In 2001, an informal agreement was reached with millers and exporters to release stocks and restrict exports (World Bank, 2014).

Lao PDR has also made use of export bans in circumstances of rising prices. Export bans are possible at both provincial and national level in Lao PDR (Eliste and Santos, 2012). Export bans were used in 2010 in response to rapidly rising prices (Durevall and van der Weide, 2014). Cambodia, on the other hand, has not imposed export restrictions, despite having experienced shortfalls in rice supply in the past (Eliste and Santos, 2012).

A number of exporting countries also use licensing arrangements to control the level and value of exports, in order to manage domestic prices and supplies in the long term. Again, the focus is primarily on rice. For example, in Viet Nam, rice exports are centrally controlled through licensing arrangements and state-owned enterprises.

Public stockholding programmes and other reserve schemes

Not all public stockholding programmes are the same. Broadly speaking, public stocks can be classified into three categories: buffer stocks, social safety net stocks and emergency stocks (Box 3.2).

Box 3.2. Types of public food stocks

Buffer stocks are used to stabilise commodity prices, with the aim of protecting producers against price drops and/or consumers against price hikes.

Social safety net stocks function as working stocks for regular food distribution programmes. These stocks aim to provide social safety nets for the impoverished and the chronically food-insecure.

Emergency stocks are maintained to provide assistance during transitory food shortages and crises which are caused by sudden supply shocks, such as natural disasters. Such stocks are generally kept at relatively low levels and thus have limited impacts on prices.

Source: Deuss (2015).

Every ASEAN country has a type of stockholding policy. These vary in purpose, scope and impacts on markets and production. The type of stockholding policy in place varies generally along importing and exporting lines. Stockholding policies in importing countries aim at both domestic price stabilisation and insurance against emergency events. For exporting countries, stocks are often established both for emergency supply management and also as insurance against rapid price spikes as opposed to price stabilisation in general. Stocks in exporting countries are generally insufficient for stabilisation purposes.

However, there are notable exceptions to these generalisations. Singapore, an importing country, applies a policy primarily aimed at emergency management, while Thailand, an exporter, has previously held stocks in order to influence world market prices and thus increase domestic prices. A brief description of the various national stockholding policies is set out below. The regional initiative, the ASEAN Plus Three Emergency Rice Reserve, is discussed in detail in Section 3.4.

- *Brunei Darussalam*: Brunei Darussalam maintains a strategic national stock equivalent to six months of demand. Given the large share of imports in domestic consumption, this stock is held for emergency and strategic purposes.
- *Cambodia*: In 2012, the Cambodian government established the Cambodian Food Reserve System, which holds stocks of rice to be distributed in case of emergencies (FAO, 2014). The reserve holds 16 000 tonnes of rice or equivalent, along with rice and vegetable seeds. The holding of rice is split between a cash equivalent of 6 000 tonnes and a physical stock of 10 000 tonnes. The reserve is only used to provide assistance during emergencies or crisis.
- *Indonesia*: Buffer stocks in Indonesia are managed by Badan Urusan Logistik (BULOG). BULOG was created in 1967 under a rice price stabilisation policy that was implemented using a price band. From 2005 onwards, the focus shifted from a price band policy to setting a floor or procurement price while selling subsidised rice to impoverished people via the Raskin programme (Deuss, 2015). The Indonesian government also controls imports to complement both its price stabilisation objectives and those of supporting producer farm gate prices. BULOG has four functions: i) it is the provider of subsidised rice to poor households under the Raskin programme (see below); ii) it intervenes in markets whenever the government deems that retail prices of rice are above tolerable levels; iii) it manages the government's emergency reserve; and iv) it sets minimum purchase prices for farmers through its purchasing responsibilities. Stocks held by BULOG are in the order of 1 million tonnes to undertake its first three functions, of which around half is for emergency management (OECD, 2012). However, over recent years, it has been reported that the Government of Indonesia has instructed BULOG to maintain a minimum secure stock of around 2 million tonnes of rice (USDA FAS, 2015).

- *Lao PDR*: In 2009, the Government of Lao PDR launched a pilot national rice reserve for emergencies. This was done through the provision of subsidised credit to millers via state-owned banks to encourage them to maintain a minimum rice stock (Eliste and Santos, 2012). This scheme, titled the National Rice Reserve, seeks to maintain stocks of around 5 000 tonnes. The reserve also includes a component for seed reserves and rice distribution as part of poverty-related safety nets (Eliste and Santos, 2012). In 2015, the Government of Lao PDR announced its intention to strengthen the scheme to hold 400 000 tonnes of paddy rice (around 3 months' supply) in its Agriculture Development Strategy to 2025 and Vision to the Year 2030 (MAF, 2015).

- *Malaysia*: Public rice stocks in Malaysia are managed by PadiBeras Nasional Berhad (BERNAS), a private company traded on the Kuala Lumpur Stock Exchange. BERNAS purchases paddy from farmers at a guaranteed minimum price, manages farm input subsidies, runs milling operations, maintains the nation's rice stockpile, and acts as the sole importer of rice. Its monopoly position was extended in 2011 for a further ten years (Deuss, 2015). BERNAS is contracted by the Malaysian government to maintain a rice stockpile for food security purposes of 292 000 tonnes – an increase from pre-2008 levels of 200 000 tonnes (Vengedasalam, Harris and MacAulay, 2011).

- *Myanmar*: The Government of Myanmar does not hold stocks in its own right, but rather makes use of a public-private partnership with the Myanmar Rice Federation to gain access to privately-held stocks (World Bank, 2014). The government co-ordinates with the federation to release stocks when prices are high. Stocks are reportedly released at market prices.

- *Philippines*: The National Food Authority (NFA) of the Philippines is tasked with stabilising rice prices by keeping farm gate prices high and retail prices reasonable for consumers. In addition, the NFA has the mandate to release rice from its stocks during emergencies and calamities in order to guarantee a stable rice supply. The authority is also responsible for the importation of rice to meet domestic production shortfalls and in distributing subsidised, mainly imported, rice to poor households. The NFA imports 35% of the import allocation of rice at duty-free rates, while 65% is imported by the private sector (Tobias et al., 2012).

- *Singapore:* In Singapore, the Rice Stockpile scheme is operated through regulations imposed on rice traders (Singapore Government, 2015). Rice is a controlled commodity under the Price Control (Rice) Act 1990, with licences required to import and trade (including re-export) rice in the wholesale market. The Rice Stockpile scheme requires importers to store two months' worth of imports in a government-designated private warehouse. The importers continue to own the stocks – however, the government can acquire the stocks subject to compensation. Importers are then responsible for rotating the stored rice, with a limit of one year imposed on the length of time that stocks can be stored under the scheme (Briones, 2014).

- *Thailand:* In 2011, the Thai government pledged to pay domestic rice producers 50% more than the market price (Permani and Vanzetti, 2014). The resulting supply response led to the accumulation of a large holding of stocks by the Thai government. Unlike in some other ASEAN member countries, the purpose of these stocks was not to stabilise prices, nor was accumulation driven by concerns over transitory food shortages or emergencies. Instead, the stocks resulted from a producer support policy and a reported belief that by withholding the supply of rice to the world market, the Thai government may be able to influence world price (discussed further in Section 3.3). The Thai government subsequently abandoned the scheme, and looked to sell much of its accumulated stocks. Nevertheless, in late 2016, in a bid to limit supply and bolster prices, the government announced that it would provide rice farmers with loans totalling over USD 1 billion, providing that they agreed to store their crop for six months (Financial Times, 2016).

- *Viet Nam*: In Viet Nam, a government-operated State Reserve holds stocks of rice and various production inputs (OECD, 2015c). The State Reserve is designed as an emergency reserve, and seeks to provide insurance against a number of risks, such as natural disasters and various epidemics. The reserve scheme is tasked with holding around 500 000 tonnes of paddy and rice (paddy equivalent), 10 000 tonnes of rice seed, 1 500 tonnes of maize seed and 130 tonnes of vegetable seeds, along with fertilisers and various pesticides and animal vaccines.

Influencing input prices

The provision of assistance to the agricultural sector via input subsidies is common the world over. Input subsidies are often provided in the belief that they will lead to increased production and food availability, and thereby increase food security. Assistance has been targeted at both physical and financial inputs. ASEAN countries vary considerably in the type and extent of their interventions.

The objectives of input subsidies are most often a combination of food security concerns and other policy objectives related to rural development and producer support. However, for countries such as Malaysia and Indonesia, strong links exist between such programmes and the objective of food security, due to the policy link between self-sufficiency and food security. As with other policy interventions, rice is the main commodity targeted.

In Malaysia, rice farmers can access subsidies for fertilisers, chemicals and for harvesting costs. Fertiliser subsidies come in the form of an allocation of free fertilisers to producers, with support for other incentives paid in monetary terms on a per hectare basis (Malaysian Department of Information, 2015; Harun, 2015).

Indonesia provides fertiliser subsidies for domestically-produced fertiliser. While having a long history dating back to 1979, the current arrangements have evolved from measures introduced in 2003, which were targeted at farmers producing on less than 2 hectares (OECD, 2012). Instead of payments to farmers, the subsidy is paid to fertiliser manufacturers (five state-owned companies) which are obliged to sell fertilisers to small-scale farmers at low prices. Fertiliser subsidies represent the largest government budgetary support programme for agriculture. In 2013, the value of this subsidy was around IDR 17.6 trillion (USD 1.7 billion), accounting for 41% of total budgetary expenditures provided to support the agricultural sector (OECD, 2015b). Indonesia has also recently provided subsidies for other inputs used in rice production, such as hand tractors and water pumps under its renewed focus on self-sufficiency.

In both Myanmar and the Philippines, fertiliser subsidies were part of government support to agriculture. However, both countries have phased out this type of support – in 2003 in Myanmar and in 2010 in the Philippines. Cambodia, Lao PDR, Singapore, and Thailand do not have explicit input subsidy arrangements. Viet Nam does provide subsidies related to irrigation, seeds and credit, amongst others (OECD, 2015c).

Influencing output prices

Beyond trade policy interventions that influence the price of food within domestic markets, a number of ASEAN governments have separate policies that seek to influence the prices received by domestic producers for key food security crops. The mechanisms used tend to vary, but most commonly, governments attempt to institute minimum prices through a system of public procurement of production, either directly or through an agent. Such policies are targeted at both maintaining returns for producers and reducing uncertainty. In doing so, it is believed that production will be more stable and investment in the sector will increase, thereby increasing the availability of food.

- *Indonesia*: Outside of trade policy, there is limited government intervention in Indonesian rice markets to influence producer prices. BULOG, which is responsible for the government's stockpile and rice for the Raskin programme, does purchase at set prices – however, the share of domestic production purchased is limited, and has ranged between a low of 4% in 2006 and a high of 9% in 2009 (OECD, 2012).
- *Lao PDR*: The Lao Agricultural Law of 1998 requires the State to maintain minimum prices and intervene in the market as a purchaser when needed. This is achieved through purchases of rice for public sector employees and military personnel, with around 25 000 tonnes purchased each year (Eliste and Santos, 2012). However, questions have been raised with regard to the contracting

arrangements used, including whether these have any influence on the prices received by farmers. Currently, contracts are formed between the government and millers (Eliste and Santos, 2012). Lao PDR has also attempted to introduce minimum prices for pigmeat by setting reference prices to which traders are supposed to adhere, but, as with rice, questions have been raised over the ability of the scheme to influence producer prices (World Bank, 2014).

- *Malaysia*: Malaysia's BERNAS, the privately-owned company responsible for the government's stockholdings, is also a purchaser of last resort for farmers (Gillson and Fouad, 2015). In this sense, it provides for some price floor, but, as in Indonesia and the Philippines, it is trade policy that is most influential on rice prices.

- *Myanmar*: In 2013, Myanmar adopted the Farmers' Rights Protection Act that sets out the possibility for the introduction of minimum prices for agricultural commodities such as rice. However, the World Bank reports that implementation details are not yet clear, and, given limited fiscal resources, a public procurement system such as that used in some other countries is unlikely to be feasible (World Bank, 2014).

- *Philippines*: Similar to Indonesia, the NFA has minimum purchase prices for rice. However, the NFA has traditionally purchased a lower share of domestic production than BULOG, meaning that much of the influence on domestic prices is driven through import controls (Gillson and Fouad, 2015).

- *Thailand*: The Thai government has variously sought to establish minimum prices for rice producers. In 2011, the government (re)instituted the paddy pledging scheme which paid domestic rice producers 50% more than the market price (Permani and Vanzetti, 2014). The scheme was abandoned due to rising costs and an accumulating stockpile of rice. However, it has been argued that the scheme was not related to food security concerns, but rather an attempt to increase world prices through stockpiling rice (Mahathanaseth and Pensupar 2014). The pre-2011 scheme provided credit (advance payment) for rice production that was "pledged" to the government in an attempt to provide market certainty for producers rather than influence prices (Chulaphan et al., 2012). More recently, in late 2016, it was announced that the government would provide rice farmers with loans totalling over USD 1 billion, providing that they agreed to store their crop for six months. On this occasion, it was clearly stipulated that the objectives of the loan were to help limit supply and support prices (Financial Times, 2016).

- *Viet Nam*: Since late 2009, Viet Nam has targeted a price for rice producers with a view to providing growers with a profit of more than 30% (OECD, 2015c). Prices are calculated regionally and based on average costs determined from surveyed farmers. The prices are "implemented" through another scheme that provides subsidised credit for the temporary storage of rice purchased at the target price during harvest time (all interest costs are paid by the government). However, this implementation of the scheme has meant it has had little influence on producer prices, with much of the benefits captured by exporters (OECD, 2015c).

Transfers to vulnerable consumers

A number of food-based safety net systems have been established across ASEAN as a means to improve access to food, either by vulnerable households or all households more generally. Programmes are either based around transfer systems or "supplementary" feeding arrangements under targeted programmes designed to increase food consumption by particular social groups.

Supplementary feeding programmes have been used in both the Philippines and Thailand. The Philippines has introduced two – a Healthy Start Feeding Program and a Food for School Program – while Thailand also provides a school-based intervention through its School Lunch Program (Hangzo, 2010).

The larger programmes are those which seek to provide subsidised food to consumers in general or to a specific set of poorer households. Most often, the focus of these programmes is on subsidised rice provision. Significant programmes of this type exist in Indonesia, Malaysia and the Philippines:

- *Indonesia*: The Raskin programme, originally called *Operasi Pasar Khusus*, first provided subsidised rice in July 1998 in order to assist households suffering from the effects of the Asian Financial Crisis (World Bank, 2012). Rice is distributed by BULOG along agreed quotas to over 50 000 regional distribution points where eligible households may purchase fixed quantities at below-market prices. Between 2005 and 2010, households could purchase a maximum of 15 kg of Raskin rice per month at 75% to 80% lower than the market price (World Bank, 2012).
- *Malaysia*: In 2009, the Malaysian government began a rice subsidy programme aimed at providing subsidised rice to impoverished households. The programme, titled SUBUR, or Rice Subsidy Programme for the People, provides cash vouchers that can be exchanged for subsidised rice (lower-quality rice). This was initially set at 30 kg a month (Malaysian Office of the Prime Minister, 2009).
- *Philippines*: The National Food Authority is tasked with providing subsidised rice to impoverished households alongside its price stabilisation responsibilities. Rice is sold to eligible households at below market prices. In 2008, for example, rice was sold at around 50% below prevailing market prices (Philippines Senate Economic Planning Office, 2010). Subsidised rice is reported to account for around 15% of total household consumption across the country.

Regulatory constraints on land use

Agricultural land is sometimes zoned in such a way that it is forbidden – or made difficult – to make use of that land for the production of other commodities, or for residential or industrial uses. In so doing, governments believe that a sufficient domestic production base can be maintained to ensure the adequate supply of staple foods.

Land use restrictions are used both in Viet Nam and Malaysia. In Viet Nam, 3.8 million hectares of land has been set aside for the production of rice only (OECD, 2015c). In Malaysia, in response to pressures posed by increasing demand for land due to urban development and to incentives to convert paddy land into palm oil production, land in granary regions has been similarly set aside exclusively for rice production.

Investments in infrastructure and other aspects of the enabling environment

A number of ASEAN governments have also made investments in the enabling environment as a means to reach either self-sufficiency targets or food security objectives. Investments are made to promote the efficiency of the sector and to spur productivity improvements. Much of the attention of governments in the region has been directed at the provision of irrigation infrastructure.

Indonesia, for example, has recently significantly expanded its investments in infrastructure as part of its renewed focus on self-sufficiency. With fiscal space created by the removal of its fuel subsidy, Indonesia has made a number of investments in irrigation infrastructure. Much of this is targeted towards rice production. In 2015, the Ministry of Agriculture committed IDR 4.2 trillion (USD 355 million) to rehabilitate irrigation canals covering an area of 1.5 million hectares, along with investments aimed at "optimising" 500 000 hectares of existing land for food production (OECD, 2015b). This increased investment is in addition to the current exemptions in place, whereby farmers are not charged for the cost of delivering water from the source to the tertiary system via primary and secondary canals. Similarly, Viet Nam allocates a large proportion of its total budgetary support for agriculture to irrigation.

Focusing on another aspect of the enabling environment, innovation, Singapore has invested in the "Food Fund". The Singapore government has invested SGD 20 million (USD 14 million) in the fund – initiated in 2009, with funds released in stages – to incentivise farms to explore new farming technologies that would ensure the resilience of the national food supply through the local production of three key food items: eggs, leafy vegetables and fish (Lim, 2013).

3.3 What have been the effects of food security policy interventions? Findings from existing research

The current set of food security policies used across ASEAN members has been examined in various levels of detail. Policy impacts differ across countries and programme types. The broad findings from these existing studies are explored in this section.

On price gaps

Overall, the effects of interventions in agricultural markets, most notably for rice, have led to higher and more stable prices in importing countries such as Indonesia, Malaysia and the Philippines (Figure 3.2). These higher prices effectively act as a tax on consumers, many of whom are food insecure, and do little to aid improvements in food security. It should be noted that the price gaps observed are unlikely to be entirely explained by policy interventions. For some ASEAN members, prices are also influenced by domestic characteristics, such as available infrastructure, short-term seasonal conditions and other market imperfections that create rigidities in the adjustment of prices. These effects mean that within each country, prices are likely to vary geographically. Particularly for less-developed countries, these "development gaps" can often result in depressed prices in rice-growing areas (an effective tax on producers) and higher prices in areas with high demand (an effective tax on consumers). Despite this, the size of the price gaps, the direction of change, and non-uniform trends relative to international markets, indicate that policy is the major driver in the observed gaps depicted in Figure 3.2.

Recent OECD analysis of Indonesia's policies demonstrates how such policies can have conflicting effects on domestic food security (OECD, 2015a). Overall, Indonesia's domestic rice prices were close to 70% higher than international prices in 2012-14 as a consequence of policy interventions, compared to just 8% higher in 2000-02 (OECD, 2015b). The current price support measures have contributed to undernourishment within poorer households by around 2.4 percentage points. Fertiliser and other input subsidies were also found to have only minor effects on decreasing rates of undernourishment, as they do not effectively decrease production costs and hence have limited effects on rice prices (Figure 3.3). In addition, as a result of current policy settings, Indonesian food security is even more susceptible to more frequent domestic economic and natural disaster risks than otherwise might be the case under more open market conditions. Policies are targeted towards decreasing the effect of international risks, but even for these, current policy tools were found to be less effective than the alternatives. For example, trade restrictions on rice can only help to avoid a surge of undernourishment in the case of a rice price spike on international markets, an event that is estimated to occur only once every 30 years.

Overall, the policies adopted by Indonesia were found to be less effective at reducing undernourishment rates in the face of a number of risks (OECD, 2015a). This, combined with the large fiscal costs of some programmes, such as fertiliser subsidies, and the efficiency costs of others, such as trade barriers and associated price support, suggests that alternative policies are likely to be more effective and efficient responses to improving the nation's food security. In exploring alternatives, OECD (2015a) highlighted the potential gains of moving towards policies that addressed food security through either food vouchers or targeted cash transfers (Figure 3.3). Both of these policies were found to have a greater potential effect on reducing undernourishment in Indonesia than the combined effects of the current policy mix. While these policies will better address the issues related to food security, to overcome a number of issues facing the agricultural sector, such as yield gaps and other issues of low productivity, other complementary policies would be required. Policies, such as those that promote sustainable productivity growth, will help improve farmer incomes and also highlight the need for a multi-faceted approach when trying to address food security.

Figure 3.2. Price gaps in ASEAN rice markets

Percentage difference to Thai wholesale price, Real prices USD, April 2001 to March 2015

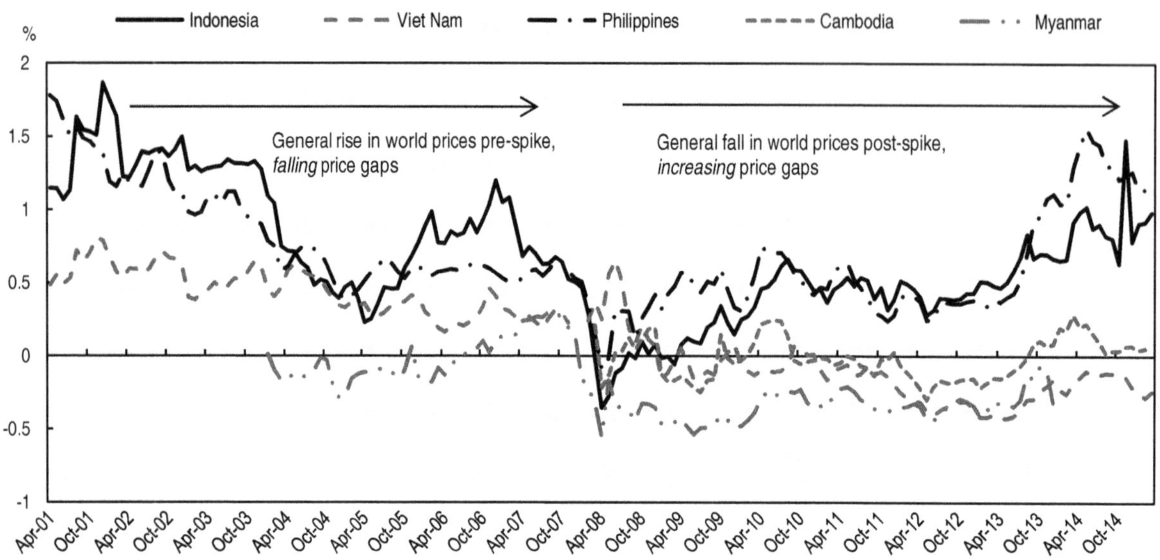

Notes: Price gap expressed as the difference in percentage terms between domestic wholesale price and Thai wholesale price. Real price series as shown in Figure 1.18 (Chapter 1).It should be noted that during the period of the paddy pledging scheme in 2011, the Thai wholesale price was artificially high, influencing the price gaps presented. This suggests that during the period from 2009 to 2011, price gaps relative to the international price are likely to be understated in high price markets and overstated in low price markets. Despite this, for the entire period, the Thai price represents a reasonable proxy for the world price. No adjustment for transportation cost has been included in the calculation. Gap for Indonesia is compared with Thai 15% broken price based on OECD (2012), for Philippines with 25% broken price based on Intal, Cu and Illescas (2012) and similarly so for Cambodia based on Maltsoglou, Dawe and Tasciotti (2010). The remainder are compared with Thai 5% price. Thai prices are based on price in Bangkok market.

Sources: Various sources: Indonesia: Indonesian Ministry of Agriculture (2015) data on rice (15% broken); Thailand: Bank of Thailand (2016) data on wholesale price (5% broken), www2.bot.or.th/statistics/BOTWEBSTAT.aspx?reportID=89&language=ENG; Viet Nam: International Grains Council (2016) data on 5% broken export rice price, www.igc.int/en/markets/marketinfo-prices.aspx; Philippines: Philippines Bureau of Agricultural Statistics (2015), regular milled rice wholesale price (25% broken based on Intal, Cu and Illescas [2012]); Cambodia: FAO (2016), *Food Security Portal*, www.foodsecurityportal.org/api/rice for rice (mix), which the FAO reports as low-quality rice (Maltsoglou, Dawe and Tasciotti [2010]); and Myanmar: Staple domestic market variety (5% broken), from World Bank (2014).

For the Philippines, Cororaton (2004) found that the controls placed on imports of rice into the country contributed to higher rates of poverty than would be observed in open markets. Given the link between poverty and food security, these results suggests that the system of supported rice prices is likely to be contributing to higher levels of food insecurity than otherwise would be the case.

Higher domestic prices for rice also create transfers between different groups, in particular from consumers (and disproportionately poor consumers) to producers (again, disproportionately large producers with a large share of production). In Thailand, for example, during the failed 2011 paddy pledging scheme which temporarily increased Thai domestic prices above world prices, it was estimated that USD 8.5 billion was transferred from consumers and government to producers (Permani and Vanzetti, 2014).

Figure 3.3. Effects of policies on food security in Indonesia

Probability-weighted impacts based on food insecurity risks

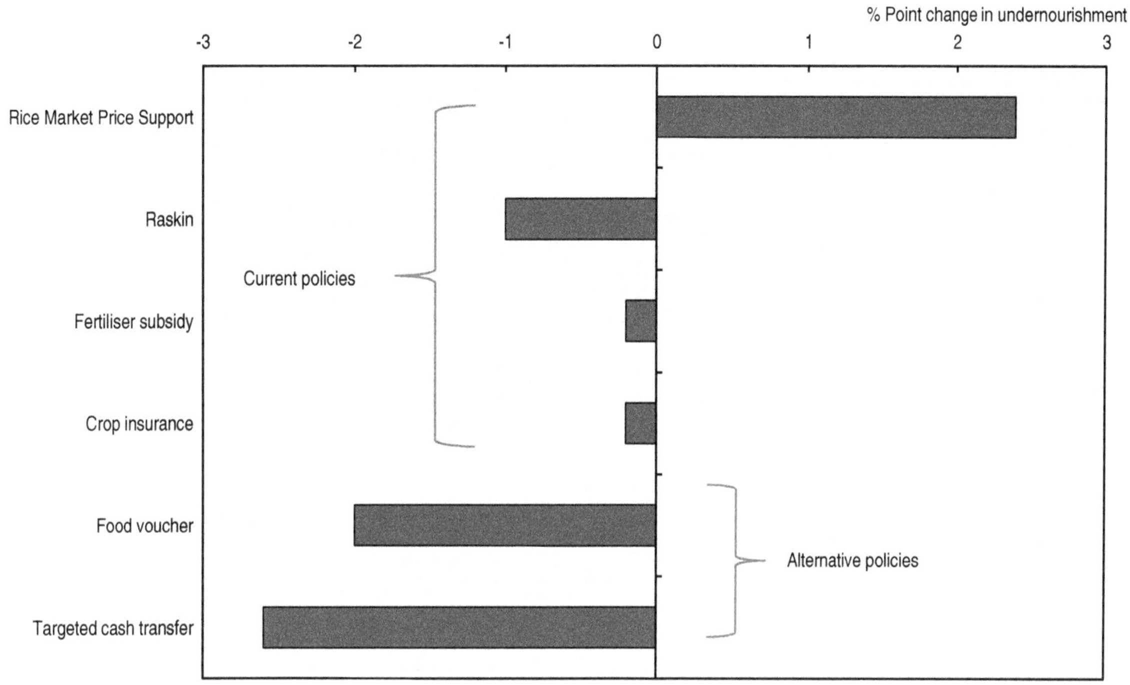

Notes: Insecurity risks include natural disasters, crop failure and a macroeconomic crisis, along with international rice and fuel price spikes. Indonesia currently has a trial crop insurance programme, an expanded version of which was analysed in the study.
Source: OECD (2015a).

On poverty rates

A large part of the motivation behind price support and related policies is that through such interventions, producer incomes will improve and thus poverty rates will fall, resulting in food security improvements in rural areas. However, analysis conducted on the distribution of the returns and the incidence of the costs of price support policies raises significant doubts about whether such policies decrease poverty and even suggests that poverty rates – and therefore food insecurity – may have increased.

In Indonesia, an examination of the impacts of rice price support policies on households has suggested that in total, around 80% of all Indonesian households are made worse off, in terms of the impact on household income, from price support policies on rice and other staples (McCulloch, 2008). And while producers are made better off, any benefits only accrue to only around 25% of all households (for some, these benefits are not sufficient to outweigh the costs from higher prices on consumption goods). Furthermore, the households that benefit are spread across the income distribution, and so benefits flow to both poor and non-poor households.[2]

In the Philippines, similar results have been found (Dawe, Moya and Casiwan, 2006). The likely negative impact on poverty of such policies in both Indonesia and the Philippines is due to the fact there are more net *buyers* of rice than there are net rice sellers, especially at the bottom of the income distribution. In addition, a large proportion of the net sellers – those who benefit from higher prices – are relatively better off. For example, in the Philippines, Dawe, Moya and Casiwan (2006) estimated that the wealthiest 40% of rice farm households account for two-thirds of the marketed surplus.

On private investment

For countries such as Myanmar, the uncertain application of export restrictions to avoid domestic price spikes has been argued to have discouraged private investment in the agricultural sector (World Bank, 2014). Despite no formal export restrictions being in place, the recent history of application of restrictions has created uncertainties. Conflicts between government objectives of export promotion and price stabilisation have been argued to have created additional uncertainty for rice millers and traders, as in the past, traders have been forced to sell privately-accumulated stocks at a loss in order to place downward pressure on domestic markets.

The added uncertainty reduces private investment in processing and storage facilities that would otherwise help to moderate price movements and also better link producers to markets (World Bank, 2014). Both developments are critical for the modernisation of agricultural production systems and for improved producer incomes – key processes that enhance food security of rural households.

Similarly, government intervention in input markets can crowd out private sector involvement. Where policies are directed to subsidies on domestically produced inputs, the incentive for investment and innovation that can flow from foreign direct investment in the sector can be muted.

On demands for fiscal resources

Some policies create large demands on fiscal resources, such as those associated with public stockholding (Deuss, 2015). For all countries, but perhaps developing countries in particular, the drain on fiscal resources, coupled with the lack of effectiveness of many of these interventions, carries with it large opportunity costs. Investments in other areas which can provide direct support to vulnerable households (for example conditional cash transfers – see Beaujeu, 2016), allowing for market-based solutions to reduce fiscal costs associated with stockholding (Beaujeu, 2016), encouraging sustainable productivity improvements, and facilitating structural adjustment, are often neglected or given less priority.

Input subsidies in a number of ASEAN countries place a significant drain on public resources. In Indonesia, for example, fertiliser subsidies remain by far the most important programme through which the government provides budgetary support to agriculture (OECD, 2015b). In 2013, the value of this subsidy was IDR 17.6 trillion (USD 1.7 billion), accounting for 41% of total budgetary expenditures provided to support agriculture – both on-farm and those paid to the agriculture sector as a whole (that is, general services such as infrastructure, research and development, amongst others).

Alongside the issues of higher domestic prices and difficulties in targeting subsidised rice distribution, stockholding and price stabilisation policies require significant resourcing. The Philippines Senate Economic Planning Office highlighted the cost of the NFA on the national budget – around PHP 27 billion in 2009 (USD 567 million). Due to its high cost and lack of effectiveness in improving food security, it was suggested that a continuation of the stockholding programme could not occur indefinitely. Instead, well-designed safety nets and policies directed at improving the productivity and sustainability of the rice sector were favourable (Philippines Senate Economic Planning Office, 2010).

On food distribution

Food-based safety nets have the potential to provide targeted assistance to food-insecure households, providing them with access to subsidised food and thereby increasing total food consumption. Systems in place in Indonesia, Malaysia and the Philippines have all delivered food to needy and vulnerable households. However, these programmes have also suffered from "leakages". Leakages occur when subsidised food flows to non-needy households or even into other uses. This decreases the effectiveness of the programme in improving food security.

In Indonesia, issues in the targeting of the Raskin programme have meant that a significant proportion of subsidised rice does not flow to poor, food-insecure households. The World Bank found that only around half of all the rice procured for the programme was delivered to households (World Bank, 2012). The remainder did not find its way to households and was unaccounted for. And of the proportion delivered to households, a significant share flowed to non-poor households.

Based on analysis of household level data, the Raskin programme has been found to only have a small impact on undernourishment overall, reducing rates by only 1.3% percentage points. This has not been enough to offset the negative impact of price support policies (OECD, 2015a).[3]

The Philippines has also reported similar issues associated with the efficacy of its programme to distribute subsidised rice (Jha and Mehta, 2008). In 2008, the programme was found to suffer from significant under-coverage and large leakages, with only 25% of poor households benefiting. Furthermore, of the rice distributed, 48% went to *non*-poor households.

Malaysia has also not been immune to issues of leakage associated with its rice distribution scheme. A recent audit by the Malaysian government's Public Accounts Committee has questioned the efficacy of the scheme, citing both large costs and high levels of leakage as issues. The impact of the programme on poorer households was also questioned (PEMANDU, 2015).

On price variability and volatility from export restrictions

At the global level, a number of analyses have shown the effects of export restrictions – and ad hoc variation in import barriers – on overall price movements. During the period of rapidly rising food prices, Anderson, Ivanic and Martin (2013) found that such policies exaggerated overall world price movements. These effects were particularly felt by net food-importing countries which already had low trade restrictions. The exaggerated price movements created by the application of insulation policies in other countries created worse outcomes globally than what would have otherwise occurred. As such, from a global perspective, the various individual country interventions that were targeted at improving food security in fact *lessened* it. Indeed, Anderson, Ivanic and Martin (2013) find that the trade-based food price insulation policies implemented in 2008 could have actually increased the number of people living in poverty around the world.

Among ASEAN countries, recent research suggests that the use of export restrictions has also contributed to price spikes and to increased volatility (Durevall and van der Weide, 2014). In Lao PDR, for example, the combinations of expectations about export restrictions and other controls encourage a surge of exports during good times when restrictions are relaxed, exacerbating the potential impact of any externally induced price spike such as that which occurred in 2010. Compared to neighbouring countries, it is argued that the removal of such controls, while beneficial on a number of efficiency grounds, would also likely reduce the incidence and severity of price spikes (Durevall and van der Weide, 2014).

On input use

Subsidies provided on inputs – and, more specifically, fertilisers – are prefaced on the belief that agricultural production has been limited by the inability of many farmers to make use of inputs and technologies that are known to be effective (Wiggins and Brooks, 2010). The lack of uptake is due to knowledge gaps, lack of functioning markets, high transport costs that result from a lack of infrastructure, or credit constraints faced by producers due to poorly functioning credit markets. Arguments have also been made from an environmental perspective.

However, despite a range of market failures put forward to justify the use of fertiliser subsidies, evidence on the functioning of such schemes show generally poor effectiveness in targeting these failures over the longer term, questions of efficiency aside. Based on studies from India, Malawi and Sri Lanka, Wiggins and Brooks (2010) suggest that input subsidies, despite promoting input use over the short to medium term, had questionable lasting effects and as such were unlikely to meet their objectives over the long term. Indeed, should the subsidies be rescinded, it is likely that input use would fall, thus having limited impacts on issues associated with knowledge gaps. Furthermore, given the considerable fiscal costs of many of these programmes which are often plagued by leakage, addressing the market failures directly, such as through investments in infrastructure and training, would often produce greater benefits at lower costs.

In the case of Indonesia, the fertiliser subsidy arrangement has not operated without difficulty. The system has experienced shortages and delays in fertiliser delivery, suffered from significant leakage, and had a limited impact on the actual price paid by farmers. Studies have found that only 10% of farmers paid the ceiling price (or below) set by the government for urea in 2007 (Osorio et al., 2011). In addition, the OECD (2012) has found that in reality, many farmers who operate more than 2 hectares also receive the subsidy by splitting land into several plots by using the names of their family members. The price disparity between subsidised and non-subsidised fertiliser on the domestic market, and between the domestic price of subsidised fertiliser and the fertiliser price in the international market, creates a strong incentive to illegally sell product to farmers who are ineligible to purchase the subsidised product, or to smuggle subsidised fertiliser abroad. As there is limited monitoring, leakage from the system is high. A second issue is that the lack of competition in the distribution system removes the incentive for manufacturers to innovate and invest in producing and distributing fertiliser more efficiently. This is compounded by the fact that while Java accounts for about 60% of demand for urea fertiliser, only about 20% of urea is produced there. Accordingly, there is a high transportation cost associated with distributing fertilisers.

3.4 Regional policy responses

Not all responses to food security issues have been of a domestic nature. Within the ASEAN co-operation framework, members have sought to also pursue regional initiatives to help collectively address issues of food security. The various policy directions related to food security are set out in the ASEAN Integrated Food Security Framework (AIFS Framework) and the Strategic Plan for ASEAN Cooperation in Food, Agriculture and Forestry (ACFAF). These documents set the platform on which ASEAN member states are pursuing policies to address long-term food security. The AIFS Framework has five components which identify the core policy areas that underpin food security in the region. Under each of these are a number of "Strategic Thrusts" which set out actions (outputs or activities) for ASEAN member states. Similarly, the ACFAF contains a number of Strategic Thrusts that also set out actions for ASEAN members. These platforms are not reviewed here, but rather the regional initiatives that have been created and implemented in response to these, along with the framework which has led to their creation – the ASEAN Economic Community Blueprint.

The ASEAN Plus Three Emergency Rice Reserve

ASEAN members have long sought to build co-operation around the public stockholding of rice to prepare for emergencies. In 1979, the Agreement of the ASEAN Food Security Reserve was signed (originally by Indonesia, Malaysia, the Philippines, Singapore and Thailand), which established the ASEAN Emergency Rice Reserve (see also Chapter 4). This agreement sets out an amount of domestically held stocks that are earmarked to meet emergency requirements in the region. The agreement does not require the holding of physical stocks, but rather a commitment to provide a set amount of rice in emergency situations (Briones, 2014). This agreement was later expanded as ASEAN grew.

In 2002, a trial of a more regional scheme was conducted which expanded beyond ASEAN members to include China, Japan and Korea. The scheme was called the East Asia Emergency Rice Reserve, and consisted of rice that would be donated in the face of acute emergency within member countries. Under this scheme, 13 000 tonnes of rice were distributed in Cambodia, Indonesia, Lao PDR, Myanmar, and the Philippines (Briones, 2014).

The East Asia Emergency Rice Reserve was later converted into the current ASEAN Plus Three Emergency Rice Reserve (APTERR) through an agreement that came into force in 2012. In total, 787 000 tonnes of rice are earmarked by members, based around historical allocations from the preceding schemes. The three non-ASEAN members account for the largest amounts of earmarked stocks, accounting for 700 000 tonnes of the total pledged amount.

Stocks can be released from APTERR under different "tiers", or types of agreements between member countries:

- Under Tier 1, stand-by arrangements between countries are negotiated which pre-specify the quantity, quality and terms and conditions of release in an effort to avoid negotiations after an emergency has occurred. The system is formalised by three-year renewable forward contracts between countries. Price does not need to be part of the contract, but if it is not, a formula for determining price must be agreed within the contract. Delivery is intended to be one month or earlier after the request.
- Under Tier 2, earmarked stocks under the scheme can be released on the basis of bilateral negotiations between members that take place after an emergency has occurred.
- Under Tier 3, stockpiled rice can be released to help meet the needs of acute emergency situations.

Members have made use of APTERR, although issues of timeliness were encountered. Following a recent natural disaster in the Philippines in late 2013, the Philippine government sought emergency supplies of rice from APTERR. In total, 6 730 tonnes of rice was delivered over an 18 month period, with rice received from China (800 tonnes in March 2014), Thailand (5 000 tonnes in April 2014), Malaysia (350 tonnes in August 2014) and Japan (580 tonnes in March 2015).

The lack of timeliness of some of these supplies has led to further reforms of the scheme. Currently, the storage of stocks outside of the donor country is being trialled, with Japan agreeing to hold some of its emergency stocks in both Cambodia and the Philippines.

ASEAN Food Security Information System

During the early stages of the development of the APTERR, ASEAN members expressed concern over the lack of information available to policy makers on factors that influence food security in the region. In response, the ASEAN Food Security Information System (AFSIS) was formed in 2002. AFSIS targets two main goals:

- Capacity development: Improve the capacity of member nations to develop, analyse and share information related to domestic and regional food security as it relates to their own country. More advanced member nations are requested to help improve the capacity of the others through capacity-building activities as necessary.
- Information network development: This includes the development of a database of relevant information so that member nations can better plan policy and its implementation.

Under the second objective, AFSIS regularly produces early warning information and a Commodity Outlook, detailing information that will help to identify possible future food security crises. The early warning information details forecasted production and yield (current year) for major food security crops of rice, maize, sugar soybean and cassava. The Commodity Outlook then provides forecasts of these crops for the following year.

ASEAN integration

A notable regional policy development that could significantly enhance food security is the push towards closer integration of ASEAN economies. The ASEAN Economic Community (AEC) initiative initially set a target for regional integration by 2015. Integration is envisaged to create:

- a single market and production base
- a highly competitive economic region
- a region of equitable economic development
- a region that is fully integrated within the global economy (ASEAN, 2015).

This initiative extends well beyond agriculture, and aims to allow for the free flow of goods, services, investment and skilled labour across the region, along with the freer flow of capital. As such, it has the potential to significantly impact growth opportunities in the region and therefore food security.

Full economic integration will take time to occur. Nevertheless, in moving down this path, and through exploiting the potential benefits of developing a single market and production base, food security could be enhanced. Bello (2005) argues that through free trade in rice and maize, enhanced through improved trade facilitation measures and the harmonisation of food regulations, food security for each of the ten ASEAN members could be improved. Such measures could exploit the natural diversity in agricultural production systems across the region to the benefit of all members. It is argued that a shift in focus of food security policy from a domestic desire for self-sufficiency to a regional perspective is required. Such a shift is possible through the pursuit of closer economic integration.

Others have explored further integration specifically in the area of rice. Currently, despite the large number of steps towards a single production base for many commodities, rice has remained a product on which differences in policy stance have remained, as outlined above. Hoang and Meyers (2015) explore the price effects of the removal of impediments to trade, mainly created through the state-trading and licencing arrangements in place, and suggest that significant falls in consumer prices are possible. For the importing countries of Indonesia, Malaysia and the Philippines, falls of around 30-40% were projected, whereas price rises on world markets were around 30% (although far less than the price spikes observed in the late 2000s). The authors stress, however, that these benefits are best realised through shared actions over time – that is, through ASEAN. In this way, the disruptions to world markets are minimised and time is allowed for adjustments in both exporting and importing countries, avoiding pressures on world markets that could lead to price surges.

3.5 Concluding comments

ASEAN food security policy can be characterised as being "rice-centric". For both large and small rice producers, an emphasis is placed on rice production. The policies chosen to pursue these objectives vary, but are in general related to attempts to spur domestic production in importing countries through the use of price support, trade barriers and input subsidies; and for exporting countries, interventions in export markets through taxes, bans and licencing arrangements, along with attempts to "lock-in" a certain quantity of rice production. At the same time, some countries have also made significant investments in the enabling environment, in irrigation infrastructure in particular. Despite this, where information on relative expenditures is available, such investments, while worthwhile, are often much smaller than the costs of interventions in output or input markets.

On the consumer side, some countries have coupled production-based policies with public stockholding regimes aimed at stabilising rice prices and providing for public distribution. These are most notable in traditionally importing countries such as Indonesia, Malaysia and the Philippines. While such regimes have stabilised prices compared to ASEAN peers, they have done so at much higher price levels, bringing into question the net impact on food security. Indeed, in 2012-14 in Indonesia, domestic prices were estimated to be 70% above comparable world prices.

There have been a number of studies that have reviewed the economic impacts of these policies. These reviews suggest that such interventions create inefficiencies in resource allocation within the economies, discourage private investment by creating greater uncertainties, and impose significant budgetary costs on governments – for which there are significant opportunity costs in terms of the alternatives to which those outlays could be better directed. Furthermore, the effects on prices for some suggest that these policies are unlikely to be effective in helping to address food security for vulnerable households. Moreover, the ineffective nature of the support on farm incomes of the poorest, and in a number of cases the incidence of support accruing to otherwise food secure households, suggests that even for poor rural producers, the long-run impacts on food security are questionable.

Alternative policies have also been suggested. In terms of food security, an emphasis on targeted support to vulnerable households that can be accessed both generally and more intensively at times of crisis has the potential to lead to significant improvements in food security. In the case of Indonesia, targeted cash transfers or food vouchers have been shown to have significant potential. For the agricultural sector, deregulated markets, increased trade and more transparent policy making are also key inputs into more stable supply, less volatile prices and increased agricultural investment. The use of trade policy in particular to maintain affordable food prices has been shown to be ineffective in both importing and exporting countries.

However, despite the significant focus of these assessments on the economic impacts of such food security policies, this review points to a gap – that of how these policies, and alternatives, perform when assessed against the risks they are intended to mitigate. It is often cited that despite the costs of such policies on the economy, they provide an effective insurance mechanism for both producers and consumers. Risks insured relate to international price movements, natural disasters and even macroeconomic events that can reduce the purchasing power of domestic consumers on international markets. Such a proposition, explored in the context of Indonesia's policies (OECD, 2015a), is questionable. For ASEAN overall, this question is explored in the following chapter.

Notes

1. The ten ASEAN member states are Brunei Darussalam, Cambodia, Indonesia, Lao People's Democratic Republic (PDR), Malaysia, Myanmar, Philippines, Singapore, Thailand and Viet Nam.
2. Relating these results to statistics on undernourishment, it would be expected that without higher prices, rates of undernourishment would have fallen faster with the overall rise in incomes that has been experienced over the relevant period.
3. As was found for the effects from McCulloch's (2008) study, relating these results to statistics on undernourishment, it would be expected that without higher prices and irrespective of the effects of Raskin, rates of undernourishment would have fallen faster with the overall rise in incomes that has been experienced over the relevant period.

References

Alavi, H., A. Htenas, R. Kopicki, A. Shepherd and R. Clarete (2012), *Trusting Trade and the Private Sector for Food Security in Southeast Asia*, World Bank, Washington, DC.

Anderson, K., M. Ivanic and W. Martin (2013), "Food price spikes, price insulation, and poverty", *Policy Research Working Paper*, No. 6535, World Bank, Washington, DC.

APEC Policy Support Unit (2012), *Food Security Policies in APEC*, Asia-Pacific Economic Cooperation Secretariat, Singapore.

ASEAN (2015), *ASEAN Economic Community* (webpage), www.asean.org/communities/asean-economic-community (accessed on 15 June 2015).

Bank of Thailand (2016), *Wholesale Price of Certain Commodities in Bangkok* Metropolis (database), Bank of Thailand, Bangkok, www2.bot.or.th/statistics/BOTWEBSTAT.aspx?reportID=89&language=ENG (accessed on 15 June 2015).

Beaujeu, R. (2016), "Alternative policies to buffer stocks for food security", in *OECD Food, Agriculture and Fisheries Papers*, OECD Publishing, Paris, http://dx.doi.org/10.1787/5jln0434qkzp-en.

Bello, A.L. (2005), "Ensuring food security – A case for ASEAN integration", *Asian Journal of Agriculture and Development*, Vol. 2/1 and 2, Southeast Asian Regional Center for Graduate Study and Research in Agriculture, Philippines, pp. 87-108,

Briones, R. (2014), *Public stockholding in Southeast Asia: Review and Prospects*, FAO Expert Meeting on Stocks, Markets and Stability, 30-31 January 2014, Food and Agriculture Organization of the United Nations, Rome.

Chulaphan, W., S. Chen, C. Jatuporn and P. Jierwiriyapant (2012), "The effect of rice price-pledging scheme on price transmission of rice markets in Thailand", *Asian Journal of Empirical Research*, Vol. 2/5, Asian Economic and Social Society, pp. 141-148.

Cororaton, C.B. (2004), "Rice reforms and poverty in the Philippines: A CGE analysis", *ADB Institute Research Paper Series*, No. 57, Asian Development Bank Institute, Manila.

Dawe, D. and T. Slayton (2010), "The world rice market crisis of 2007-2008", in *The Rice Crisis: Markets, Policies and Food Security*, Food and Agriculture Organization of the United Nations, and Earthscan, London and Washington, DC.

Dawe D., P. Moya and C. Casiwan, eds. (2006), *Why Does the Philippines Import Rice? Meeting the Challenge of Trade Liberalization,* International Rice Research Institute and Philippine Rice Research Institute, Manila.

Deuss, A. (2015), "Review of the performance and impacts of recent stockholding policies", in *Issues in Agricultural Trade Policy: Proceedings of the 2014 OECD Global Forum on Agriculture*, OECD Publishing, Paris, http://dx.doi.org/10.1787/9789264233911-5-en.

Durevall, D. and R. van der Weide (2014), "Importing high food prices by exporting: Rice prices in Lao PDR", *Working Papers in Economics*, No. 607, University of Gothenburg, Sweden.

Eliste, P. and N. Santos (2012), *Lao People's Democratic Republic Rice Policy Study,* International Rice Research Institute, World Bank and Food and Agriculture Organization of the United Nations, Rome.

FAO (2016), "Rice Prices", *Food Security Portal* (database), Food and Agriculture Organization of the United Nations, Rome, www.foodsecurityportal.org/api/rice (accessed on 27 April 2015).

FAO (2014), "Cambodia", *Country Fact Sheet on Food and Agriculture Policy Trends*, April 2014, Food and Agriculture Organization of the United Nations, Rome.

Financial Times (2016), "Thailand offers $1bn loan to rice farmers in push to boost prices" (webpage), *Financial Times*, www.ft.com/content/5cab20c8-9f64-11e6-891e-abe238dee8e2 (accessed on 15 March 2017).

Gillson, I., and A. Fouad (eds.) (2015), *Trade Policy and Food Security: Improving Access to Food in Developing Countries in the Wake of High World Prices,* Directions in Development, World Bank, Washington, DC.

Hangzo, P.K.K. (2010), "Comprehensive food security: An approach to sustainably address food insecurity", *NTS Perspectives*, Issue No. 3, Centre for Non-Traditional Security Studies, S. Rajaratnam School of International Studies, Singapore.

Harun, R. (2015), *Policies and Economic Development of Rice Production in Malaysia*, Asia-Pacific Information Platform on Agricultural Policy.

Headey, D. (2011), "Rethinking the global food crisis: The role of trade shocks", *Food Policy*, Vol. 36, Elsevier, Amsterdam, pp. 136-146.

Hoang, H.K. and W.H. Meyers (2015), "Price stabilization and impacts of trade liberalization in the Southeast Asian rice market", *Food Policy,* Vol. 57, Elsevier, Amsterdam, pp. 26-39.

Indonesian Ministry of Agriculture (2015), data received from Indonesian Ministry, Bangkok.

Intal, P.S., L.F. Cu and J.A. Illescas (2012), "Rice prices and the National Food Authority", *Philippine Institute for Development Studies Discussion Paper Series*, No. 2012-27, Makati City, Philippines.

International Grains Council (2016), "Export Prices", *International Grains Council* (database), London, www.igc.int/en/markets/marketinfo-prices.aspx (accessed on 7 May 2015).

Jha, S. and A. Mehta. (2008), "Effectiveness of public spending: The case of rice subsidies in the Philippines", *ADB Economics Working Paper* No. 138, Asian Development Bank, Manila.

Jones, D. and A. Kwieciński (2010), "Policy responses in emerging economies to international agricultural commodity price surges", *OECD Food, Agriculture and Fisheries Papers*, No. 34, OECD Publishing, http://dx.doi.org/10.1787/5km6c61fv40w-en.

Lim, G. (2013), "Singapore's food security strategy" (webpage), *New Mandala*, Australian National University College of Asia and the Pacific, Canberra, http://asiapacific.anu.edu.au/newmandala/2013/02/07/singapores-food-security-strategy/ (accessed on 15 May 2015).

MAF (2015), *Agriculture Development Strategy to 2025 and Vision to the Year 2030*, Lao PDR Ministry of Agriculture and Forestry, Vientiane.

Mahathanaseth, I. and K. Pensupar (2014), *Thai Agricultural Policies: The Rice Pledging Scheme*, Department of Agricultural and Resource Economics, Kasetsart University, Bangkok.

Malaysian Department of Information (2015), *Subsidi, Bantuan dan Insentif Pertanian* (website), http://pmr.penerangan.gov.my/index.php/penstrukturan-subsidi/9893-subsidi-bantuan-dan-insentif-pertanian.html (accessed on 18 May 2015).

Malaysian Office of the Prime Minister (2009), "NAJIB: 200,000 families to benefit from rice subsidy programme", *News* (webpage), www.pmo.gov.my/home.php?menu=newslist&news_id=706&news_cat=13&cl=1&page=1731&sort_year=2009&sort_month (accessed on 18 May 2015).

Maltsoglou, I., D. Dawe and L. Tasciotti (2010), "Household level impacts of increasing food prices in Cambodia", *Environment and Natural Resources Management Working Paper*, No. 37, Food and Agriculture Organization, Rome.

McCulloch, N. (2008), "Rice prices and poverty in Indonesia", *Bulletin of Indonesian Economic Studies*, Vol. 44/1, Taylor and Francis, pp. 45-64.

MOA (2015), *Bahagian Industri Padi Dan Beras* (website), Malaysian Ministry of Agriculture, www.moa.gov.my/bahagian-industri-padi-dan-beras (accessed on 18 May 2015).

Naylor, R.L. and W.P. Falcon (2010), "Food security in an era of economic volatility", *Population and Development Review*, Vol. 36/4, New Jersey, United States, pp. 693-723.

OECD (2017), *Agricultural Policies in the Philippines*, OECD Publishing, Paris, http://dx.doi.org/10.1787/9789264269088-en.

OECD (2015a), *Managing Food Insecurity Risk: Analytical Framework and Application to Indonesia*, OECD Publishing, Paris, http://dx.doi.org/10.1787/9789264233874-en.

OECD (2015b), *Agricultural Policy Monitoring and Evaluation 2015*, OECD Publishing, Paris, http://dx.doi.org/10.1787/agr_pol-2015-en.

OECD (2015c), *Agricultural Policies in Viet Nam 2015*, OECD Publishing, Paris, http://dx.doi.org/10.1787/9789264235151-en.

OECD (2012), *OECD Review of Agricultural Policies: Indonesia 2012*, OECD Publishing, Paris, http://dx.doi.org/10.1787/9789264179011-en.

OECD (2008), "Rising food prices: Causes and consequences", *OECD Policy Brief*, OECD Publishing, Paris, www.oecd.org/trade/agricultural-trade/40847088.pdf.

Osorio, C.G., D.E. Abriningrum, E.B. Armas and M. Firdaus (2011), "Who is benefiting from fertilizer subsidies in Indonesia?", *World Bank Policy Research Working Papers*, Washington, DC.

PEMANDU (2015), "Tajuddin: Government Not Planning To Stop ST15 Rice Subsidy Programme" (webpage), Performance Management and Delivery Unit of the Malaysian Government, http://gtp.pemandu.gov.my/gtp/Media_Coverage-@-Tajuddin-;_Government_Not_Planning_To_Stop_ST15_Rice_Subsidy_Programme.aspx (accessed on 5 September 2015).

Permani, R. and D. Vanzetti (2014), *Rice Mountain: An Assessment of the Thai Rice Pledging Program*, Contributed paper prepared for presentation at the 58th AARES Annual Conference, Port Macquarie, New South Wales, 4-7 February 2014.

Philippines Bureau of Agricultural Statistics (2015), "Rice" (database), *Bureau of Agricultural Statistics*, Manila.

Philippines Government (2011), "Briefer on the Food Staples Self-Sufficiency Roadmap 2011-2016" (webpage), *Official Gazette*, Manila.

Philippines Senate Economic Planning Office (2010), "Subsidizing the National Food Authority: Is it a good policy?", *SEPO Policy Brief*, Senate Economic Planning Office.

Piesse, J. and C. Thirtle (2009), "Three bubbles and a panic: An explanatory review of recent food commodity price events", *Food Policy*, Vol. 34, Elsevier, Amsterdam, pp. 119-129.

Singapore Government (2015), *Rice Stockpile Scheme* (website).

Tobias A, I. Molina, H.G. Valera, K.A. Mottaleb, and S. Mohanty (2012), *Handbook on Rice Policy for Asia,* International Rice Research Institute, Philippines.

USDA FAS (2015), *Indonesia Grain and Feed Annual Report 2015*, USDA Foreign Agricultural Service GAIN Report No. ID1512, USDA, Washington, DC.

Vengedasalam, D., M. Harris and G. MacAulay (2011), *Malaysian Rice Trade and Government Interventions*, Contributed paper presented to the 55th Annual Conference of the Australian Agricultural and Resource Economics Society, Melbourne, 8-11 February 2011.

Wiggins, S. and J. Brooks (2010), *The Use of Input Subsidies in Developing Countries*, Background paper to the OECD Global Forum on Agriculture: Policies for Agricultural Development, Poverty Reduction and Food Security, 29-30 November 2010, Paris.

WITS (2016), *World Integrated Trade Solution* (database), The World Bank, Washington, DC, http://wits.worldbank.org/default.aspx (accessed on 15 November 2016).

World Bank (2014), *Myanmar: Rice Price Volatility and Poverty Reduction*, Economic and Sector Work, Report No. 89687-MM, World Bank, Washington, DC.

World Bank (2012), *Raskin Subsidized Rice Delivery: Social Assistance and Public Expenditure Review 3*, World Bank, Jakarta.

Chapter 4

Current approaches and alternatives for managing food insecurity risk in ASEAN

This chapter examines food security at household level within selected ASEAN countries, and assesses the effectiveness and efficiency of current policy approaches in addressing transitory food insecurity risks. First, household survey data for five ASEAN countries is evaluated in order to shed light on the incidence of undernourishment at household level, including details on the characteristics of undernourished households. Second, the impact of three major food insecurity risks on rice prices and the prevalence of undernourishment in the selected countries are simulated. Finally, the performance of current and alternative trade and domestic safety net policies under the risk scenarios is compared.

Key points

- Poverty is strongly linked with the *risk* of undernourishment. Due the high share of food costs in lower-income household expenditures, they are particularly vulnerable to price and income shocks.
- The importance of rice in regional diets means that rice prices are particularly important for households, including low-income farm households in certain Association of Southeast Asian Nations (ASEAN) countries, as they tend to purchase a significant proportion of the rice that they consume.
- High rice prices resulting from market interventions and trade restrictions are persistently increasing the rates of undernourishment, and are ineffective in managing key food insecurity risks such as regional *El Niño* events and domestic crop failures.
- Of the alternative policy options explored, integrated rice markets and targeted social safety net measures are the most effective means to manage such risks. Simulations indicate that the regional integration of rice markets would both reduce the undernourished population by 5% in five ASEAN member states, and mitigate the otherwise large food insecurity impact of weather risks.
- Alternative policies such as cash transfers and food voucher programmes perform better than existing rice distribution programmes in reducing undernourishment across the risk scenarios analysed. Nonetheless, of the policies explored, ASEAN rice market integration is found to be the most effective in reducing the risk of food insecurity across the scenarios in both countries.
- The analysis also suggests that the gradual integration of regional rice markets would prevent a sharp increase in rice prices in exporting countries. The reforms should start with greater private sector involvement in regional trade. In addition, in response to the expected higher prices in rice-exporting countries, the development of targeted safety net measures will be an important complementary measure.

4.1 Introduction[1]

Notwithstanding the rapid economic growth experienced by a number of countries in the Southeast Asian region, food security remains one of the most important policy issues for the Association of Southeast Asian Nations (ASEAN).[2] While economic growth has yielded significant declines in populations suffering from chronic poverty, the *nature* of food insecurity in some instances has shifted from permanent to transitory, whereby some populations are currently food secure, but are nevertheless at risk of becoming food insecure as a result of temporary shocks, be they economic, natural, domestic or international.

Chapter 1 of this publication provides an overview of the food security status of ASEAN countries, while Chapter 3 reviews the current agriculture-related policy approaches to address food security. The review indicates that the current policy approach in ASEAN countries is largely oriented towards relatively *short-term* risk management through measures such as rice price stabilisation and some public distribution of food, rather than addressing the long-term food security objective.

This chapter assesses specific food insecurity situations in selected ASEAN countries by means of household survey data, and evaluates current and alternative policy measures by applying the OECD analytical framework for transitory food insecurity (OECD, 2015a). The use of household-level data has the advantage of enabling the identification of the profiles of populations at risk of undernourishment and the assessment of impacts of different policies. Consultation with ASEAN on the initial assessment results was conducted through a series of OECD-ASEAN events on food security (Seoul in October 2015 and NayPyiTaw in June 2016).

The application process of the transitory food insecurity framework includes three steps: preparatory work, risk assessment and policy assessment. This chapter first assesses the food security situation by examining food consumption patterns and nutrition outcomes at the household level, followed by an estimation of households' consumption response to price and income shocks. In the second part of the chapter, the potential food insecurity risk scenarios in the ASEAN region are identified, and the potential impact on the prevalence of undernourishment in selected ASEAN member states is quantified. The food security impacts of different trade policy measures under different food insecurity scenarios are also subsequently quantified. Similarly, the impacts of current and alternative safety net policy measures – such as subsidised rice distribution, cash transfer and food voucher programmes – are compared in two food-importing ASEAN countries: Indonesia and the Philippines. The chapter concludes with policy recommendations derived from the results of the risk assessment.

4.2 Household-level assessment of food security in selected ASEAN countries

A snapshot of food security in the ASEAN region, presented in Chapter 1, highlights the impressive progress that has been made in the reduction of undernourishment in Southeast Asia in comparison with other regions worldwide. In the early 1990s, undernourishment rates in the region were the highest in the world. Since then, however, the rate of undernourishment has fallen below those observed in a number of other regions. Nevertheless, a great number of households may still be at risk of undernourishment as a result of temporary shocks. To be food secure, a population, household or individual must have access to adequate food at all times, and should not risk losing access to food as a consequence of sudden shock or cyclical events (FAO, 2006). In addition, there are indications of high rates of stunting among children under five years of age across Southeast Asia.

While a snapshot of the food security situation in the ASEAN region shows an overall improvement of food security in recent decades, household survey data provides information on the incidence of undernourishment at the household level.[3] The survey data enables the characteristics of undernourished households – such as income, location and professional choice – to be assessed. It also provides information on food consumption patterns, which is necessary to estimate responses to different risk scenarios and policies.

The data show that the incidence of undernourishment is concentrated among low-income households in the countries examined, with the exception of Thailand. The poorest 20% of households account for more than 40% of the incidence of undernourishment in Indonesia, Myanmar, Philippines and Viet Nam (Figure 4.1). However, access to food is not *solely* determined by the purchasing power of households. A comparison of the prevalence of undernourishment between farm and non-farm households in Indonesia, Myanmar, the Philippines and Viet Nam shows higher incidence of undernourishment among *non*-farm households, despite higher poverty rates in farm households (Figure 4.2). The reason for this is that farm households that produce food crops are likely to gain better access to food by adjusting the proportion of their production that is sold to the market, in accordance with their own consumption needs.

Non-farm households purchase nearly 90% of the rice that they consume. However, in Indonesia, Myanmar and the Philippines, even *farm* households purchase the majority of rice that they consume (Figure 4.3). A large number of farm households in these three countries are small-scale subsistence producers or landless farm workers whose production does not satisfy their own needs.

Policy makers should therefore take into consideration the fact that vulnerable farm households tend to be small-scale producers that source a large part of the household's rice consumption from the market. The survey data show that the dependence on purchased rice is particularly high for *poor* farm households in all five countries. Farm households in the poorest quintile class in Indonesia, Myanmar and the Philippines tend to source a higher share of rice from the market than average farm households. In particular, the poorest farm households in the Philippines were found to purchase 77% of rice that they consume (OECD calculations, based on Philippine Statistics Authority, 2014). For those farm households that source a high share of rice from the market, higher rice prices are likely to have a negative impact on their food security, notwithstanding higher revenue from rice production.

Food accounts for a larger share of the expenditure of poorer households (Figure 4.4). The poorest 20% of households in Myanmar, Viet Nam and Indonesia spend on average more than 60% of total expenditure on food and 20% on rice. These households are more vulnerable to food insecurity shocks such as high rice price and income loss, as they have limited capacity to reallocate expenditure to staple food items in response to shocks.

Figure 4.1. Composition of undernourished households by expenditure quintile

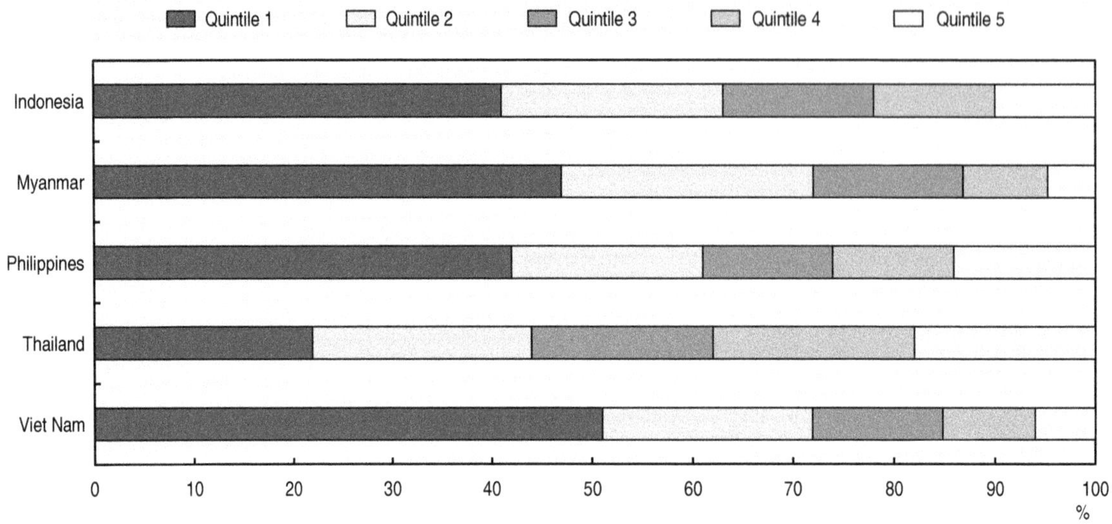

Sources: OECD calculations based on data from: Indonesia: BPS-Statistics Indonesia (2010), Myanmar: MNPED (2010), Philippines: Philippine Statistics Authority (2014), Thailand: TNSO (2014), and Viet Nam: General Statistics Office of Viet Nam (2010).

Figure 4.2. Prevalence of undernourishment by household type

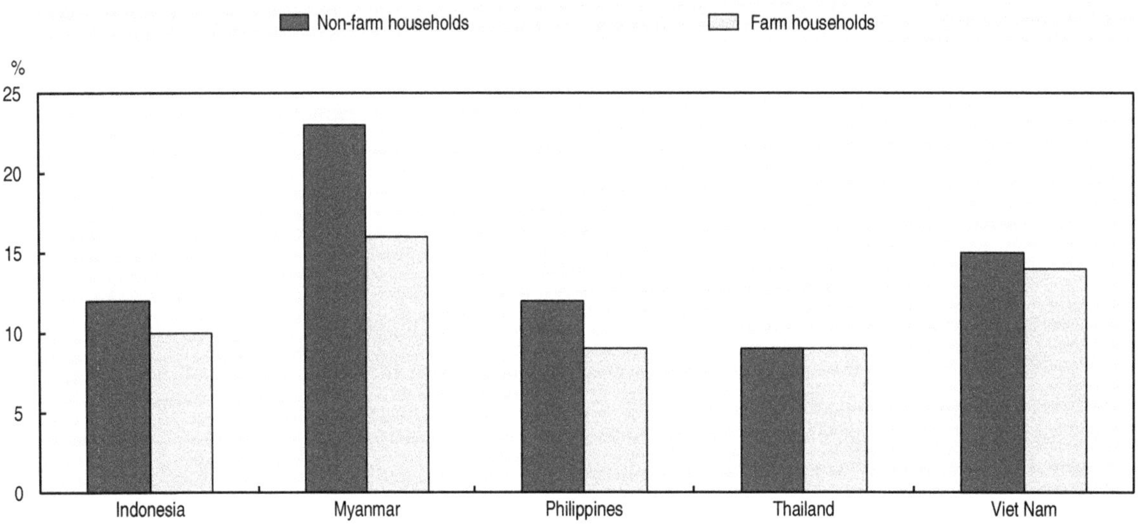

Sources: OECD calculations based on data from: Indonesia: BPS-Statistics Indonesia (2010), Myanmar: MNPED (2010), Philippines: Philippine Statistics Authority (2014), Thailand: TNSO (2014), and Viet Nam: General Statistics Office of Viet Nam (2010).

Figure 4.3. Source of rice consumption, by household type

Percentage of rice consumed

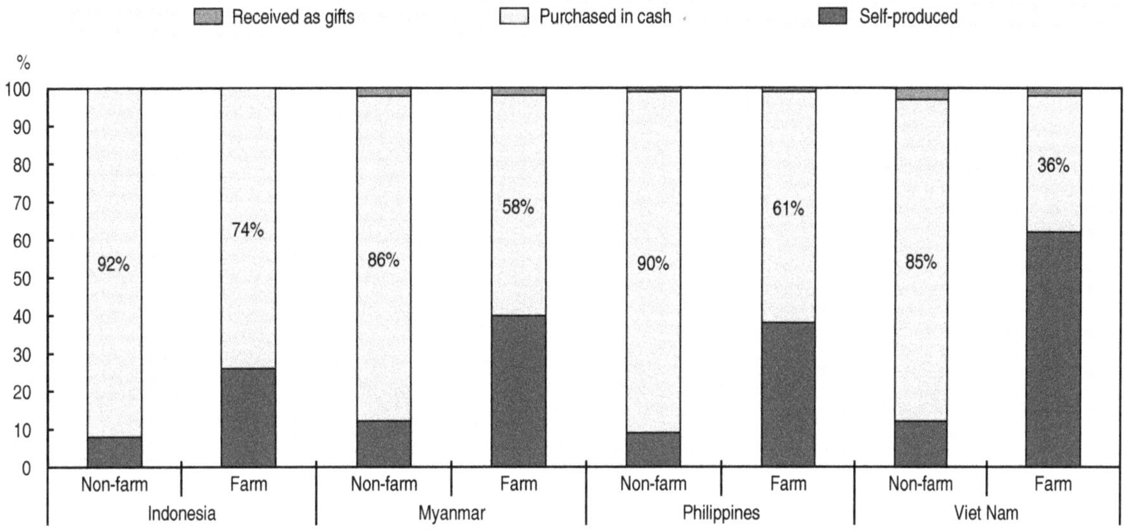

Note: In the absence of household income sources in some surveys, farm and non-farm households are classified based on the profession of household head.

Sources: OECD calculations based on data from: Indonesia: BPS-Statistics Indonesia (2010), Myanmar: MNPED (2010), Philippines: Philippine Statistics Authority (2014), and Viet Nam: General Statistics Office of Viet Nam (2010).

Figure 4.4. Percentage share of food and rice in household expenditure, by expenditure quintile

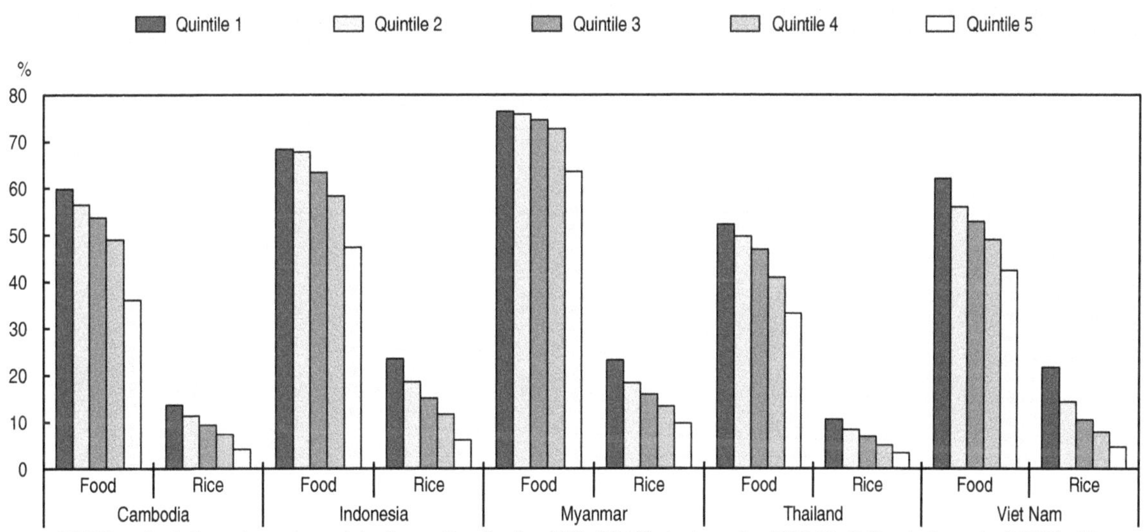

Sources: OECD calculations based on data from: Cambodia: NIS (2013), Indonesia: BPS-Statistics Indonesia (2010), Myanmar: MNPED (2010), Thailand: TNSO (2014), and Viet Nam: General Statistics Office of Viet Nam (2010).

Rice is a staple in ASEAN countries, accounting for a large part of the energy and nutrition intake. The average share of rice in calorie intake is the highest in Myanmar (63%), followed by Viet Nam (54%). The dependence of calorie intake on rice is, on average, less in Indonesia (46%) and the Philippines (48%). In Thailand, the average share of rice in calorie intake has already declined to less than 40%. Rice is also a significant source of protein intake in these countries, accounting for 30-50% of a household's protein intake on average. A comparison of the share of rice in calorie and protein intake reveals that in all five countries, poorer households source a larger proportion of their energy and nutrition intake from rice than more affluent households (Figure 4.5). The poorest 20% of households in Myanmar and Viet Nam obtain around 70% of calories from rice, while the wealthiest 20% of households in Indonesia, the Philippines and Thailand obtain only one-third of their calories from rice. High shares of protein intake from rice among poor households are an indication of strongly staple-based nutrition, which often leads to micronutrient deficiency.

Figure 4.5. Percentage share of rice in calorie and protein intake, by expenditure quintile

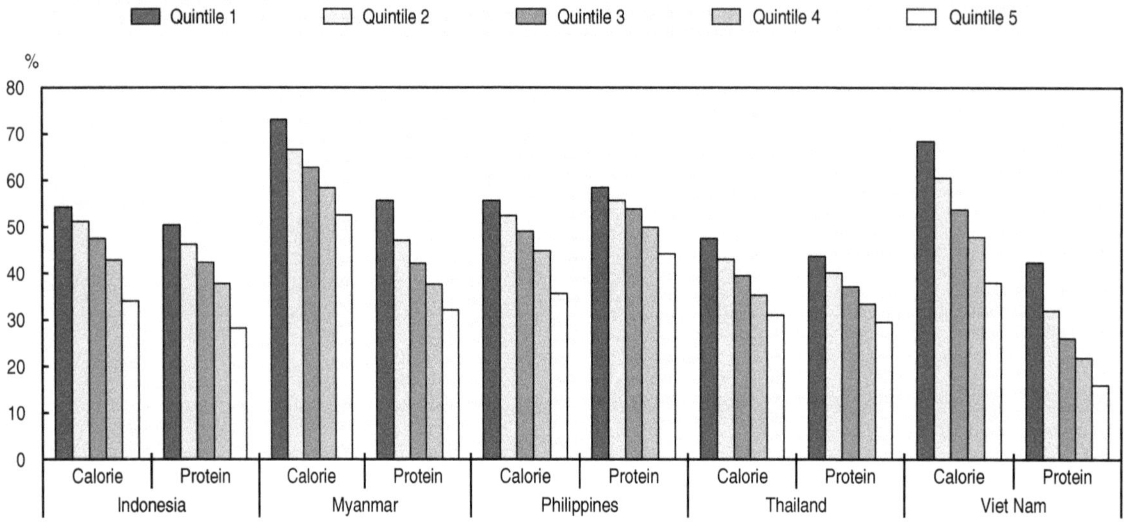

Sources: OECD calculations based on data from: Indonesia: BPS-Statistics Indonesia (2010), Myanmar: MNPED (2010), Philippines: Philippine Statistics Authority (2014), Thailand: TNSO (2014), and Viet Nam: General Statistics Office of Viet Nam (2010).

Figure 4.6. Per capita rice consumption, by expenditure class

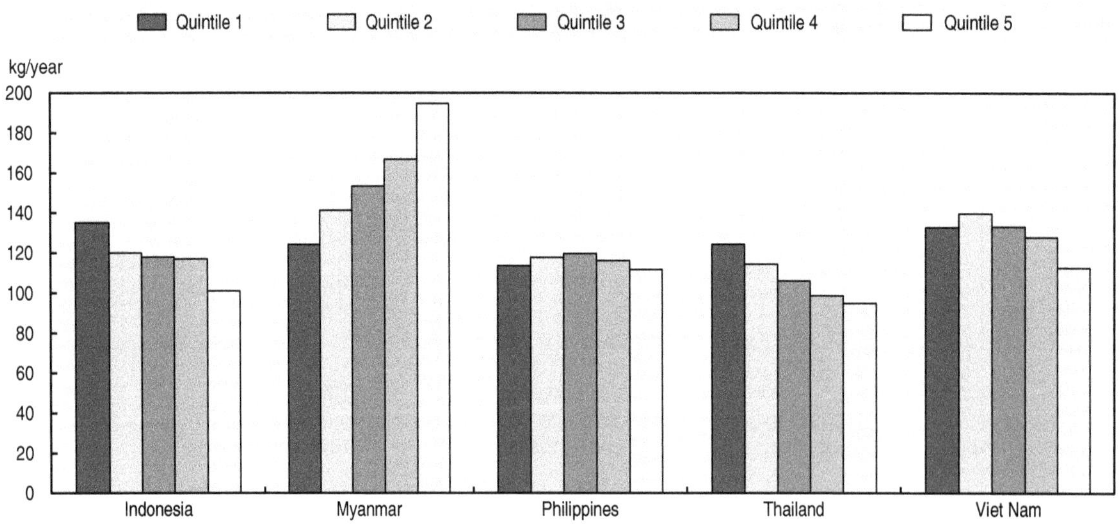

Sources: OECD calculations based on data from: Indonesia: BPS-Statistics Indonesia (2010), Myanmar: MNPED (2010), Philippines: Philippine Statistics Authority (2014), Thailand: TNSO (2014), and Viet Nam: General Statistics Office of Viet Nam (2010).

The lower calorie dependence of richer households on rice indicates that households diversify their diets away from rice as their incomes increase. The household survey data in ASEAN member countries suggest a future decline in per capita rice consumption as per capita income increases. In Indonesia, Thailand and Viet Nam, higher-income households consume less rice per capita (Figure 4.6). Per capita rice consumption has already begun to decline in these countries. However, the inverse relationship between per capita rice consumption and levels of income is less clear in the Philippines, and, in Myanmar, richer households consume *more* rice per capita.

On a broader scale, this trend is consistent with the experience of other countries in East Asia, whereby per capita rice consumption increases at the initial stages of economic development, but starts to decline as the per capita gross domestic product (GDP) of a country reaches a certain level, after which rice changes from a normal to an inferior good (Box 4.1).[4] It is likely that, while rice is currently a normal good in Myanmar and the Philippines, these countries will follow the path of other East Asian nations in the future.

Box 4.1. Economic development and rice consumption in East Asia

East Asian countries share a similar trend of economic development and per capita rice consumption (Figure 4.7). The relationship between per capita gross domestic product (GDP) and per capita consumption of rice shows that rice consumption increases at the early stages of economic development. At this stage, rice is a "normal" good, meaning that it is consumed in greater quantities as households obtain higher income. At a higher income stage, however, the population starts to diversify its diet away from rice. For example, per capita rice consumption in Japan halved between 1961-63 and 2008-10, while the share of rice in calorie supply fell from 44% to 23% during the same period. Ito, Peterson and Grant (1989) find that rice became an inferior good in advanced economies in Asia such as Japan, Chinese Taipei, Malaysia and Singapore in the mid-1980s. It is likely that ASEAN countries will follow a similar path of economic growth and declining per capita consumption of rice.

Figure 4.7. Economic development and per capita supply of rice in East Asia

Three-year moving average

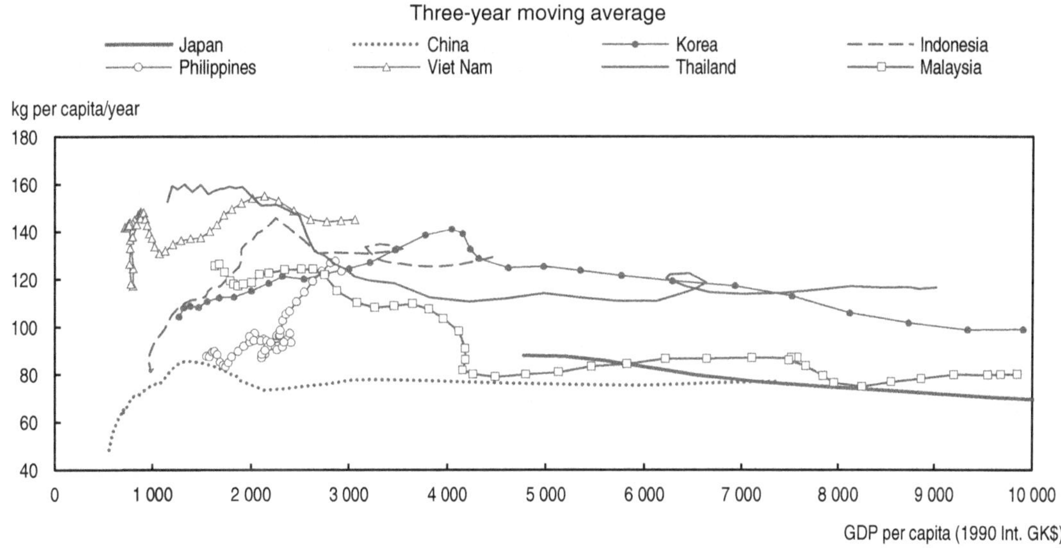

Sources: Per capita supply of rice: FAO (2016), *FAOSTAT*, http://www.fao.org/faostat/en/#home; gross domestic product (GDP) per capita: Maddison-Project (2013), *Maddison Project*, http://www.ggdc.net/maddison/maddison-project/home.htm.

Although the review of current policies (Chapter 3 of this publication) finds that the current policy approach of ASEAN countries could be characterised as "rice centric", the importance of a stable rice supply for food security may reduce in future, as diets shift away from rice. Moreover, an emphasis on providing policy support to rice producers may delay the structural adjustment of agricultural production that is necessary to meet increasing demand for diverse high-value added products, and may therefore reduce income opportunities for producers.

Household response to price and income shock

In order to simulate the consumption response to an external shock, household demand systems are estimated from household survey data by applying the Almost Ideal Demand System (AIDS). The estimation of AIDS enables the calculation of the price and expenditure elasticity of food items at the household level, which are then used to simulate the response to price and income shocks. Assuming that food consumption behaviour differs according to different income levels, the AIDS is estimated by expenditure quintile class. Both the model specifications and the detailed estimated results for Myanmar, Thailand and Viet Nam are presented in Annex 4.A2.[5]

The estimated own price elasticity of rice demand indicates that the households in the lower expenditure quintile are likely to have larger price elasticity than those in lower quintiles (Figure 4.8). This means that the percentage reduction in rice consumption in response to an increase in the price of rice would be larger for poorer households. For example, the poorest 20% of households in Myanmar, Thailand and Viet Nam reduce rice consumption by more than 6% in response to a 10% increase in rice price. The higher price elasticity of rice demand for poorer households indicates that they are forced to reduce the consumption of rice in response to a positive price shock. This behavioural response shows a vulnerability of the poorer households to price risk. The estimated own price elasticity of rice demand turns to positive for the highest quintile in Thailand and Viet Nam. This group of households also tends to consume high-quality rice, which is often associated with higher prices.

The estimated expenditure elasticity of rice demand shows higher demand elasticity for the households in lower expenditure quintiles, meaning that the rice consumption of poorer households is likely to reduce more than that of more affluent households, when faced with the same relative magnitude of income loss (Figure 4.9). The estimated expenditure elasticity is particularly high for the majority of households in Myanmar and Viet Nam, revealing their vulnerability to income shocks. The estimation also indicates that households in the highest expenditure quintile class in Indonesia, Thailand and Viet Nam have *negative* expenditure elasticity, meaning that increases in income reduce rice consumption. For these households, rice could be an inferior good. This result also implies that per capita rice consumption is likely to reduce further as the level of income increases in these countries.

Figure 4.8. Own price elasticity of rice demand

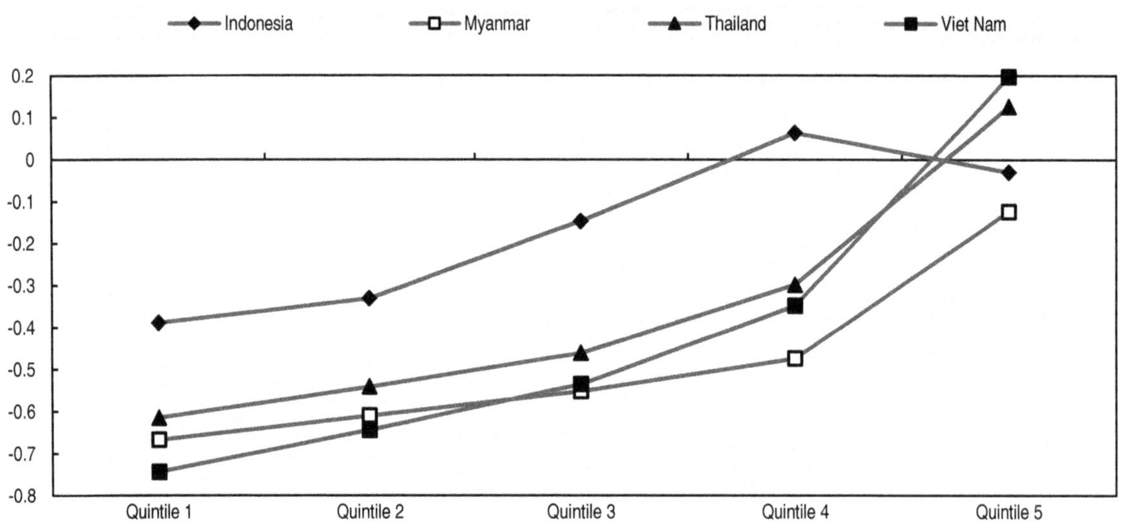

Note: Some estimations of own price elasticity are not statistically significant, particularly in the case of Myanmar.
Sources: OECD estimations based on: Indonesia: BPS-Statistics Indonesia (2010), Myanmar: MNPED (2010), Thailand: TNSO (2014), and Viet Nam: General Statistics Office of Viet Nam (2010).

Figure 4.9. Expenditure elasticity of rice demand

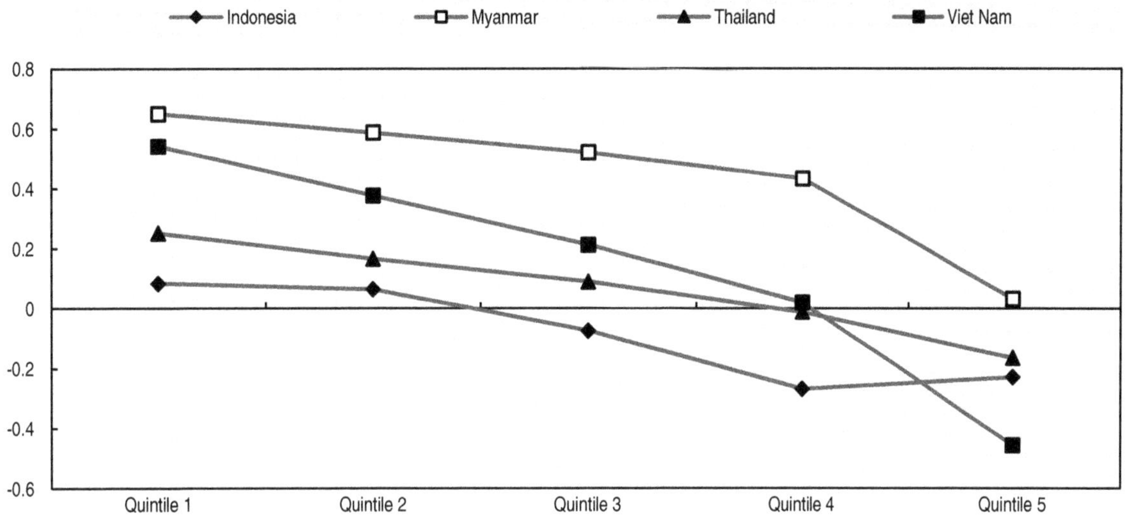

Note: Some estimations of expenditure elasticity of rice demand are not statistically significant, particularly in the case of Myanmar.
Sources: OECD estimations based on: Indonesia: BPS-Statistics Indonesia (2010), Myanmar: MNPED (2010), Thailand: TNSO (2014), and Viet Nam: General Statistics Office of Viet Nam (2010).

4.3 Assessment of food insecurity risks in selected ASEAN countries

The second step of the analytical framework of transitory food insecurity is a risk assessment. This step requires the identification of possible risks and shocks in consultation with experts, stakeholders and policy makers. This process should lead to a small set of risk scenarios, reflecting diversity in both the scale of impacts (regional and national), as well as the sources of shock (economic, natural, domestic and international). The risk identification is followed by risk evaluation to quantify the impacts of risk scenarios, both at country and household level. Based on the risk assessment, the following section assesses the performance of alternative trade and domestic safety net policies under different risk scenarios.

Identification of food insecurity risk scenarios

Significant risks and threats to the ASEAN rice market and regional food security as a whole still remain, despite the implementation of several initiatives in the ASEAN region to improve food security through a permanent regional rice reserve, better policy co-ordination, and data and information-sharing. The most important threat results from severe crop loss in one or more countries in the region due to extreme weather, such as floods, drought and typhoons.

In addition to natural disasters, the economic recessions that occasionally occur in Southeast Asia also pose important threats to food insecurity. Three stylised risk scenarios are analysed here.[6] The first scenario is the systemic risk of *El Niño* at the regional level, which simultaneously affects ASEAN member states. The second and third scenarios are country-specific risks, including both natural disasters and economic downturn.

Scenario 1: Regional El Niño risk scenario

OECD estimates, based on the IMPACT model and presented in Chapter 2, indicate that climate change will negatively affect agriculture for the ASEAN region as a whole. The findings suggest that climate change will lead to as much as 17% lower yield for rain-fed rice (16% for irrigated rice) by 2050, compared with what would have been the case without climate change. Prices worldwide are projected to rise for all agricultural products examined. However, for Southeast Asia, prices are expected to rise by a greater amount for staple crops of rice, maize and cassava. The price of rice is on average between 45% and over 55% higher when climate effects are included in the projections.

In addition to long-term impacts on the production and price of agricultural products, climate change is also expected to increase the frequency of extreme weather events. One of the potential systemic risks in the region is a strong *El Niño* event. The distribution of regional shocks to agricultural production is calculated based on Iizumi et al. (2014), who estimate the average effects of *El Niño* events on agriculture worldwide. However, the assumption of the average effect is not appropriate for this extreme scenario. The magnitude of the production shocks on rice were assumed based on a series of scenarios run by the International Food Policy Research Institute's IMPACT team as inputs to a Lloyd's Risk Report assessing the potential of *El Niño* events to disrupt agricultural markets (Lloyd's, 2015). The impact of the regional *El Niño* on rice production is assumed to be 10% in Indonesia, Thailand and the Philippines, and 20% in Viet Nam.

The market impacts of a rice yield shock in the regional *El Niño* risk scenario are quantified using the AGLINK-COSIMO model. Following the *OECD-FAO Agricultural Outlook 2016-2025* (OECD-FAO, 2016), the OECD (Furuhashi, 2017 [forthcoming]) established a baseline scenario of rice markets in Southeast Asia in 2016-25, assuming the existing rice trade policy regime (tariffs and other measures that restrict trade). The yield shocks are added to this baseline scenario to simulate a short-term impact on the regional commodity market. The impact on the domestic rice price is measured as a percentage difference between the baseline and the shock scenarios.[7]

The impacts of a regional *El Niño* event on domestic rice price are by far the largest in Indonesia due to its restrictive rice trade regime, followed by the Philippines (Figure 4.10). In the absence of representation of farm income in the AGLINK-COSIMO model, the elasticity of the farm household income with respect to rice price is assumed to be 0.25 for the farm household, based on the previous OECD study on income transfer efficiency (OECD, 2001).

Figure 4.10. Impact of regional *El Niño* on rice prices

Percentage change

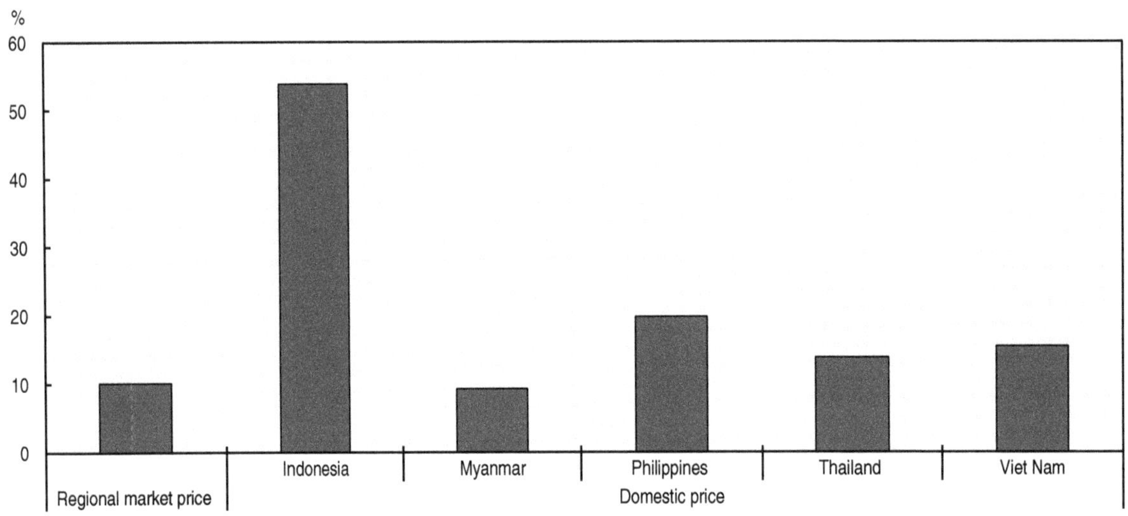

Source: Shocks introduced in the baseline scenario of Furuhashi (2017, forthcoming) based on OECD-FAO (2016).

Scenario 2: Crop failure

Extreme weather events frequently cause damage to the standing rice crop in the Southeast Asian region. On average, the Philippines is hit by 20 typhoons and Viet Nam by six to nine typhoons a year. In 2013, Typhoon Haiyan destroyed 170 000 metric tonnes of rice that was ready for harvest, and another 117 000 tonnes of rice that was already planted. In 2011, floods in Thailand damaged about 700 000 ha of rice crops, while flooding in Myanmar in 2013 following cyclone Nargis affected 24% of the country's rice area. In 1997, drought in Southeast Asia caused significant crop loss in Indonesia, Lao PDR, and the Philippines.

Similarly, the 2009 drought in the Philippines reduced its rice production by 3.31% (Redfern, Azzu and Binamira, 2012).

Of the various abiotic stresses, drought is the largest constraint, affecting nearly a fifth of the total rice area in Asia and causing millions of dollars in economic losses to poor rice producers (Pandey, Bhandari and Hardy [eds.], 2007). Similarly, 10–15 million hectares of rice fields are affected by floods every year in South and Southeast Asia, causing crop losses of up to USD 1 billion (Dey and Upadhyaya, 1996). Salinity, although not as severe as submergence and drought, is an increasing problem in the coastal areas of Southeast Asia. In northeast Thailand alone, salinity affects about 3 million hectares of surface area (Clermont-Dauphin et al., 2010). Similarly, salinity affects 1.8 million hectares of rice fields in the Mekong Delta in the dry season (MRC, 2010).

The frequency and duration of extreme weather such as droughts, floods, typhoons and heat waves are expected to rise in Southeast Asia in the future, along with higher temperatures and a rise in sea level (Lian and Bhullar, 2011). The rising night-time temperature of the past three decades is expected to continue, and will also have a significant negative impact on rice yield (Mohanty et al., 2013). According to Peng et al. (2004) and Welch et al. (2010), a 1°C increase in temperature above the critical level of 24°C is linked to a 10% reduction in grain yield and biomass. The rice-growing countries in the region also frequently face typhoons in the wet season, resulting in significant damage to the rice crop.

Rice is grown as a monoculture and is therefore particularly susceptible to disease and pest infestations. Pests such as the brown planthopper (BPH), known locally as wereng coklat, have been a serious problem in the past. In Indonesia, major outbreaks occurred during the 1974-75 planting season, and an especially severe outbreak occurred in 1985-86. More recently, outbreaks occurred in 1998 and 2011. The historical annual rice yield growth data also show that Myanmar, Thailand and Viet Nam have all experienced more than 10% of yield loss at least once in the last 55 years (Figure 4.11).

Figure 4.11. Annual growth rate of paddy rice yield

Percentage change

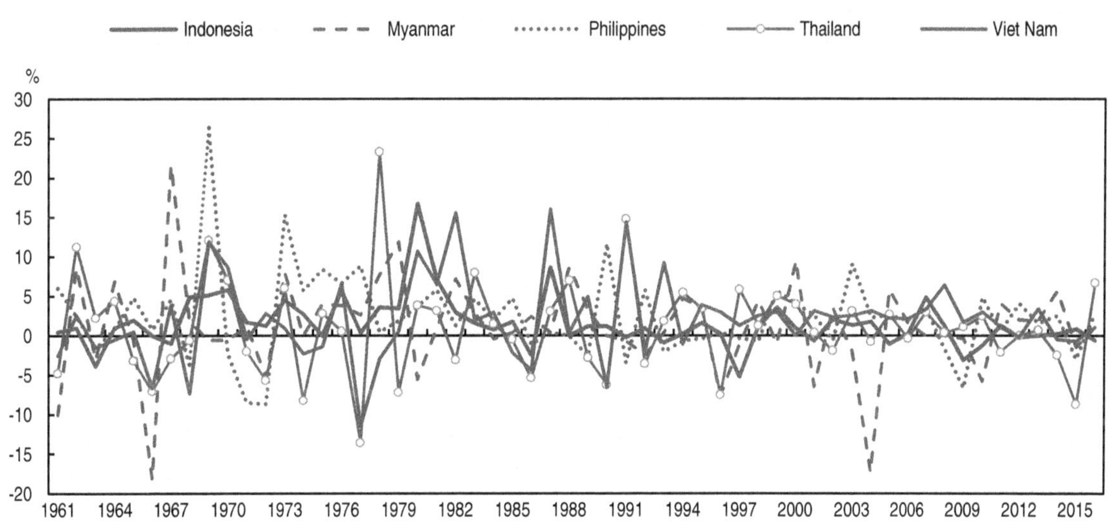

Source: USDA (2016), *Production, Supply and Distribution Online*, http://apps.fas.usda.gov/psdonline/psdquery.aspx.

Considering the historical yield data, this scenario assumes a 15% loss of rice yield in five ASEAN countries (Indonesia, Myanmar, the Philippines, Thailand and Viet Nam). Unlike the previous scenario, the yield loss is country-specific, meaning that the yield loss occurs only in one country while the yield of the four other countries remains constant. The simulation introducing country-specific rice yield loss to the baseline scenario (Furuhashi, 2017 [forthcoming]) shows that the impact of domestic crop failure on rice prices is particularly large in Indonesia (Figure 4.12). Under the restrictive rice trade regime in Indonesia, the domestic rice price is formed predominantly by domestic supply and demand, leading to a surge in domestic price when production contracts. A crop failure in one country also has an impact on the domestic rice price in other countries through a change in the regional market price. The impact on regional prices is particularly high for Viet Nam, where a 15% rice yield loss leads to a 7% increase in the regional rice price.

Figure 4.12. Impact of country-specific crop failure on domestic rice price

Percentage change

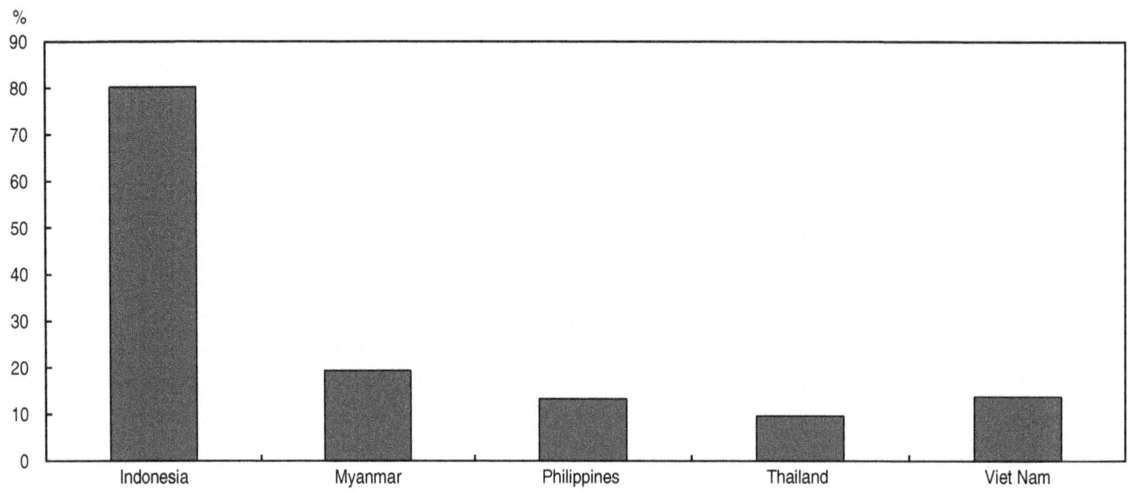

Source: Shocks introduced in the baseline scenario of Furuhashi (2017, forthcoming), based on OECD-FAO (2016).

Scenario 3: Economic downturn

Economic recessions that occasionally occur in Southeast Asia are posing important threats to food insecurity. In some households just above the poverty or undernourishment thresholds, an economic slowdown may have significant effects on poverty incidence and undernourishment. The most severe economic crisis in recent years was the one that accompanied the 1997-99 Asian financial crisis. The economic impacts were particularly large in Indonesia and Thailand. Real consumption in Indonesia and Thailand decreased by 13.1% and 7.6% respectively in 1998 (Figure 4.13).

Economic recessions are not necessarily caused by international crises, however. The recessions of 1981-82 and 1985 in Indonesia, 1984 in the Philippines and 1988 in Myanmar, for example, occurred for domestic reasons. Given the improvement in macroeconomic management in ASEAN countries in recent years, experts and policy makers tend to agree that there is unlikely to be another massive regional-scale economic crisis similar to the Asian financial crisis. Nevertheless, the major economic risk remains an economic downturn at the country level.

The "economic downturn" scenario considers a broad-based macroeconomic slowdown caused by an economy-wide collapse in production. A domestic macroeconomic slowdown also affects incomes, and not necessarily predominantly in investment-related industries as is usually the case with a financial crisis. Instead, the income effects are likely to be widely distributed across industries, although agricultural incomes tend to be less sensitive to domestic macroeconomic conditions. Following the estimation by OECD (2015a), this scenario assumes that real income declines by 15% for non-farm households and 11% for farm households.

The impact of macroeconomic slowdown on rice prices is quantified by the AGLINK-COSIMO model, assuming the contraction of GDP by 10%. However, the impact on domestic rice price is between -0.3% and -0.6% in five countries. The main food insecurity shock in this scenario is a reduction of household income.

Figure 4.13. Annual growth rate of final consumption expenditure, 1961-2013

Percentage

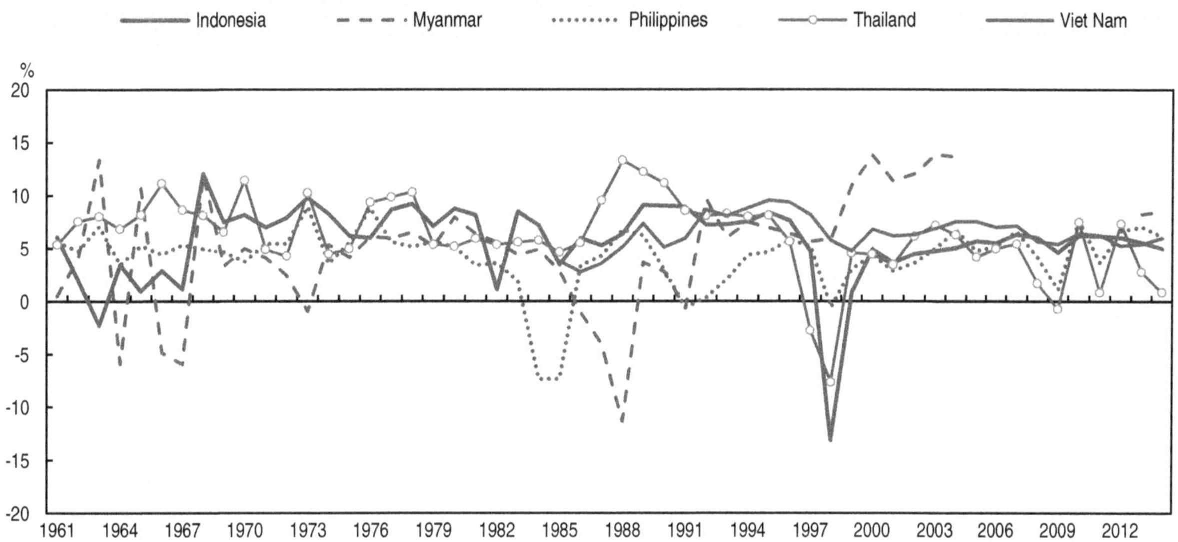

Source: World Bank (2016), *World Development Indicators*, http://data.worldbank.org/indicator.

Impact assessment

The impact of three risk scenarios on the prevalence of undernourishment in five ASEAN member states is quantified by introducing the shocks in rice prices and household income to the household demand system and deriving the change in household calorie consumption. The change in consumption pattern is then translated into the change in household calorie consumption to assess the incidence of undernourishment.

The comparison of the impact of the three risk scenarios shows that an economic crisis leads to the largest impact on undernourishment in Myanmar, Thailand and Viet Nam. However, the regional *El Niño* event has the largest impact on undernourishment in the Philippines (Table 4.1). The increase in the undernourished population in Indonesia is the largest under the domestic crop failure scenario. This is because of the large impact of crop failure on domestic rice prices under the current restrictive rice trade regime. In Thailand, the impacts of all three scenarios are found to be marginal both on farm and non-farm households. Due to its lower expenditure share of food, particularly rice, households in Thailand are much more resilient under the risk scenarios than the other four ASEAN member states.

The adverse impact of high rice prices in Scenarios 1 and 2 is larger for non-farm households in Indonesia, Myanmar, the Philippines and Viet Nam (Figure 4.14). Although high domestic rice prices boost the income of those farm households that are unaffected by crop failure, the simulation result shows that high rice prices increase the incidence of undernourishment among farm households in Indonesia, Myanmar and the Philippines. The farm households in these countries are dominated by small-size subsistent farms or farm workers who purchase a large part of the rice they consume (Figure 4.3). On the other hand, a positive net impact of high price scenarios on the prevalence of undernourishment is found for farm households in Viet Nam, where an increase in farm income due to high rice prices offsets the negative impact of high rice prices as a consumer (Figure 4.14).

An overall risk assessment requires the weighting of different risk scenarios considering the probability of each scenario. The regional *El Niño* event is assumed to occur once in 20 years, considering the historical record of a very strong *El Niño* (1982-83, 1997-98 and 2015-16). Although the probability of domestic risks such as crop failure and economic crisis is likely to be different across countries, this exercise assumed a uniform probability of domestic crop failure (once in every 15 years) and economic downturn (once in every 25 years), following the previous assessment made for Indonesia (OECD, 2015a). The reference scenario is the situation without any shock (once in every two years).

Table 4.1. Frequency of scenarios and prevalence of undernourishment

	0. Without shock	1. Regional *El Niño*	2. Domestic crop failure	3. Economic downturn
A. Average number of years per occurrence	2	20	15	25
B. Prevalence of undernourishment				
Indonesia	13.7	23.6	29.5	19.6
Myanmar	20.2	24.1	26.8	28.0
Philippines	14.0	23.6	20.0	23.1
Thailand	8.0	8.2	8.0	8.9
Viet Nam	14.5	17.8	17.5	23.2
C. Expected increase in undernourishment				
Indonesia	0.0	0.50	**1.05**	0.24
Myanmar	0.0	0.20	**0.44**	0.31
Philippines	0.0	**0.48**	0.40	0.36
Thailand	0.0	0.01	0.00	**0.04**
Viet Nam	0.0	0.16	0.20	**0.35**

Sources: OECD estimates based on data from: Indonesia: BPS-Statistics Indonesia (2010), Myanmar: MNPED (2010), Philippines: Philippine Statistics Authority (2014), Thailand: TNSO (2014), and Viet Nam: General Statistics Office of Viet Nam (2010).

Figure 4.14. Impact of three risk scenarios on the prevalence of undernourishment

Percentage point change in share of population that is undernourished

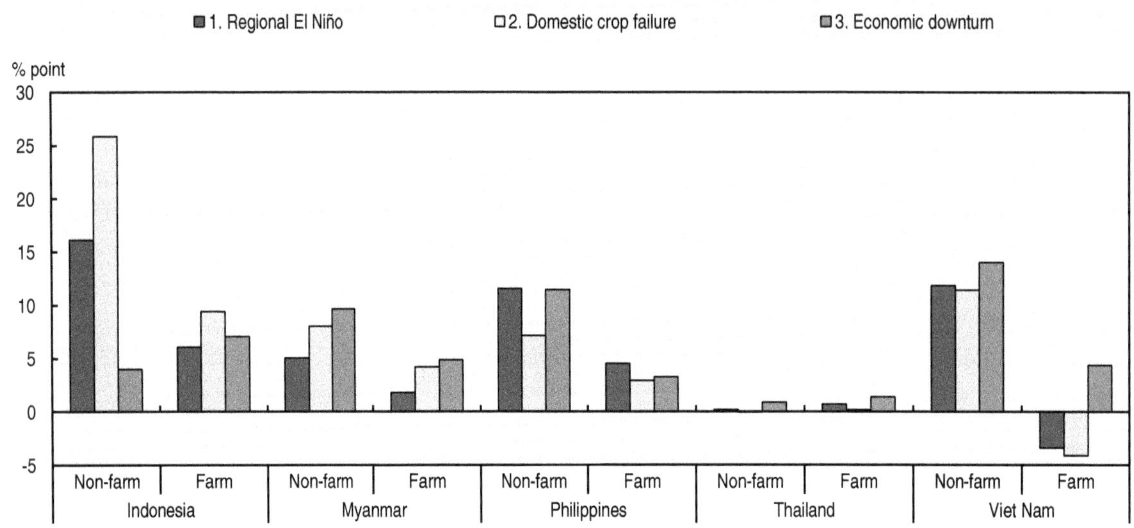

Sources: OECD estimates based on data from: Indonesia: BPS-Statistics Indonesia (2010), Myanmar: MNPED (2010), Philippines: Philippine Statistics Authority (2014), Thailand: TNSO (2014), and Viet Nam: General Statistics Office of Viet Nam (2010).

Based on the assumption of the probability of three risk scenarios, the expected increase in the rate of undernourishment is calculated for each risk scenario (Table 4.1). While the economic downturn scenario leads to the largest increase in the rate of undernourishment, domestic crop failure is found to have a larger expected impact in Myanmar. Even after considering the lower probability of the scenario, the economic downturn still has the largest expected impact on the prevalence of undernourishment in Thailand and Viet Nam. Similarly, domestic crop failure leads to by far the largest expected increase in the rate of undernourishment in Indonesia.

The sum of the expected increase in undernourishment in the three risk scenarios indicates the country's overall exposure to risk. The sum of the expected increase in undernourishment is the largest in Indonesia (1.78), followed by the Philippines (1.24), Myanmar (0.94), Viet Nam (0.71) and Thailand (0.05). This exercise shows that Indonesia is exposed to the largest potential food insecurity risk among the five countries. The Philippines and Myanmar are also exposed to relatively larger food insecurity risks, but the potential food insecurity risk is marginal in Thailand.

4.4 Policy assessment

ASEAN rice market integration policy

The impacts of ASEAN rice market integration on regional and domestic rice prices are taken from a projection for the increased integration scenario, carried out in a recent OECD study (Furuhashi, 2017 [forthcoming]). This scenario assumes a phase-out of tariff and other domestic policy barriers within ASEAN member states, leading to reduced price gaps between domestic and border prices. Regional *El Niño* and domestic crop failure shocks are added to the increased integration scenario to estimate the impact of ASEAN rice market integration under the risk scenarios (Box 4.2).

Box 4.2. ASEAN rice market integration scenario

The scenario analysis assesses the impacts of rice market integration within ASEAN member countries with the use of the AGLINK-COSIMO model.[8] The increased integration scenario chosen for this analysis gradually eliminates the differences between domestic prices and border prices – adjusted for transport costs and quality differences – for all rice trades within the region. This scenario not only involves tariff removal among ASEAN member countries, but also the removal of *non*-tariff trade barriers, including state trading agency involvement. It implicitly assumes that rice exporting countries would not impose export restrictive measures.

In ASEAN rice-trading nations, state trading agencies often have the exclusive right to import rice. For example, the National Food Authority (NFA) of the Philippines, the Bureau of Logistics (BULOG) of Indonesia, and BERNAS of Malaysia are solely in charge of importing rice into their respective countries. NFA accounted for more than 90% of total rice imports into the Philippines in 2014, and the majority of imports were procured through government-to-government (G-to-G) contracts. In the case of Indonesia, BULOG has the monopoly in importing medium-grain rice, and can import only if instructed by the Ministry of Trade. Private traders certified by the Ministry of Trade can however import 100% broken rice as a raw material for the food manufacturing industry. Hoang and Meyers (2015) reveal that – due to the involvement of state trading enterprises, in addition to other domestic policies – *implicit* tariffs on rice in these three countries are much higher than the official tariffs committed to within the ASEAN Free Trade Area.

Under the reference scenario without shocks, domestic rice prices in Indonesia and the Philippines decline by 39% and 45%, respectively, due to the reduction of tariff and non-tariff trade restrictions. The domestic prices in rice-exporting countries (Myanmar, Thailand and Viet Nam) increase by 9-17% due to increased exports to Indonesia and the Philippines (Figure 4.15). Consequently, the overall prevalence of undernourishment in Indonesia and the Philippines is reduced by 1.3 and 7.5 percentage points, respectively (Figure 4.16). As lower domestic rice prices reduce the income of rice farms, the share of undernourished farm households increases in Indonesia by 1.1 percentage points, but that of *non*-farm households falls by 5.2 percentage points. In the Philippines, the prevalence of undernourishment reduces both among farm and non-farm households by 3.5 and 9.2 percentage points, respectively. This is because small-scale subsistence farm households benefit from lower rice prices as a consumer, fully offsetting the negative impacts of lower rice prices as a producer.

The net impact of higher domestic rice prices on overall undernourishment in Viet Nam under ASEAN rice market integration policy is small, mainly due to the reduced incidence of undernourishment among farm households associated with higher farm income. However, high domestic prices increase the undernourished population in Myanmar by 5.6 percentage points, with a particularly large impact on non-farm households (7.2 points). Overall, ASEAN rice market integration reduces the rate of undernourishment in five ASEAN member states by 0.7 percentage points, which is equivalent to a 4.9% reduction.

Figure 4.15. Impact of ASEAN rice market integration on domestic rice price

Percentage change

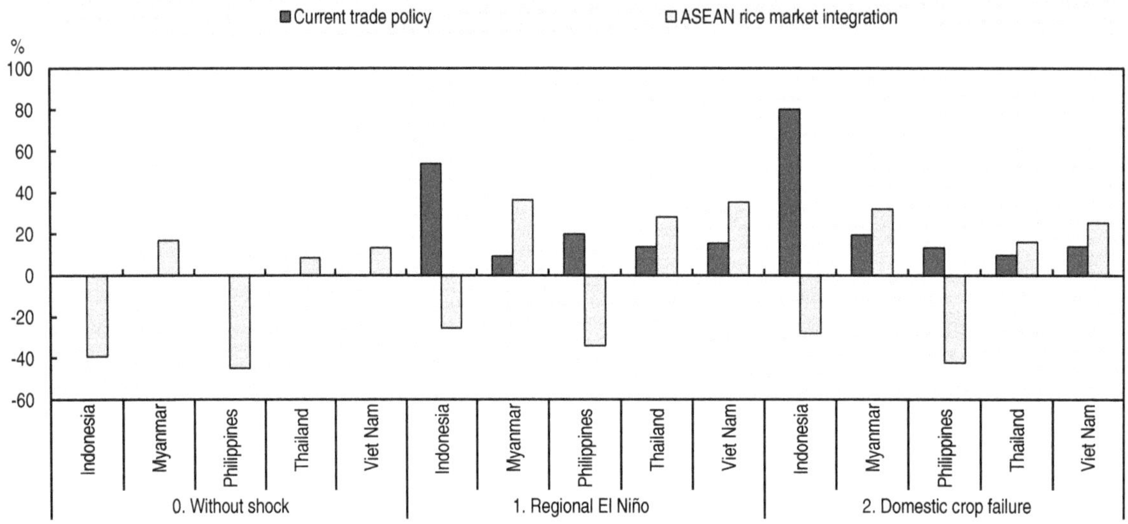

Source: OECD-FAO AGLINK-COSIMO model.

With access to the integrated ASEAN rice market, the domestic price in Indonesia and the Philippines would be lower than the reference level (under current trade policy), even in both the regional *El Niño* and domestic crop failure scenarios. ASEAN rice market integration can thereby offset the negative impact of the two high price risk scenarios on the prevalence of undernourishment in these two rice-importing countries.

On the other hand, the integrated ASEAN rice market leads to higher domestic prices in rice-exporting countries in the production loss scenarios compared with the current trade regime. In particular, the prevalence of undernourishment in Myanmar and Viet Nam increases by 6.9 and 6.3 percentage points more in the regional *El Niño* scenario, and by 3.1 and 3.5 percentage points more in the domestic crop failure scenario, respectively.

Under the current rice trade regime, the regional *El Niño* scenario increases the prevalence of undernourishment in five ASEAN member states by 6.6 percentage points (equivalent to a 49% increase). However, integrating the rice market in ASEAN could mitigate the impact to an increase of 1.5 percentage points in the prevalence of undernourishment (equivalent to an 11% increase). The integration of the rice market in ASEAN mitigates the impact of systemic regional production failure. When the production loss is domestic, the potential risk of surge both in domestic price and the undernourished population can be mitigated by imports from regional rice markets. While the benefit is particularly large for rice-importing countries, exporting countries can also rely on imports from regional markets in the case of emergency. The integrated rice market in ASEAN therefore provides an opportunity to share production risk across countries in the region.

Figure 4.16. Impact of ASEAN rice market integration on the prevalence of undernourishment

Percentage point change in the share of population that is undernourished

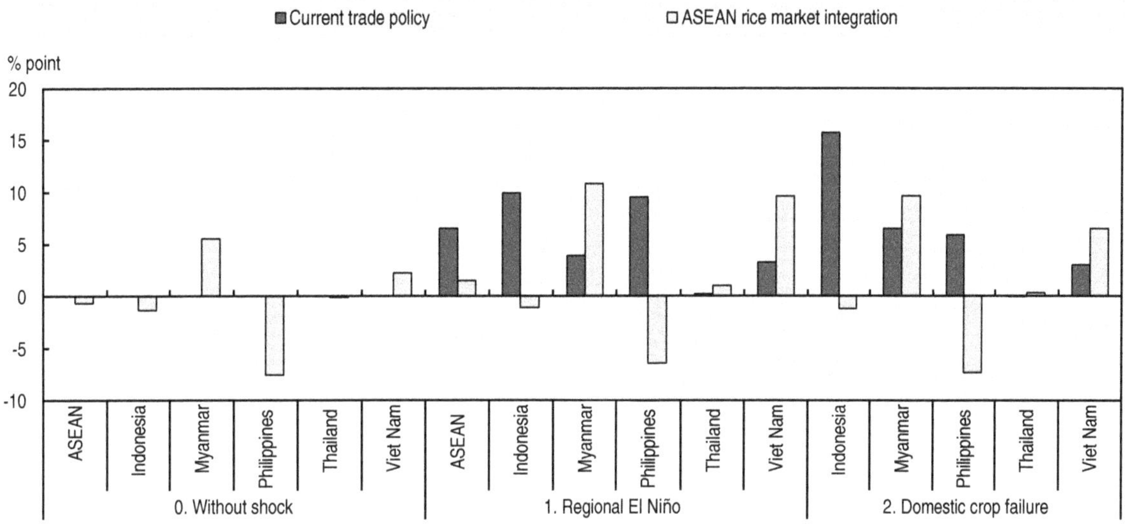

Note: ASEAN total includes only five countries, but these countries account for more than 90% of total population as well as of undernourished population in ASEAN region.

Sources: OECD estimations based on data from: Indonesia: BPS-Statistics Indonesia (2010), Myanmar: MNPED (2010), Philippines: Philippine Statistics Authority (2014), Thailand: TNSO (2014), and Viet Nam: General Statistics Office of Viet Nam (2010).

Safety net policies

Most ASEAN member states implement some form of safety net policies to protect vulnerable households. The majority of countries maintain emergency or buffer stocks of rice to ensure a supply of rice in the event of an emergency (Chapter 3 and Deuss, 2017 [forthcoming]). Rice-importing countries such as Indonesia and the Philippines manage larger stocks, and implement rice distribution programmes which deliver rice at a subsidised price. Self-sufficiency policies to protect domestic rice producers through border protection and domestic support are also considered to be a means to prevent a volatile world price from transmitting to the domestic market price. This section assesses the impacts of current and alternative social safety net policies in two rice-importing countries: Indonesia and the Philippines.

Rice distribution programmes in Indonesia (Raskin) and the Philippines (NFA rice) account for approximately 8% and 13% of domestic rice consumption, respectively (OECD, 2015a; OECD, 2017). Indonesian National Socioeconomic Survey (SUSENAS) data show that the unit price of Raskin rice is on average 16% lower than the market price in Indonesia. Similarly, the wholesale and retail prices of National Food Authority (NFA) rice were more than 20% lower, on average, in 1990-2014, compared with the respective domestic market prices (OECD, 2017). In addition to the large costs involved in both stock management and the physical delivery of rice to consumers, these rice distribution programmes are known to have difficulties in targeting vulnerable households. While Raskin prioritises its delivery to impoverished or almost-impoverished households, roughly 50% of the Indonesian household population purchases Raskin rice (OECD, 2015a). Under the NFA programme, anyone can buy NFA rice sold in the accredited retail stores without needing to be pre-qualified (Fernandez and Velarde, 2012). Jha and Mehta (2008) find that while the programme reaches 16% of the population, only 25% of impoverished people actually benefit from it.

Alternative safety net measures include food vouchers and cash transfer programmes. Both Indonesia and the Philippines have already implemented the latter. Indonesia introduced the unconditional cash transfer programme, *Bantuan Langsung Tunai*, as an *ad hoc* social assistance programme in 2005 (OECD, 2015a), while the Philippines launched the conditional cash transfer programme, *Pantawid Pamilya*, in 2007. The Philippines programme is considered to be highly successful, and currently provides financial grants to

4.4 million poor families with children aged 0-18 years across the country, subject to compliance with education and health requirements related to the children.

A food voucher programme could also be an alternative safety net policy which enables households to purchase a diverse set of food items depending on food consumption patterns in the region. Unlike the rice distribution programme, a voucher programme has the advantage of not requiring governments to purchase and physically deliver rice to consumers. It also has the benefit of allowing consumption to diversify away from rice.

The policy simulation compared the performance of the existing rice distribution programme and alternative targeted cash transfer and food voucher programmes in Indonesia and the Philippines under current rice trade policy. While the recipients of Raskin rice are simulated according to the actual household survey data, those of NFA rice are randomly selected based on the probability of 25% and 11% of poor and non-poor households purchasing the NFA rice. The cash transfer and food voucher programmes are assumed to be targeted at poor populations, and the size of payments are set according to the cost of rice distribution programmes, to make the impact of these programmes comparable.[9]

Existing rice distribution programmes in these countries reduce the prevalence of undernourished people by 1.1 and 1.0 percentage points, respectively, under the reference scenario (Figure 4.17). However, the policy impacts will be lower in both high price scenarios (Scenarios 1 and 2) in Indonesia, and in Scenario 2 in the case of the Philippines. Under the high price scenarios, the relative size of rice price subsidy would be reduced as the base price becomes higher. Figure 4.17 compares the impact of three programmes on the proportion of undernourished populations in Indonesia and the Philippines under four scenarios.

Alternative safety net policies are found to be more effective at reducing the number of undernourished people. The targeted cash transfer programmes reduce undernourished populations by 2.1 and 1.3 points in Indonesia and the Philippines, respectively. The cash transfer programme performs well across all of the risk scenarios, as it directly supports the household income, and the impact will be larger in the risk scenarios when households have reduced purchasing power. It shows a higher performance than rice distribution programmes, or than food vouchers in most scenarios, as it is more targeted at the poor population.

Food voucher programmes perform better than cash transfers – with reductions of 2.6 percentage points in Indonesia and 1.3 points in the Philippines – in the reference scenario without shock. This is due to less leakage to the consumption of non-staple food and non-food items compared with the cash transfer programme. However, the performance of the food voucher programme is lower than the cash transfer programme in the high price scenarios (Scenarios 1 and 2), because the relative size of the subsidy is reduced, as is the case of the rice distribution programme. The simulation shows that in Indonesia, the effectiveness of the rice distribution and food voucher programmes is reduced under the high price scenarios. In contrast, the cash transfer programme is counter-cyclical by design, with the result that its effectiveness increases when households suffer from income shock.

The expected performance of both the ASEAN rice market integration policy and safety net programmes are compared in Indonesia and the Philippines, weighted by the probability of each scenario. In the absence of either policy, the four risk scenarios lead to an increase in the undernourished populations by 1.8 and 1.2 percentage points, respectively (Figure 4.18). The simulation indicates that the integration of the ASEAN rice market is most effective at reducing undernourished populations both in Indonesia (2.4 points) and the Philippines (5.8 points). It also shows that rice market integration has the capacity to fully offset the expected adverse impact on undernourishment across all four risk scenarios. The rice distribution programme, on the other hand, has the lowest expected performance and does not offset the potential food insecurity impact across risk scenarios, either in Indonesia or the Philippines.

Figure 4.17. Impact of safety net policies on the prevalence of food insecurity: Indonesia and the Philippines

Percentage point change in the share of population that is undernourished

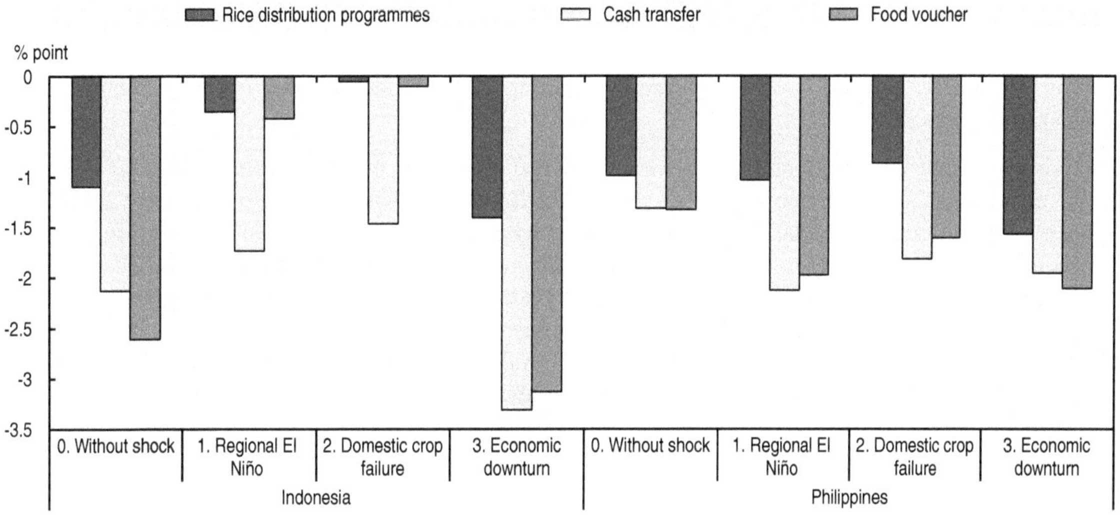

Sources: OECD estimations based on data from: Indonesia: BPS-Statistics Indonesia (2010), and Philippines: Philippine Statistics Authority (2014).

Figure 4.18. Expected policy impacts under risk scenarios in Indonesia and the Philippines

Percentage point change in the share of population that is undernourished

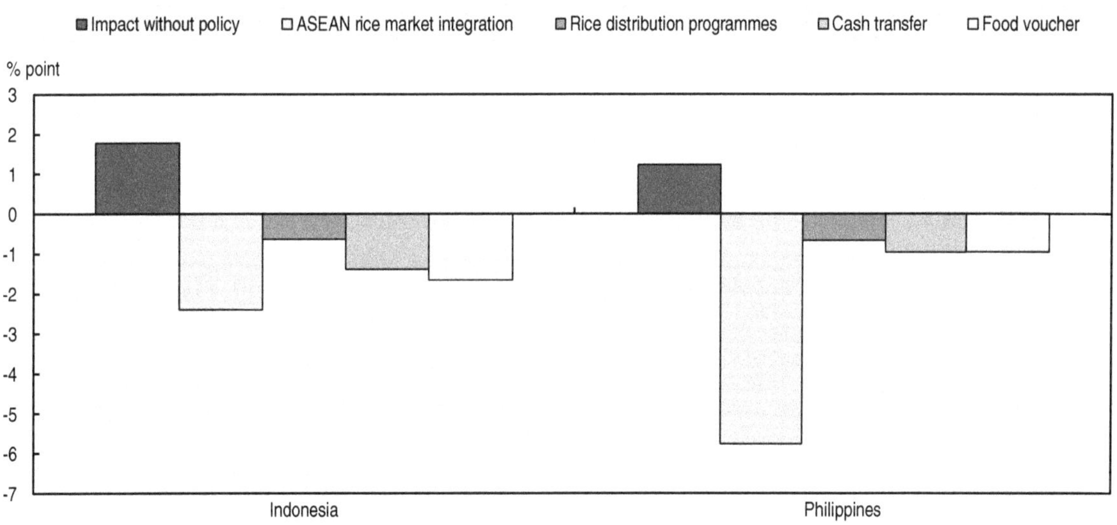

Sources: OECD estimations based on data from: Indonesia: BPS-Statistics Indonesia (2010), and Philippines: Philippine Statistics Authority (2014).

While ASEAN rice market integration provides an opportunity for two rice-importing countries to reduce the impacts of food insecurity risk, it increases regional and domestic rice prices in three rice-exporting countries (Myanmar, Thailand and Viet Nam). The impact can be particularly large for non-farm households in Myanmar and Viet Nam. The development of effective safety net measures such as targeted cash transfer and food voucher programmes in these exporting countries would mitigate the potential adverse impact of integrating regional rice markets, and is therefore an important policy agenda to complement rice market integration in the region.

Exporting countries occasionally avail of export restrictions in order to restrain increases in domestic price. Given that the top four rice exporters (India, Thailand, Viet Nam and Pakistan) have a 70% market share in the global rice market, the use of such policies by exporting countries can be a significant source of food insecurity in the region. This is exactly what occurred in May 2007, when India imposed an export ban on non-basmati exports to ensure domestic supply. Viet Nam also implemented similar export restrictions on rice (Jones and Kwieciński, 2010). While it is a fiscally low-cost policy measure which also has an immediate effect, it reduces both the incentive for investment by industries and also growth potential (Chapter 3 and World Bank, 2014). In addition, it induces importing countries to seek self-sufficiency.

The simulation result also implies that the gradual integration of regional rice markets would avoid a sharp increase in rice prices in exporting countries. While ASEAN member states are building a policy co-ordination and information exchange mechanism to facilitate more reliable regional rice markets, including the establishment of emergency stocks (Box 4.3), the integration of the ASEAN rice market could take several steps. The first step is to increase the involvement of the private sector in rice trade (FAO, 2011). In the ASEAN region, the private sector plays a smaller role than state trading agencies in most rice-trading nations, with the exception of Thailand. The majority of rice imported by large Southeast Asian importers such as the Philippines, Indonesia and Malaysia is through government-to-government (G-to-G) deals. The mandates of NFA in the Philippines and BULOG in Indonesia could be reduced to neutral managers of emergency stocks so that private traders could be more involved in rice imports.

Similarly, on the export side, the Vietnam Food Association (VFA) controls rice exports from Viet Nam. Two export agencies of VFA, Vinafood 1 and 2, account for nearly half of total rice exports, while other smaller joint stock companies – of which the government is a shareholder – and private traders account for the rest (Hoang, 2009). Following the Thai example, Viet Nam could allow its private exporters to play a significant role in the export market and reduce the role of Vinafood 1 and 2 in rice exports.

Box 4.3. Policy co-ordination and information exchange mechanisms in Southeast Asia

Since the 2007-08 rice crisis, ASEAN member states have revamped their efforts to improve regional food security through policy co-ordination and information exchange among ASEAN countries. In 2003, the member states adopted the ASEAN Integrated Food Security (AIFS) framework. The AIFS framework provides measures and activities to facilitate co-operation in the implementation of an integrated food security information system. In 2012, the first Asian Rice Trade Forum was held in Siem Reap, Cambodia, with the aim of promoting coherent policy actions on rice trade for the improvement of food security in the ASEAN region. The food security information system is designed to provide an annual and semi-annual commodity outlook for rice and other crops, and additionally provides estimates of crop losses caused by extreme weather events and market information.

The aftermath of the rice crisis in 2007-08 witnessed a flurry of debates among eminent experts on the need for a rice reserve to stabilise market prices and avoid a repeat of the 2007 rice spike. Within Southeast Asia, debate has been extensive on the need for the ASEAN Plus Three emergency rice reserve (APTERR) as a tool for market stabilisation. The APTERR began in 2002 as East Asia Emergency Rice Reserve, replacing the ASEAN Emergency Rice Reserve (AERR), with the addition of three East Asian countries (Japan, China, and Korea) to revive regional co-operation for strengthening food security and reducing poverty among ASEAN countries. The AERR, which operated between 1979 and 2002, with voluntary contributions of 87 000 metric tonnes (mt) of rice stocks, was never used by the member countries. Even in 1997, during the Asian financial crisis, Indonesia preferred to borrow money from the International Monetary Fund to purchase rice from the global market rather than access the rice from the AERR. The reluctance of the member countries to access rice from the AERR was partly because the reserves were too small and also because of the complexity of the release mechanism. The release process involves a direct request by a member state to another member country, and the terms of trade would be negotiated between the two countries concerned. The release mechanism, which was similar to purchasing rice from the open market or through a Government-to-Government (G-to-G) agreement, dissuaded countries from accessing the reserve.

The APTERR has two types of reserves: earmarks and stockpiles. The earmarked stocks are released through special commercial transactions between two member countries, and the APTERR secretariat acts as market intermediary or broker (Briones, 2011, see also Chapter 3). The size of the earmarked stocks is 837 000 mt, with most of this obtained from the Plus Three countries (Briones, 2011). Release from the stockpile is designed for emergency and humanitarian purposes. Since its establishment in 2004, rice from the stockpile has been released in ten instances under the emergency relief and poverty alleviation programmes.

Transparency and price formation in the ASEAN market could be vastly improved by the establishment of a futures market in the region. A pre-feasibility study by McKenzie (2012) of a rice futures market in ASEAN concluded that two major economic benefits of a futures market are price discovery and price risk management. However, the author concluded that the futures market will have a limited role in stabilising prices across multiple years.

The success of a rice futures market largely depends on liquidity (Mohanty, 2011). Without adequate liquidity, a futures contract cannot be used as an effective tool for hedging, and the market will be exposed to manipulation. In addition, the price discovery role of a futures market will be compromised without adequate liquidity. The current environment of the ASEAN market, in which G-to-G deals account for the bulk of the trade, is not ideal for a futures market to thrive. Apart from private sector participation in rice trade, several other conditions such as regional co-operation on policies, harmonisation of quality standards, the formation of an independent regulatory body, and others, are necessary elements for the successful creation of an ASEAN rice futures market (McKenzie, 2012).

4.5 Concluding comments

While economic growth has yielded significant declines in populations suffering from chronic poverty within ASEAN, the *nature* of food insecurity has shifted in a significant way from permanent to transitory. This means that while some households currently have adequate access to food, they are nevertheless at risk of losing this access as a result of temporary shocks that alter incomes and prices, such as those related to economic events, natural disasters or domestic or international shocks to markets.

Chapter 1 of this publication provides an overview of the food security status of the ASEAN member states, while Chapter 3 reviews the current agriculture-related policy approaches to address food security. The review indicates that the current policy approach in ASEAN countries is largely oriented towards relatively short-term risk management through measures such as rice price stabilisation and social safety nets rather than addressing the objective of long-term food security.

Based on the preceding work, this chapter assesses the major risks of undernourishment in the ASEAN region and evaluates the performance of current and alternative policy approaches, applying the OECD's analytical framework of transitory food insecurity (OECD, 2015a). The analysis takes into account specific food insecurity situations in five ASEAN member countries, which account for more than 90% of the population in the region: Indonesia, Myanmar, the Philippines, Thailand and Viet Nam.

Some key findings related to investigating food security at the household level in five ASEAN countries include:

- A strong link is found between poverty and undernourishment. Low-income households are also found to be more vulnerable to market or income shocks, indicated by a larger reduction of rice consumption in response to an increase in rice price or a loss of income.
- Farm households producing food crops sometimes have better access to food, even if their income levels are lower. Farm households often consume their own food products, while non-farm households purchase nearly 90% of rice that they consume. However, farm households also purchase a significant part of the rice they consume from the market, particularly low-income farm households in the Philippines, Indonesia and Myanmar, where small-scale subsistence farms or landless farm workers dominate the sector. For example, the poorest 20% farm households in the Philippines purchase on average 77% of rice that they consume.
- Rice is a staple among ASEAN member states, accounting for a large part of energy and nutrition intake, particularly for low-income households. There is a clear trend that richer households have a lower share of rice in both their expenditure and calorie intake.

- Per capita quantity of rice consumption is lower for higher-income households in Thailand, Viet Nam and Indonesia. In these countries, negative income elasticity for rice consumption was estimated among high-income households, indicating that rice may become an inferior good as incomes rise. In Myanmar, high-income households consume more rice per capita. These trends are consistent with the experience of economic development and rice consumption in East Asia. Per capita rice consumption in these ASEAN member states is therefore likely to decline in future as per capita income levels continue to grow.

Three major risk scenarios were identified in the region through several consultation events with ASEAN, and their impact on the prevalence of undernourishment in five ASEAN member states was quantified. Three risk scenarios identified through this exercise include: 1) a region-wide *El Niño* event which leads to yield losses in major rice-producing countries in the region, 2) country-specific crop failure due to natural disaster, and 3) country-specific economic downturn. The risk assessment shows that both the regional *El Niño* event and domestic crop failure lead to high rice prices and have particularly high adverse effects on undernourishment among non-farm households, but that these events also increase the undernourished population among *farm* households in Indonesia, Myanmar and the Philippines. On the other hand, the economic downturn scenario increases the incidence of undernourishment for both farm and non-farm households in all five countries.

Considering the probability of risk scenarios, domestic crop failure has the largest expected impact on undernourishment in Indonesia and Myanmar, while the regional *El Niño* event was identified as the largest risk for the Philippines. An economic downturn was found to be the largest risk to food security in Thailand and Viet Nam. Overall, the risk assessment shows that Indonesia is exposed to the largest risk of increasing undernourishment under the scenarios considered, followed by the Philippines and Myanmar. On the other hand, these risks only posed marginal threats to food security in Thailand.

The analysis found that there are alternative policies to those currently in place that can significantly *reduce* both current levels of undernourishment and *mitigate* the potentially large impacts of risk scenarios on undernourishment. In particular, among the alternative policies examined, ASEAN rice market integration had the largest positive effects. While such moves could increase the regional price of rice, they lead to reductions in the domestic price of rice in importing countries such as Indonesia and the Philippines. These changes lead to significant reductions in undernourishment: overall, ASEAN rice market integration would reduce the undernourished population by 5% in the five ASEAN members examined. The integration of regional rice markets was found to be an effective policy alternative under the production risk scenarios. In particular, the impact of a regional *El Niño*, which has the potential to increase the number of undernourished by up to 49% in the five ASEAN members examined, could be mitigated to an 11% increase if rice markets are integrated.

Across the five countries examined, rice market integration reduces undernourished populations in Indonesia and the Philippines by 10% and 54% respectively by improving access to rice through lower domestic prices. The analysis indicates that the respective rice price stabilisation policies in these two countries that place restrictions on trade comes at the cost of higher price levels in domestic markets, leaving the households more vulnerable to production risks. The rice market integration policy can also offset the impact of the regional *El Niño* and domestic crop failure scenarios on undernourishment, which are identified as the largest risks to food insecurity in these two countries. While the benefit is particularly large for rice-importing countries, exporting countries can also rely on imports from the regional market in the event of emergency. The integrated rice market in ASEAN therefore provides an opportunity to share production risks across countries in the region and, as a result, is a significant means to manage risk of food insecurity in the region. Although the ASEAN rice market integration scenario does not explicitly define the removal of restrictive measures on rice exports, the analysis implies that rice export restrictions imposed by rice exporting countries would increase the regional rice price, leading to higher rates of undernourishment in rice importing countries as well as in exporting countries which do not impose such measures. The scope of the analysis is limited to the rice market – however, the results also imply the potential benefit of a more integrated regional market to manage production risk of *other* commodities such as maize and soybean.

In addition to rice market integration, the impacts of current and alternative safety net policies were compared in two rice-importing countries: Indonesia and the Philippines. The analysis shows that existing programmes related to subsidised rice distribution in both Indonesia and the Philippines do not offset the adverse impacts on undernourishment of either the higher prices generated by the current policies used to stabilise prices and stimulate rice production, nor the potential impacts of the risk scenarios examined. Alternative safety net policies such as cash transfers and food voucher programmes show a higher performance in reducing undernourishment across scenarios. However, the targeting of social safety net policies at vulnerable households is a challenging task for governments. One way of improving targeting is to link these programmes with general social security policy through identification cards. In 2013, for example, the Indonesian government introduced a Social Protection Card to identify low income households that were eligible for social programmes.

Of the alternative policies explored, ASEAN rice market integration is, nonetheless, found to be more effective in reducing the risk of food insecurity across the scenarios in Indonesia and the Philippines. However, it should be noted that this chapter examines only a subset of risk management policies. For example, farm risk management tools such as crop insurance are not examined. Production risk in agriculture could also be mitigated by medium- to long-term investment in innovation – for example, by the development of a drought-tolerant rice variety, as highlighted in Chapter 2. It is also important to note that improving agricultural productivity and increasing income opportunities for vulnerable households through a variety of policies are essential pre-requisites for long-term food security. With respect to consumption, the diversification of diets away from the traditional dominance of staple rice is necessary to improve existing micronutrient deficiency.

Higher regional rice prices resulting from regional market integration increase the incidence of undernourishment among low-income non-farm households in two rice-exporting countries: Myanmar and Viet Nam. The development of targeted safety net measures in these exporting countries is also an important policy agenda to complement rice market integration in the region. The analysis also implies that the gradual integration of regional rice markets would enable a sharp increase in rice prices in exporting countries to be avoided, and make the process of rice trade liberalisation more politically feasible. Tariffs on rice should be further reduced beyond the reductions committed to by ASEAN Economic Community members in 2015. The reforms should also start with the greater involvement of the private sector in regional rice trade. In particular, Viet Nam could allow its private exporters to play a significant role in the export market, reducing the role of state-owned enterprises in rice exports. Finally, the role of state agencies in managing rice imports and buffer stocks (such as NFA in the Philippines and BULOG in Indonesia) could be reduced to neutral managers of emergency stocks, thus enabling private traders to become more involved in rice imports.

Notes

1. This chapter benefited greatly from input from a number of contributors. In particular, the estimations of the food demand system based on household expenditure survey data for Myanmar, Thailand and Viet Nam were performed by Hoa Hoang, Research Associate, Food and Agricultural Policy Research Institute (FAPRI), University of Missouri-Columbia. In addition, a review of the rice market structure in Southeast Asia was provided by Samarendu Mohanty, International Rice Research Institute, and Roderick Rejesus, North Carolina State University.

2. ASEAN is comprised of Brunei Darussalam, Cambodia, Indonesia, Lao PDR, Malaysia, Myanmar, the Philippines, Singapore, Thailand and Viet Nam.

3. The threshold of undernourishment based on per capita calorie consumption is set to replicate the prevalence of undernourishment published in FAO food security indicators for the corresponding years. The quantity of consumption is converted to calorie consumption based on the ASEAN Food

Composition Tables in Thailand, the Philippines and Viet Nam. For Indonesia and Myanmar, a calorie conversion table attached to the household survey data was used to make such calculations.

4. The definition of an inferior good is based solely on the inverse relationship between demand for the commodity and the income level of consumers, holding everything else constant.

5. Annex 3.A in OECD (2015a) provides the documentation for the AIDS estimation in Indonesia. The simulation in the Philippines applied the estimated parameters in Lantican, Sombilla and Quilloy (2013).

6. The stylised scenarios were identified in a series of consultations with ASEAN member countries through two OECD-ASEAN events co-organised with the ASEAN Secretariat and regional organisations.

7. The impact on domestic rice price is approximated by change in producer price in the AGINK-COSIMO model. Since Myanmar is not included as a single country in the AGINK-COSIMO model, the impact in Myanmar is approximated by that of the group of least developed Asian countries, including Cambodia and Lao PDR, in which Myanmar accounts for 60% of rice production in 2013, according to FAOSTAT (FAO, 2016).

8. The AGLINK-COSIMO model represents Indonesia, Malaysia, the Philippines, Thailand and Viet Nam individually, but Myanmar, Cambodia and Lao PDR are aggregated within the group of least developed Asian countries. This scenario analysis assesses rice market integration between Indonesia, Malaysia, the Philippines, Thailand, Viet Nam and the group of least developed Asian countries, some of which are members of ASEAN. Furuhashi (2017, forthcoming) documents more technical details of implementing ASEAN rice market integration, using the AGLINK-COSIMO model.

9. In the absence of budgetary data, the cost includes only the rice price subsidy component of the programme, excluding stock, logistics and other programme management costs. The potential impacts of alternative programmes are likely to be larger once the larger cost of administering the rice distribution programme is considered – this is not necessarily reflected in the simulation.

References

BPS-Statistics Indonesia (2010), *National Socioeconomic Survey,* Central Bureau of Statistics of Indonesia.

Banks, J., R. Blundell and A. Lewbel (1997), "Quadratic Engel curves and consumer demand", *Review of Economics and Statistics*, Vol. 79/4, pp. 527–539.

Briones, R.M. (2011), "Regional cooperation for food security: The case of emergency rice reserves in the ASEAN Plus Three", *ADB Sustainable Development Working Paper Series*, No. 18, Asian Development Bank, Manila.

Clermont-Dauphin, C., N. Sunnawang, O. Grünberger, C. Hammecker and J.L. Maeght (2010), "Yield of rice under water and soil salinity risks in farmers' fields in northeast Thailand", *Field Crops Research,* Vol. 118/3, Elsevier, Amsterdam, pp. 289–296.

Cox, T. and M. Wohlgenant (1986), "Prices and quality effects in cross-sectional demand analysis", *American Journal of Agricultural Economics,* Vol. 68/4, Blackwell Publishing, pp. 908-919.

Deaton, A. (1988), "Quality, quantity, and spatial variation of price", *The American Economic Review*, Vol. 78/3, American Economic Association, pp. 418–430.

Deaton, A. and J. Muellbauer (1980), "An almost ideal demand system", *American Economic Review*, 70/3, pp. 312–326.

Deuss, A. (2017), *Market Impacts of Asian Public Stockholding Policies Related to Rice*, OECD publications, forthcoming.

Dey, M.M. and H.K. Upadhyaya (1996), "Yield loss due to drought, cold, and submergence in Asia", in *Rice Research in Asia: Progress and Priorities*, CAB International, Oxon, United Kingdom, and the International Rice Research Institute, Manila.

FAO (2016), *FAOSTAT* (database), www.fao.org/faostat/en/#home (accessed on 2 September 2016).

FAO (2011), "The 2007-08 rice price crisis: How policies drove up prices… and how they can help stabilize the market", *Economic and Social Perspectives Policy Brief 13*, Food and Agriculture Organization, Rome.

FAO (2006), "Food security", *Policy Brief*, Issue 2, June 2006, Food and Agriculture Organization, Rome.

Fernandez, L. and R. Velarde. (2012), "Who benefits from social assistance in the Philippines? Evidence from the latest national household surveys", *Philippine Social Protection Note,* No. 4, The World Bank, Washington, DC.

Furuhashi, G. (2017), *Market Implications of Integration of Asian Rice Markets*, OECD publications, forthcoming.

General Statistics Office of Viet Nam (2010), *Household Living Standard Survey*, Ha Noi.

Gibson, J. and S. Rozelle (2011), "The effects of price on household demand for food and calories in poor countries: Are our databases giving reliable estimates?", *Applied Economics*, Vol. 43/27, pp. 4021–4031, Taylor and Francis, London.

Hoang, L.V. (2009), "Estimation of food demand from household survey data in Vietnam", *Working Paper,* No. 12, Development and Policies Research Center, Viet Nam.

Hoang, H.K. and W.H. Meyers (2015), "Price stabilization and impacts of trade liberalization in the Southeast Asian rice market", *Food Policy*, Vol. 57, Elsevier, Amsterdam, pp. 26-39.

Iizumi, T., J-J. Luo, A.J. Challinor, G. Sakurai, M. Yokozawa, H. Sakuma, M.E. Brown and T. Yamagata (2014), "Impacts of *El Niño* Southern Oscillation on the global yields of major crops", *Nature Communications,* Vol. 5, Springer Nature.

Ito, S., E.W.F. Peterson and W.R. Grant (1989), "Rice in Asia: Is it becoming an inferior good?", *American Journal of Agricultural Economics*, Vol. 71/1, Oxford University Press, pp. 32-42.

Jha, S. and A. Mehta (2008), "Effectiveness of public spending: The case of rice subsidies in the Philippines", *ADB Economics Working Paper Series,* No. 138, Asian Development Bank, Manila.

Jones, D. and A. Kwieciński (2010), "Policy responses in emerging economies to international agricultural commodity price surges", *OECD Food, Agriculture and Fisheries Papers*, No. 34, OECD Publishing, Paris, http://dx.doi.org/10.1787/5km6c61fv40w-en.

Lantican, F.A., M.A. Sombilla and K.P. Quilloy (2013), *Estimating the Demand Elasticities of Rice in the Philippines,* Southeast Asian Regional Center for Graduate Study and Research in Agriculture, Laguna, Philippines.

Lecocq, S. and J.-M. Robin (2015), "Estimating almost-ideal demand systems with endogenous regressors", *The Stata Journal*, Vol. 15/2, pp. 554–573.

Lian, K.K. and L. Bhullar (2011), "Governance on adaptation to climate change in the Asean region", *Carbon and Climate Law Review*, Volume 5/1, pp. 82-90.

Lloyd's (2015), "Food system shock: The insurance impacts of acute disruption to global food supply". *Emerging Risk Report: 2015 Innovation Series*, London.

Maddison-Project (2013), *Maddison Project* (database), www.ggdc.net/maddison/maddison-project/home.htm (accessed on 2 September 2016).

Majumder, A., R. Ray and K. Sinha (2012), "Calculating rural-urban food price differentials from unit values in household expenditure surveys: A comparison with existing methods and a new procedure", *American Journal of Agricultural Economics*, Vol. 94/5, Oxford University Press.

McKenzie, A. (2012), "Prefeasibility study of an ASEAN rice futures market", *ADB Sustainable Development Working Paper Series,* No. 19, Asian Development Bank, Manila.

Mergenthaler, M., K. Weinberger and M. Qaim (2009), "The food system transformation in developing countries: A disaggregate demand analysis for fruits and vegetables in Vietnam", *Food Policy*, Vol. 34/5, Elsevier, Amsterdam, pp. 426–436.

MNPED (2010), *Integrated Household Living Conditions Survey in Myanmar (2009-2010),* Myanmar Ministry of National Planning and Economic Development, United Nations Development Programme, United Nations Children's Fund, and the Swedish International Development Cooperation Agency, Yangon, Myanmar.

Mohanty, S. (2011), *Singapore: Home of an International Futures Market*, International Rice Research Institute, Manila.

Mohanty, S., R. Wassmann, A. Nelson, P. Moya and S.V.K. Jagadish (2013), "Rice and climate change: Significance for food security and vulnerability", *IRRI Discussion Paper Series,* No. 49, International Rice Research Institute, Manila.

MRC (2010), "Impacts of changes in salinity intrusion, assessment of basin-wide development", *Technical Note 8*, Mekong River Commission, Phnom Penh and Vientiane.

Niimi, Y. (2005), "An analysis of household responses to price shocks in Vietnam: Can unit values substitute for market prices?", *PRUS Working Paper,* No. 30, Poverty Research Unit at Sussex, University of Sussex, United Kingdom.

NIS (2013), *Cambodia Socio-Economic Survey 2013*, National Institute of Statistics, Ministry of Planning, Phnom Penh.

OECD (2017), *Agricultural Policies in the Philippines*, OECD Publishing, Paris, http://dx.doi.org/10.1787/9789264269088-en.

OECD-FAO (2016), *OECD-FAO Agricultural Outlook 2016-2025*, OECD Publishing, Paris, http://dx.doi.org/10.1787/agr_outlook-2016-en.

OECD (2015a), *Managing Food Insecurity Risk: Analytical Framework and Application to Indonesia*, OECD Publishing, Paris, http://dx.doi.org/10.1787/9789264233874-en.

OECD (2015b), *Agricultural Policy Monitoring and Evaluation 2015*, OECD Publishing, Paris, http://dx.doi.org/10.1787/agr_pol-2015-en.

OECD (2001), *Market Effects of Crop Support Measures*, OECD Publishing, Paris, http://dx.doi.org/10.1787/9789264195011-en.

Pandey S., H. Bhandari and B. Hardy (eds.) (2007), *Economic Costs of Drought and Rice Farmers' Coping Mechanisms: A Cross-Country Comparative Analysis*, International Rice Research Institute, Manila.

Peng, S., J. Huang, J.E. Sheehy, R.C. Lanza, R.M. Visperas, X. Zhong, G.S. Centeno, G.S. Khush and K.G. Cassman (2004), "Rice yields decline with higher night temperature from global warming", *PNAS*, Vol. 101/27, Proceedings of the National Academy of Sciences of the United States of America, Washington, DC.

Philippine Statistics Authority (2014), *Survey of Food Demand for Agricultural Commodities in the Philippines,* Department of Agriculture Bureau of Agricultural Statistics, Quezon City, Philippines.

Poi, B. (2013), "Easy demand system estimation with quaids", *The Stata Journal*, Vol. 12/3, pp. 433–446.

Ray, R. (1983), "Measuring the costs of children : An alternative approach", *Journal of Public Economics*, Vol. 22/1, Elsevier, pp. 89–102.

Redfern, S.K., N. Azzu and J.S. Binamira (2012), "Rice in Southeast Asia: Facing risks and vulnerabilities to respond to climate change", in *Building Resilience for Adaptation to Climate Change in the Agriculture Sector: Proceedings of a Joint FAO/OECD Workshop*, Food and Agriculture Organization, Rome.

TNSO (2014), *Household Socio-Economic Survey*, National Statistics Office of Thailand, Bangkok.

USDA (2016), *Production, Supply and Distribution Online* (database), United States Department of Agriculture, https://apps.fas.usda.gov/psdonline/app/index.html#/app/home (accessed on 2 September 2016).

Vu, L. (2009), *Estimation of Food Demand from Household Survey Data in Vietnam*, *Working Paper,* No. 12, Development and Policies Research Center, Vietnam.

Welch, J.R., J.R. Vincent, M. Auffhammer, P.F. Moya, A. Dobermann and D. Dawe (2010), "Rice yields in tropical/subtropical Asia exhibit large but opposing sensitivities to minimum and maximum temperatures", *PNAS*, Vol. 107/33, Proceedings of the National Academy of Sciences of the United States of America, Washington, DC.

World Bank (2016), *World Development Indicators* (database), http://data.worldbank.org/indicator (accessed on 2 September 2016).

World Bank (2014), "Myanmar: Rice price volatility and poverty", *Economic and Sector Work*, Report No. 89687-MM, World Bank, Washington, DC.

Annex 4.A1

Descriptions of household survey data

Myanmar

The Integrated Household Living Conditions Survey (IHLCS) conducted by the UNDP, UNICEF, the Swedish International Development Cooperation Agency (SIDA), and the Myanmar Ministry of National Planning and Economic Development in 2009 and 2010, is used for the analysis. IHLCS is a nationwide quantitative survey of 18 660 households, with two rounds of data collection (December 2009/January 2010 and May 2010). The second round, conducted in May 2010, is the follow-up of the first round, with the number of households unchanged. The Technical Report associated with the 2010 IHLCS shows that both rounds provide similar statistics, although the average of both rounds was used for reporting. For the purpose of this analysis, we only use data provided from the first round of the survey.

The survey generally comprises nine modules, including:

Module 1: Household Basic Characteristics
Module 2: Housing
Module 3: Education
Module 4: Health
Module 5: Consumption Expenditures
Module 6: Household Assets
Module 7: Labour and Employment
Module 8: Business
Module 9: Finance and Savings

In this analysis, we mainly draw information from Modules 1, 3 and 7 for household demographic characteristics and Module 5 for food and non-food consumption expenditure.

For food consumption expenditure, the survey asks for information on three different types of consumption, based on the household's frequency of purchasing these items on a regular basis (food consumption expenditure in the last seven days, last 30 days, and implicit food expenditure). There are more than 200 different food items in total.

Food consumption expenditures in the last seven days:

- pulses, beans, nuts and seeds
- meat, dairy products, eggs
- fish and other seafood
- roots and tubers
- vegetables
- fruits
- spices and condiments
- other food products
- alcoholic beverages
- food and beverages consumed outside the home.

Food consumption expenditures in the last 30 days:

- rice and cereals
- oil and fats
- milk products
- other food items (tea, coffee, sugar, etc.).

Specific information collected includes:

- the quantity and value of each food item purchased in cash
- the quantity of each food item obtained in kind through barter or received as gifts, loans, wages or payment
- the quantity of each home-produced food item consumed.

Similarly, for non-food consumption expenditures, the survey asks for two types of consumption based on the frequency of purchasing these items on a regular basis (non-food consumption expenditure in the last 30 days and 6 months)

Non-food consumption expenditure in the last 30 days:

- energy for household use
- water
- personal apparel
- medicines/drugs (including traditional medicine)
- local transport (daily travel)
- other non-food items (telephone services, cigarettes, entertainment, etc.).

Non-food consumption expenditures in the last six months:

- clothing and other apparel
- home equipment
- house rent and repair
- health (including traditional medicine)
- education
- travel/trips (overnight travel)
- other (household worker services, etc.).

Specific information collected includes:

- the value of each non-food item purchased in cash
- the value of each non-food item obtained in kind through barter or received as gifts, loans, wages or payment.

Thailand

The Household Socio-Economic Survey (HSES) conducted by the National Statistics Office of Thailand in 2014 is used for the analysis. HSES is a nationwide quantitative survey of 44 342 households and is conducted on an annual basis. The data provided, however, are already aggregated. For example, food consumption is reported in 13 broad food groups, while in the original survey, each broad food group includes 5-8 different food items. Aggregate data, in fact, restricts the separation of rice consumption from the remainder of staple foods. The share of rice and other cereal consumption that comes from home production also cannot be estimated, nor how much rice is produced at the household level, since this information is not available in the data set.

For food consumption, the survey asks for two types of food: a) only dried food/canned food consumed regularly by households (often stored for over one week), and b) all food items consumed by households during the previous week. Total consumption is the sum a) and b). For each food type, the survey also asks for three different types of consumption based on how the food is acquired: 1) purchase in cash, 2) payment in kind, and 3) from home production. However, in the aggregate dataset, this valuable information is not provided.

The 13 broad aggregate food groups include:

1. Grain, cereal, and carbohydrate products, including rice, rice flour, wheat products such as breads
2. Meats and poultry (dried and processed food)
3. Fishes and seafood
4. Milk, cheese and egg products
5. Oil and fat products
6. Fruits and nuts
7. Vegetables (dried/pickled/salted)
8. Sugar and sweets
9. Spices and condiments (dried/canned/preserved)
10. Non-alcoholic beverages (consumed at home)
11. Prepared food (consumed at home)
12. Alcoholic beverages (consumed at home)
13. Tobacco products *(included in non-food consumption instead).*

Specific information reported includes the quantity and monetary value of each broad food group for types a) and b).

The estimation of non-food expenditure includes a similar list of consumption as mentioned above in the Myanmar section.

Viet Nam

The Viet Nam Household Living Standard Survey (VHLSS), conducted by the General Statistics Office of Viet Nam in 2010, is used for the analysis. VHLSS is a nationwide quantitative survey of 9 393 households, conducted every two years. The 2010 VHLSS was conducted in May, September and December of 2010 in which each month was round targeted to one third of the survey size. Thus, unlike the Myanmar's IHLCS, the final survey is the sum of three surveys conducted at three different times of the year.

For food consumption, the survey asks for two types of expenditure: a) expenditures on foods and drinks on festive occasions, and b) recurrent (monthly) expenditures on food and drinks. For each food item of type b), the survey also asks for three different types of consumption based on how the food is acquired: 1) purchased in cash, 2) from home production; and 3) received as gift or in-kind. For food items of type b, self-subsidy and in-kind consumption are grouped together in one category.

The total number of individual food items in the survey is 154, which covers:

1. Rice, including white rice and sticky rice
2. Other staples such as cassava, maize, wheat and wheat products
3. Meats including pork, beef, chicken, duck etc.
4. Seafood
5. Beans and vegetables
6. Fruits
7. Sugar and confectionery
8. Milk and eggs
9. Alcoholic and non-alcoholic beverages
10. Tobacco and cigarettes *(included in non-food expenditure instead)*
11. Food consumed outside the home and other miscellaneous foods.

Specific information includes:

- the quantity and monetary value of food items that are purchased in cash
- the value of food items that are received in kind, as gifts or sourced from home production.

The estimation of non-food expenditure includes a similar list of consumption as mentioned above in the Myanmar section.

Annex 4.A2

Almost Ideal Demand System (AIDS) model specifications and estimation results

QUAIDS model specifications and estimation

In this analysis, the Quadratic AIDS (QUAIDS), developed by Banks, Blundell, & Lewbel (1997) is used, which is an extended version of the Almost Ideal Demand System (AIDS) developed by Deaton and Muellbauer (1980) as the theoretical basis. One of the advantages of QUAIDS is that it allows for a quadratic relationship between expenditure and budget shares, which appears to be a better fit for a demand system that includes non-food consumption.

There is a potential source for error, however. In this analysis, only cross-sectional data of a particular year for each country was used. Thus, the difference in income, prices and demand patterns over time was not taken into account. The estimated elasticities are deemed to be helpful, though it should be noted that the relationships between prices and demand or expenditure and demand are estimated at one point in time.

Another issue that is popular in household analysis is outliers. There are two main sources of outliers. One comes from raw data. For this type of outlier, mostly values on the right-hand side of the distribution, the exact causes are unknown, whether it is human error or processor error. The other type of outlier is the values predicted by the model or calculated from the predicted values. For some households, the results could be a very big negative value for predicted consumption. This is admittedly a disadvantage of the model that ensures some condition is preserved for post-estimation. For QUAIDS, since the total budget share of a household must add up to 1, some predicted values could go beyond the [0,1] range.

Unfortunately, there is no hard and fast rule to treat outliers. An effort is made to be consistent in ruling out outliers across the three surveys, but also to be case-specific in making adjustments as long as the values look reasonable to the overall distribution.

Another issue is that quality differences within food groups was not considered, which means by assumption composite food groups are homogeneous. In reality, assuming that there is no constraint in market access, consumers – especially middle and high-income – could switch to lower-quality food without sacrificing the consumption quantity if there is a hike in prices. In this analysis, it is assumed there is no such quality adjustment. Thus, the impacts on consumption simulated by the models could be overestimated.

The lack of market prices also creates sources for error, since proxies for prices are used. In this analysis, the unit price approach is applied with adjustment for quality biases. Since unit price is expenditure divided by the corresponding quantity, there is a potential for collinearity since (total) expenditure also enters the demand system. While fulfilling the adding-up condition (budget share must sum to 1) after estimations is a significant advantage of the models, it is not tested whether other assumptions still hold, as this is not the focus of the analysis.

Based on an indirect utility function, the QUAIDS model has a form as follows:

$$w_i = \alpha_i + \sum_{j=1}^{n} \gamma_{ij}\, ln p_j + \beta_j ln\left[\frac{m}{a(p)}\right] + \frac{\lambda_i}{b(p)}\left\{ln\left[\frac{m}{a(p)}\right]\right\}^2 \quad (1)$$

where w_i is the budget share of household i derived from price, quantity and total expenditure, $w_i = p_i q_i / m$, and satisfies the constraint $\sum_{i=1}^{n} w_i = 1$, n is the number of goods in the system (in this analysis include different food groups and non-food), p_j is the price of good j, m is per capita total expenditure of food and non-food, $a(p)$ and $b(p)$ are the price indices, p is the vector of prices and α, β, γ, and λ are parameters to be estimated. Price indices are defined below:

$$lna(p) = a_0 + \sum_{n=1}^{n} \alpha_i lnp_i + \frac{1}{2}\sum_{i=1}^{n}\sum_{j=1}^{n} \gamma_{ij} lnp_i lnp_j \quad (2)$$

$$b(p) = \prod_{i=1}^{n} p_i^{\beta_i} \quad (3)$$

All parameters need to satisfy the adding-up condition, homogeneity condition, and Slutsky symmetry restriction:

Adding-up: $\sum_{i=1}^{n} \alpha_i = 1, \sum_{i=1}^{n} \beta_i = \sum_{i=1}^{n} \gamma_{ij} = 0,$

Homogeneity: $\sum_{j=1}^{n} \gamma_{ij} = 0 \; \forall \, j$

Symmetry: $\gamma_{ij} = \gamma_{ji}$

Expenditure elasticities are obtained from

$$\eta_i = {\mu_i}/{w_i} + 1 \text{ where } \mu_i = \beta_i + \frac{2\lambda_i}{b(p)}\left\{ln\left[\frac{m}{a(p)}\right]\right\} \quad (4)$$

Uncompensated price elasticities are given by

$$e_{ij}^{u} = {\mu_{ij}}/{w_i} - \delta_{ij} \text{ where } \mu_{ij} = \gamma_{ij} - \mu_i\left(\alpha_j + \sum_k \gamma_{jk} lnp_k\right) - \frac{\lambda_i \beta_i}{b(p)}\left\{ln\left[\frac{m}{a(p)}\right]\right\}^2 \quad (5)$$

Compensated price elasticities are derived from the Slutsky equation:

$$e_{ij}^{c} = e_{ij}^{u} + \eta_i w_i \quad (6)$$

In addition, to account for demographic characteristics of a household, Poi (2013) extended equation 1 using the scaling technique proposed by Ray (1983). Assuming a utility maximizing household with s demographic characteristics, represented by vector z, the scaled expenditure function has the form:

$$m_0\,(p,z,u) = \overline{m_0}\,(z).\,\phi(p,z,u) \quad (7)$$

in which $\overline{m_0}\,(z)$ measures the change in a household's expenditure with respect to demographic characteristics holding consumption patterns constant. The second term, $\phi(p,z,u)$, on the other hand, accounts for actual prices and quantities consumed by a household. It is defined by:

$$ln\phi(p,z,u) = \frac{\prod_{j=1}^{k} p_j^{\beta_j}\left(\prod_{j=1}^{k} p_j^{\eta'_j z} - 1\right)}{\frac{1}{u} - \sum_{j=1}^{k} \lambda_j lnp_j} \quad (8)$$

QUAIDS with a vector of demographic variables z now has the form:

$$w_i = \alpha_i + \sum_{j=1}^{n} \gamma_{ij}\, lnp_j + (\beta_j + \eta_i z)\, ln\left[\frac{m}{m_0\,(z) a(p)}\right] + \frac{\lambda_i}{b(p)c(p,z)}\left\{ln\left[\frac{m}{m_0\,(z) a(p)}\right]\right\}^2 \quad (9)$$

where $m_0\,(z) = 1 + \rho' z$ and $c(p,z) = \prod_{j=1}^{k} p_j^{\eta'_j z}$ with $\sum_{j=1}^{k} \eta_{rj} = 0$ (r=1…s) to satisfy adding-up condition. Two additional vectors of demographic parameters ρ and η are to be estimated.

It is noted that when $\lambda_i = 0$ equation 1 becomes the original AIDS model. With a quadratic term λ_i in the expenditure *m*, QUAIDS allows a good to change from luxury (expenditure elasticity>1) to necessity (expenditure elasticity<1) as expenditure increases.

The estimation is conducted using Stata version 14 with the help of *aidsills* command developed by Lecocq and Robin (2015). *Aidsills* estimates AIDS and its extensions with endogenous regressors using iterated linear least squares, which is proved to be much faster than other similar procedures such as *quaids*.

Expenditure and price elasticities are estimated by the model at the sample means, reported along with standard deviations.

Data treatment issues

Household characteristics

For all three surveys, household demographic characteristics are represented by the following variables:

- age of household head
- household size
- if household head is a male
- if household head is married
- if household head works in agriculture, hunting or forestry
- if the household lives in urban area
- educational level of the household head
  - elementary school or less
  - middle school
  - high school
  - college
  - graduate (post-graduate)
- per cent of households with children aged less than 5 years old
- per cent of households with children aged between 5 and 14
- per cent of households with adults aged 60 and above

The categorisation of educational level remains the same across three countries, even though the name of each category could vary depending on how it is called in each country. For example, in Thailand, elementary school is called "primary", while middle school is called "secondary".

Convert to annual basis

To convert to annual consumption, weekly consumption is multiplied by a factor of 52 and monthly consumption is multiplied by a factor of 12. Annual quantity, monetary value or calorie intake are divided by household size to convert to a per capita basis. The size of the household is defined as the number of household members without adjustments for adult equivalence.

Treatment of outliers

Outliers are popular in household surveys. It is not easy to come up with a solid validation for the criteria to define and eliminate outliers. For all three surveys, there are no problems of very small values but of very big values on the right tail of the distribution. Thus, values that exceed the sample's 99th percentile are considered as outliers. These outliers are not eliminated, as that would cause great detrition of the sample size. In particular, values are replaced, either reported as quantity or expenditure that exceed the sample's 99th percentile by the 99th percentile value. Since the 99th percentile value is still very big, this replacement, in fact, does not affect the household's position in the sample distribution, but it helps to make the distribution much less skewed to the right.

Occasionally in the analysis, another criteria is used to treat outliers and missing values. Missing values or values that are larger than 3 to 5 standard deviations from the sample mean are replaced by the mean value. This approach is used particularly for unit prices, which are discussed below. Since the prices that the household might have paid for a particular food are unknown, it is assumed that the mean values represent the prevailing price in a market that is closest to where the household lives, which could be at a village or a province level.

Treatment of missing, unreported values and quality bias

Household surveys often do not ask for information on paid prices or market prices of foods, but monetary value of cash purchases instead – which turns out to be true for three surveys used in this analysis.[1] Following the literature, unit price is used as a proxy for missing prices. Unit price is calculated as value of food consumption purchased in cash divided by its quantity.

This approach, however, suffers from quality effects and measurement errors, which are common in household data analysis (Deaton, 1988). Consumers choose quality which is reflected by the price (unit value). When prices change, however, consumers react by changing both quality and quantity. Measurement errors in reported quantities and expenditures also cause inaccuracy in enumerated unit prices. To account for these potential biases, the communal mean price method originally developed by Cox and Wohlgenant (1986) and later modified by Vu (2009) in his food demand study using VHLSS 2006 is employed. Several studies have affirmed the usefulness of this method in eliminating spatial and quality variations in price data (Gibson and Rozelle, 2011; Majumder, Ray and Sinha, 2012; Niimi, 2005).

First, prices are adjusted for quality differences. The equation takes the following form:

$$p_i = \alpha\, p_i^c + \beta f_i + \gamma\, x_i + \sum_n \eta_{in} z_{in} + e_i \qquad (1)$$

where i denotes the household i in the dataset, p_i is the unit price of an individual food faced by household i, p_i^c is the mean of unit prices at communal level[2], $f_i$is the share of food away from home, $x_i$is the household food expenditure per cap and e_i is the error terms. Household characteristics z_{in} include household size, urban and region dummy variables, the sex, education and age of the household head.

The residual for every household i in equation 10 is added to the communal mean unit price p_i^c to obtain the quality-adjusted prices p_i^a at the household level.

$$p_i^a = p_i^c + \hat{e_i} \qquad (2)$$

According to Deaton (1988), household surveys normally collect data from households in the same village at the same time. Thus, it is plausible that these households should face the same price. Taking this insight into consideration, this study assumes that households in the same commune (the smallest geographic unit in the dataset) face the same prices. This communal mean quality-adjusted price of the individual food item is the mean of p_i^a calculated at the communal level.

$$p_i^{c*} = \overline{p_i^a} \qquad (3)$$

With the exception of the group of miscellaneous foods, the composite price of the food group is also computed at the communal level, i.e. households in the same commune face the same unit prices for these composite food groups. Following Niimi (2005), the commune mean budget shares are used as weights.

$$p_g^c = \sum_{i=1}^{k} p_i^{c*} u_i^c \Big/ \sum_{i=1}^{k} u_i^c \qquad (4)$$

where $u_i^c$is the mean budget share at the communal level of individual food item i, k is the number of food item i in the group, p_g^c is the price of the composite food group g at the communal level.

Generally, for each country, consumption is grouped into seven composite food groups covering rice, meats, vegetables, sugar, oils and fats, drinks and miscellaneous (food consumed outside the home is included in this last group). In the case of Thailand, due to the fact that data are already aggregated, instead of categorising rice as a single group to be consistent with other countries, a group was created, named "rice and cereals", which includes the consumption of rice and other cereals such as wheat noodles and breads.

The miscellaneous food group is a combination of disparate food items with different quantity units. While the IHLCS and HES generally report both value and quantity for food consumed outside the home and other miscellaneous food items, the VHLSS does not. Thus, to generate a proxy for prices of this group in VHLSS, Vu (2009) is followed to use the regional 2010 CPIs published by the General Statistics Office of

Viet Nam. The quality-adjusted unit prices are then multiplied with the total quantity consumed for a particular food item to derive the total expenditure on food.

Food and non-food consumption

While VHLSS and HSES include tobacco and cigarettes (and betel leaves in particular for VHLSS) in food consumption, IHLCS does not. To be consistent, the consumption of these items is considered as non-food expenditure for all three countries.

Non-food expenditure will enter the demand system as one single variable along with other composite food groups. Non-food expenditure generally includes the household's expenses on education, health care, personal care, housing, transportation, utilities, communication, other activities and the consumption of tobacco and cigarettes, as mentioned earlier. Non-food expenditure, however, does not include the household's financial status, such as debts or savings. Since it is a combination of many disparate consumption categories and quantity is often not reported, it is impossible to use the unit price approach as for food consumption. Thus, different approaches are used to generate a proxy for the price of non-food using the existing data. For Myanmar and Thailand, the inverse of the unit price of alcoholic beverages is used. For Viet Nam, the average of the unit prices of several disparate food items in the miscellaneous food group is used. Total expenditure is the sum of food and non-food expenditures.

QUAIDS estimation results

Table 4.A2.1. Expenditure elasticity by expenditure quintile class: Myanmar

	Expenditure	Quintile 1	Quintile 2	Quintile 3	Quintile 4	Quintile 5
Rice	0.508	0.65	0.587	0.521	0.433	0.032
	(0.664)	(0.34)	(0.47)	(0.627)	(0.875)	(2.525)
Pulses, beans, nuts and seeds	0.896*	0.971**	0.929**	0.901*	0.870*	0.794
	(0.366)	(0.345)	(0.346)	(0.358)	(0.384)	(0.483)
Meats	1.041**	1.142*	1.083**	1.046**	1.009*	0.928*
	(0.396)	(0.451)	(0.41)	(0.398)	(0.397)	(0.429)
Fish and seafood	1.117*	1.271	1.176*	1.124*	1.077*	0.992*
	(0.485)	(0.717)	(0.56)	(0.495)	(0.451)	(0.412)
Vegetables and fruits	0.921**	0.995***	0.953**	0.924**	0.894**	0.82
	(0.322)	(0.295)	(0.304)	(0.32)	(0.342)	(0.428)
Oils and fats	0.857**	0.914***	0.884***	0.860**	0.836**	0.767
	(0.294)	(0.247)	(0.267)	(0.289)	(0.316)	(0.421)
Other	0.993***	1.070***	1.024***	0.996***	0.969***	0.912**
	(0.293)	(0.323)	(0.301)	(0.294)	(0.293)	(0.312)
Non-food	1.400**	1.236	1.349*	1.399**	1.434***	1.454***
	(0.436)	(0.881)	(0.634)	(0.46)	(0.302)	(0.072)

Notes: Estimated at sample means. Standard errors are in bracket. * $p<0.05$, ** $p<0.01$, *** $p<0.001$.
Number of observations: 18 544.
Source: OECD calculation based on QUAIDS.

Table 4.A2.2. Uncompensated own price elasticity by expenditure quintile class: Myanmar

	Own-price	Quintile 1	Quintile 2	Quintile 3	Quintile 4	Quintile 5
Rice	-0.541	-0.667*	-0.61	-0.552	-0.474	-0.124
	(0.567)	(0.287)	(0.4)	(0.536)	(0.751)	(2.179)
Pulses, beans, nuts and seeds	-0.842***	-0.845***	-0.846***	-0.844***	-0.839***	-0.822***
	(0.108)	(0.076)	(0.092)	(0.104)	(0.12)	(0.17)
Meats	-0.806***	-0.822***	-0.814***	-0.807***	-0.798***	-0.773***
	(0.098)	(0.05)	(0.076)	(0.095)	(0.117)	(0.179)
Fish and seafood	-0.930***	-0.939***	-0.933***	-0.930***	-0.926***	-0.915***
	(0.081)	(0.07)	(0.075)	(0.08)	(0.086)	(0.103)
Vegetables and fruits	-0.818***	-0.840***	-0.829***	-0.820***	-0.810***	-0.779***
	(0.101)	(0.077)	(0.089)	(0.099)	(0.112)	(0.156)
Oils and fats	-0.663***	-0.685***	-0.676***	-0.665***	-0.654***	-0.608*
	(0.164)	(0.105)	(0.135)	(0.159)	(0.187)	(0.287)
Other	-1.067***	-1.098***	-1.080***	-1.068***	-1.058***	-1.037***
	(0.118)	(0.13)	(0.121)	(0.119)	(0.119)	(0.126)
Non-food	-0.948***	-0.818**	-0.893**	-0.942***	-0.985***	-1.061***
	(0.253)	(0.254)	(0.275)	(0.256)	(0.223)	(0.145)

Notes: Estimated at sample means. Standard errors are in bracket. * p<0.05, ** p<0.01, *** p<0.001.
Number of observations: 18 544.
Source: OECD calculation based on QUAIDS.

Table 4.A2.3. Uncompensated cross price elasticity: Myanmar

	p1	p2	p3	p4	p5	p6	p7	p8
Rice	-0.541	-0.01	-0.071	-0.171	0.001	0.014	0.153	0.114
	(0.567)	(0.011)	(0.043)	(0.206)	(0.044)	(0.016)	(0.258)	(0.136)
Pulses, beans, nuts and seeds	-0.132*	-0.842***	-0.285**	0.476*	-0.165***	-0.019	0.004	0.012
	(0.054)	(0.108)	(0.104)	(0.237)	(0.028)	(0.05)	(0.225)	(0.075)
Meats	-0.218	-0.088	-0.806***	0.126	0.032	-0.109	0.106	-0.155
	(0.146)	(0.053)	(0.098)	(0.077)	(0.089)	(0.059)	(0.151)	(0.134)
Fish and seafood	-0.45	0.161***	0.144	-0.930***	0.051	0.004	-0.023	-0.154
	(0.281)	(0.023)	(0.087)	(0.081)	(0.096)	(0.054)	(0.187)	(0.177)
Vegetables and fruits	-0.079	-0.043	0.043	0.057	-0.818***	-0.074***	-0.037	-0.023
	(0.073)	(0.027)	(0.105)	(0.081)	(0.101)	(0.02)	(0.125)	(0.092)
Oils and fats	-0.017	-0.011	-0.226***	0.035	-0.183**	-0.663***	-0.067	0.237***
	(0.056)	(0.053)	(0.061)	(0.093)	(0.06)	(0.164)	(0.144)	(0.044)
Other	0.067	-0.002	0.077	0.001	-0.036	-0.026	-1.067***	-0.058
	(0.092)	(0.031)	(0.087)	(0.065)	(0.066)	(0.033)	(0.118)	(0.096)
Non-food	-0.052	-0.008	-0.088	-0.063	-0.047	0.031	-0.076	-0.948***
	(0.173)	(0.09)	(0.215)	(0.141)	(0.094)	(0.133)	(0.335)	(0.253)

Notes: Estimated at sample means. Standard errors are in bracket. * p<0.05, ** p<0.01, *** p<0.001.
Number of observations: 18 544.
Source: OECD calculation based on QUAIDS.

Table 4.A2.4. Estimation results by household type: Myanmar

	Expenditure		Own price	
	Farm	Non-Farm	Farm	Non-Farm
Rice	0.563	0.466	-0.587	-0.505
	(0.528)	(0.778)	(0.451)	(0.664)
Pulses, beans, nuts and seeds	0.919**	0.880*	-0.854***	-0.833***
	(0.332)	(0.392)	(0.092)	(0.12)
Meats	1.057**	1.030*	-0.831***	-0.787***
	(0.361)	(0.424)	(0.088)	(0.106)
Fish and seafood	1.155*	1.097*	-0.927***	-0.931***
	(0.553)	(0.452)	(0.076)	(0.083)
Vegetables and fruits	0.940**	0.907**	-0.830***	-0.810***
	(0.302)	(0.338)	(0.093)	(0.107)
Oils and fats	0.878***	0.842**	-0.686***	-0.646***
	(0.263)	(0.317)	(0.137)	(0.186)
Other	1.010**	0.983***	-1.077***	-1.061***
	(0.315)	(0.282)	(0.121)	(0.117)
Non-food	1.397*	1.400***	-0.913***	-0.969***
	(0.549)	(0.377)	(0.268)	(0.242)

Notes: Estimated at group means. Standard errors are in bracket. * p<0.05, ** p<0.01, *** p<0.001.
Number of observations: 18 544.
Source: OECD calculation based on QUAIDS.

Table 4.A2.5. Expenditure elasticity by expenditure quintile class: Thailand

	All	Quintile 1	Quintile 2	Quintile 3	Quintile 4	Quintile 5
Rice	**0.078**	**0.252*** **	**0.166** **	**0.09**	**-0.012**	**-0.165*** **
	(0.054)	(0.045)	(0.055)	(0.061)	(0.054)	(0.041)
Meats and fish	**0.452*** **	**0.610*** **	**0.542*** **	**0.475*** **	**0.351** **	**0.041**
	(0.086)	(0.047)	(0.062)	(0.08)	(0.118)	(0.248)
Vegetables and fruits	**0.548*** **	**0.702*** **	**0.628*** **	**0.562*** **	**0.448*** **	**0.168**
	(0.074)	(0.04)	(0.056)	(0.072)	(0.103)	(0.209)
Milk and sugar	**0.557*** **	**0.608*** **	**0.592*** **	**0.578*** **	**0.531*** **	**0.453*** **
	(0.056)	(0.048)	(0.051)	(0.053)	(0.061)	(0.079)
Oils and fats	**0.484*** **	**0.672*** **	**0.586*** **	**0.507*** **	**0.354***	**-0.046**
	(0.127)	(0.086)	(0.107)	(0.127)	(0.166)	(0.322)
Drinks	**1.001*** **	**1.276*** **	**1.151*** **	**1.051*** **	**0.892*** **	**0.624** **
	(0.146)	(0.172)	(0.153)	(0.145)	(0.151)	(0.203)
Other	**0.895*** **	**1.005*** **	**0.958*** **	**0.918*** **	**0.853*** **	**0.750*** **
	(0.061)	(0.063)	(0.059)	(0.059)	(0.064)	(0.077)
Non-food	**1.292*** **	**1.316*** **	**1.304*** **	**1.295*** **	**1.284*** **	**1.270*** **
	(0.009)	(0.009)	(0.009)	(0.009)	(0.009)	(0.01)

Notes: Estimated at sample means. Standard errors are in bracket. * p<0.05, ** p<0.01, *** p<0.001.
Source: OECD calculation based on QUAIDS.

Table 4.A2.6. Uncompensated own price elasticity by expenditure quintile class: Thailand

	All	Quintile 1	Quintile 2	Quintile 3	Quintile 4	Quintile 5
Rice	**-0.428*** **	**-0.615*** **	**-0.541*** **	**-0.461*** **	**-0.298**	**0.126**
	(0.115)	(0.056)	(0.08)	(0.106)	(0.16)	(0.308)
Meats and fish	**-0.451*** **	**-0.595*** **	**-0.533*** **	**-0.472*** **	**-0.360** **	**-0.079**
	(0.101)	(0.058)	(0.075)	(0.094)	(0.134)	(0.267)
Vegetables and fruits	**-0.674*** **	**-0.766*** **	**-0.720*** **	**-0.680*** **	**-0.615*** **	**-0.456*** **
	(0.038)	(0.019)	(0.027)	(0.036)	(0.054)	(0.114)
Milk and sugar	**-0.511*** **	**-0.564*** **	**-0.548*** **	**-0.533*** **	**-0.484*** **	**-0.402*** **
	(0.057)	(0.044)	(0.049)	(0.052)	(0.064)	(0.085)
Oils and fats	**-0.199**	**-0.433* **	**-0.318**	**-0.219**	**-0.039**	**0.434**
	(0.301)	(0.177)	(0.236)	(0.292)	(0.397)	(0.733)
Drinks	**-0.876*** **	**-0.884*** **	**-0.883*** **	**-0.881*** **	**-0.869*** **	**-0.839*** **
	(0.008)	(0.005)	(0.006)	(0.007)	(0.011)	(0.02)
Other	**-0.480*** **	**-0.468*** **	**-0.497*** **	**-0.495*** **	**-0.474*** **	**-0.418*** **
	(0.029)	(0.018)	(0.022)	(0.026)	(0.034)	(0.049)
Non-food	**-1.064*** **	**-1.046*** **	**-1.055*** **	**-1.062*** **	**-1.071*** **	**-1.082*** **
	(0.023)	(0.024)	(0.023)	(0.023)	(0.023)	(0.023)

Notes: Estimated at sample means. Standard errors are in bracket. * p<0.05, ** p<0.01, *** p<0.001.
Number of observations: 40 250.
Source: OECD calculation based on QUAIDS.

Table 4.A2.7. Uncompensated cross price elasticity: Thailand

	p1	p2	p3	p4	p5	p6	p7	p8
Rice	**-0.428*** **	**0.043* **	**0.086*** **	**0.002**	**-0.004**	**-0.015**	**-0.131**	**0.178**
	(0.115)	(0.017)	(0.008)	(0.016)	(0.028)	(0.017)	(0.097)	(0.168)
Meats and fish	**0.02**	**-0.451*** **	**0.046*** **	**-0.012* **	**0**	**0.012*** **	**-0.173*** **	**0.117*** **
	(0.012)	(0.101)	(0.008)	(0.005)	(0.025)	(0.002)	(0.033)	(0.013)
Vegetables and fruits	**0.078*** **	**0.069**	**-0.674*** **	**0.002**	**0.013**	**0.008*** **	**-0.150*** **	**0.131*** **
	(0.02)	(0.042)	(0.038)	(0.006)	(0.03)	(0.002)	(0.024)	(0.014)
Milk and sugar	**-0.012**	**-0.037**	**0.001**	**-0.511*** **	**-0.01**	**0.021*** **	**-0.147*** **	**0.141** **
	(0.016)	(0.039)	(0.01)	(0.057)	(0.039)	(0.003)	(0.033)	(0.048)
Oils and fats	**-0.049**	**0.007**	**0.117*** **	**-0.060** **	**-0.199**	**0.027*** **	**-0.192* **	**-0.1**
	(0.054)	(0.214)	(0.03)	(0.019)	(0.301)	(0.006)	(0.084)	(0.38)
Drinks	**-0.061**	**0.007**	**-0.001**	**0.019**	**0.004**	**-0.876*** **	**-0.049**	**0.127**
	(0.036)	(0.078)	(0.022)	(0.02)	(0.052)	(0.008)	(0.035)	(0.081)
Other	**-0.078*** **	**-0.130*** **	**-0.067*** **	**-0.048*** **	**-0.009**	**-0.009** **	**-0.480*** **	**-0.002**
	(0.013)	(0.024)	(0.009)	(0.008)	(0.019)	(0.003)	(0.029)	(0.047)
Non-food	**-0.043*** **	**-0.059** **	**-0.030*** **	**-0.020*** **	**-0.006**	**-0.007*** **	**-0.077*** **	**-1.064*** **
	(0.004)	(0.019)	(0.003)	(0.002)	(0.018)	(0.002)	(0.011)	(0.023)

Notes: Estimated at sample means. Standard errors are in bracket. * p<0.05, ** p<0.01, *** p<0.001.
Number of observations: 40 250.
Source: OECD calculation based on QUAIDS.

Table 4.A2.8. Estimation results by household type: Thailand

	Expenditure		Own price	
	Farm	Non-Farm	Farm	Non-Farm
Rice	0.174**	0.076	-0.583***	-0.423***
	(0.067)	(0.053)	(0.072)	(0.116)
Meats and fish	0.590***	0.447***	-0.577***	-0.447***
	(0.053)	(0.088)	(0.064)	(0.102)
Vegetables and fruits	0.702***	0.542***	-0.766***	-0.671***
	(0.042)	(0.076)	(0.02)	(0.039)
Milk and sugar	0.628***	0.555***	-0.586***	-0.509***
	(0.047)	(0.057)	(0.043)	(0.058)
Oils and fats	0.683***	0.476***	-0.451*	-0.19
	(0.089)	(0.129)	(0.178)	(0.306)
Drinks	1.191***	0.994***	-0.924***	-0.875***
	(0.117)	(0.147)	(0.008)	(0.008)
Other	1.007***	0.892***	-0.498***	-0.480***
	(0.062)	(0.061)	(0.022)	(0.029)
Non-food	1.322***	1.292***	-1.044***	-1.065***
	(0.01)	(0.009)	(0.025)	(0.023)

Notes: Estimated at group means. Standard errors are in bracket. * p<0.05, ** p<0.01, *** p<0.001.
Number of observations: 40 250.
Source: OECD calculation based on QUAIDS.

Table 4.A2.9. Expenditure elasticity by expenditure quintile class: Viet Nam

	All	Quintile 1	Quintile 2	Quintile 3	Quintile 4	Quintile 5
Rice	**0.283*** **	**0.542*** **	**0.377*** **	**0.213*** **	**0.02**	**-0.455**
	(0.017)	(0.054)	(0.025)	(0.027)	(0.093)	(0.347)
Pork	**0.836*** **	**0.902*** **	**0.871*** **	**0.845*** **	**0.802*** **	**0.667*** **
	(0.056)	(0.057)	(0.051)	(0.052)	(0.064)	(0.129)
Meats and fish	**1.009*** **	**1.093*** **	**1.039*** **	**1.011*** **	**0.975*** **	**0.907*** **
	(0.047)	(0.059)	(0.054)	(0.046)	(0.041)	(0.055)
Vegetables and fruits	**0.783*** **	**0.849*** **	**0.810*** **	**0.785*** **	**0.751*** **	**0.673*** **
	(0.087)	(0.028)	(0.11)	(0.085)	(0.075)	(0.133)
Sugar	**0.928*** **	**1.041* **	**0.976*** **	**0.934*** **	**0.868*** **	**0.711**
	(0.245)	(0.089)	(0.281)	(0.235)	(0.248)	(0.51)
Drinks	**0.968*** **	**0.975* **	**0.971*** **	**0.969*** **	**0.965*** **	**0.956** **
	(0.192)	(0.038)	(0.253)	(0.185)	(0.174)	(0.305)
Other	**0.990*** **	**1.066*** **	**1.014*** **	**0.991*** **	**0.968*** **	**0.938*** **
	(0.043)	(0.159)	(0.059)	(0.042)	(0.034)	(0.036)
Non-food	**1.226*** **	**1.273*** **	**1.241*** **	**1.227*** **	**1.214*** **	**1.194*** **
	(0.042)	(0.567)	(0.057)	(0.043)	(0.041)	(0.057)

Notes: Estimated at sample means. Standard errors are in bracket. * p<0.05, ** p<0.01, *** p<0.001.
Number of observations: 9 383.
Source: OECD calculation based on QUAIDS.

Table 4.A2.10. Uncompensated own price elasticity by expenditure quintile class: Viet Nam

	All	Quintile 1	Quintile 2	Quintile 3	Quintile 4	Quintile 5
Rice	-0.554***	-0.743***	-0.644***	-0.535***	-0.348*	0.197
	(0.101)	(0.048)	(0.084)	(0.12)	(0.165)	(0.333)
Pork	-0.815***	-0.851***	-0.840***	-0.825***	-0.799***	-0.705***
	(0.069)	(0.053)	(0.06)	(0.065)	(0.075)	(0.119)
Meats and fish	-0.788***	-0.810***	-0.806***	-0.796***	-0.783***	-0.725***
	(0.05)	(0.037)	(0.049)	(0.048)	(0.046)	(0.06)
Vegetables and fruits	-0.883***	-0.902***	-0.891***	-0.884***	-0.874***	-0.850***
	(0.028)	(0.022)	(0.027)	(0.027)	(0.028)	(0.034)
Sugar	-0.701***	-0.734***	-0.744***	-0.715***	-0.667***	-0.547***
	(0.092)	(0.084)	(0.069)	(0.077)	(0.093)	(0.159)
Drinks	-0.899***	-0.898***	-0.901***	-0.904***	-0.901***	-0.891***
	(0.037)	(0.036)	(0.035)	(0.033)	(0.032)	(0.032)
Other	-1.331***	-1.439***	-1.365***	-1.326***	-1.303***	-1.275***
	(0.094)	(0.176)	(0.11)	(0.092)	(0.082)	(0.074)
Non-food	-0.935*	-0.902	-0.925*	-0.935*	-0.944**	-0.960***
	(0.371)	(0.489)	(0.434)	(0.373)	(0.31)	(0.228)

Notes: Estimated at sample means. Standard errors are in bracket. * $p<0.05$, ** $p<0.01$, *** $p<0.001$.
Number of observations: 9 383.
Source: OECD calculation based on QUAIDS.

Table 4.A2.11. Uncompensated cross price elasticity: Viet Nam

	p1	p2	p3	p4	p5	p6	p7	p8
Rice	-0.554***	-0.073	-0.116*	-0.03	-0.015	-0.003	0.351***	1.174***
	(0.101)	(0.047)	(0.054)	(0.073)	(0.042)	(0.075)	(0.088)	(0.317)
Pork	-0.171*	-0.815***	0.082*	-0.049	-0.060*	-0.031	0.265	-0.339
	(0.078)	(0.069)	(0.032)	(0.026)	(0.025)	(0.024)	(0.149)	(0.332)
Meats and fish	-0.219*	0.048	-0.788***	0.037	-0.016	0.021	0.01	-0.612*
	(0.089)	(0.054)	(0.05)	(0.024)	(0.022)	(0.021)	(0.096)	(0.264)
Vegetables and fruits	-0.062	-0.051	0.063**	-0.883***	0.036	0.025	0.155	-0.31
	(0.126)	(0.078)	(0.023)	(0.028)	(0.024)	(0.039)	(0.208)	(0.844)
Sugar	-0.127	-0.316	-0.168	0.180**	-0.701***	-0.004	0.388	-0.889
	(0.395)	(0.23)	(0.093)	(0.067)	(0.092)	(0.051)	(0.647)	(2.575)
Drinks	-0.187	-0.072	0.132*	0.062	0.004	-0.899***	0.017	-0.07
	(0.265)	(0.135)	(0.052)	(0.068)	(0.032)	(0.037)	(0.415)	(2.09)
Other	0.170**	0.104	0.019	0.059*	0.033	-0.004	-1.331***	-0.403
	(0.059)	(0.058)	(0.025)	(0.026)	(0.023)	(0.02)	(0.094)	(0.284)
Non-food	-0.079	-0.03	-0.040***	-0.034*	-0.008	-0.006	-0.044	-0.935*
	(0.054)	(0.032)	(0.011)	(0.014)	(0.012)	(0.025)	(0.096)	(0.371)

Notes: Estimated at sample means. Standard errors are in bracket. * $p<0.05$, ** $p<0.01$, *** $p<0.001$.
Number of observations: 9 383.
Source: OECD calculation based on QUAIDS.

Table 4.A2.12. Estimation results by household type: Viet Nam

	Expenditure		Own price, uncompensated	
	Farm	Non-Farm	Farm	Non-Farm
Rice	0.387***	0.068	-0.618***	-0.418**
	(0.027)	(0.06)	(0.079)	(0.151)
Pork	0.859***	0.798***	-0.834***	-0.785***
	(0.05)	(0.068)	(0.062)	(0.082)
Meats and fish	1.024***	0.988***	-0.797***	-0.776***
	(0.051)	(0.044)	(0.05)	(0.051)
Vegetables and fruits	0.793***	0.770***	-0.885***	-0.881***
	(0.101)	(0.074)	(0.029)	(0.028)
Sugar	0.953***	0.893***	-0.716***	-0.678***
	(0.267)	(0.238)	(0.087)	(0.099)
Drinks	0.971***	0.962***	-0.906***	-0.888***
	(0.205)	(0.192)	(0.037)	(0.037)
Other	1.001***	0.979***	-1.378***	-1.288***
	(0.056)	(0.033)	(0.114)	(0.077)
Non-food	1.233***	1.218***	-0.930*	-0.941**
	(0.048)	(0.04)	(0.402)	(0.335)

Notes: Estimated at group means. Standard errors are in bracket. * $p<0.05$, ** $p<0.01$, *** $p<0.001$ Number of observations: 9 383.
Source: OECD calculation based on QUAIDS.

Notes

1. There is a market price survey, the Spatial Cost-of-Living Index (SCOLI), conducted in conjunction with VHLSS 2010, but it only followed the second and third rounds and is thus not complete for the analysis.
2. Commune is the subgroup stratum above village in the context of VHLSS. The definition and name could be varied across household surveys, depending on the sampling method of each survey. In this chapter, township, area and commune levels are used for Myanmar, Thailand and Viet Nam, respectively.

Chapter 5

Improving the enabling environment for agriculture in ASEAN

This chapter assesses the enabling environment for sustainable agricultural productivity growth in the countries that comprise the Association of Southeast Asian Nations (ASEAN). It first outlines the key challenges and opportunities facing the agricultural sectors of ASEAN members that motivate the need for sustainable agricultural productivity growth. Second, it presents an updated and revised Agricultural Growth Enabling Index (AGEI). Third, it applies the AGEI to a subset of ASEAN members: Cambodia, Indonesia, Lao PDR, Malaysia, Myanmar, the Philippines, Thailand and Viet Nam. Finally, it presents country-specific profiles of the performance of each of the ASEAN members across all components of the AGEI.

Key points

- In the long term, a positive enabling environment contributes to eliminating food insecurity by creating conditions that support investment in innovation, and by providing incentives for the sustainable use of natural resources.
- Aspects of the enabling environment that are relatively strong in Association of Southeast Asian Nations (ASEAN) countries include stable macroeconomic environments, labour market operations, human capital, and relatively abundant water resources.
- However, agriculture-related aspects of the enabling environment are relatively weak, including public investments in agricultural research and development (R&D) and the quality and presence of agricultural infrastructure. Agriculture in ASEAN may be underprovided with public goods and other economic services relative to other sectors.
- Sustainability aspects of the enabling environment are also relatively weak in ASEAN. Agricultural land is scarce and environmental governance and regulations are weak and/or poorly enforced.
- Future sustainable productivity growth in ASEAN may be constrained by low investment in agricultural sector capacities and a lack of policies to protect the natural resource base for agricultural production.

5.1 Introduction[1]

Ensuring food security is a policy priority for Southeast Asia. As mentioned in Chapter 1, the region has made significant progress in reducing poverty and undernourishment, and the proportion of undernourished in the population has fallen by almost 70% for the region as a whole,[2] from 30.6% for the period 1990-92 to 9.6% for the period 2014-16 (FAO, 2015a). Nevertheless, despite this progress, Southeast Asia still contains around 60 million of the world's undernourished populations.

To address food security issues, some governments in the region have focused on increasing domestic production with a view to achieving self-sufficiency in rice or across a number of staple products, through price and input incentives for producers, supported by trade policies. However, past reviews of agriculture-related food security policies used in the region have pointed to shortcomings with such policy approaches. In addition to the shortcomings already discussed in Chapter 3, these policies divert attention away from opportunities to promote growth in food production across the region by improving the enabling environment for sustainable agricultural productivity growth. In the long term, growth in food production, including by small-scale farmers, and agricultural growth more broadly, depends on improvements in agricultural *productivity*. By creating conditions that promote environmentally sustainable agricultural productivity growth, the enabling environment plays an important role in determining food security outcomes over the long term.

It is therefore useful to also consider potential opportunities to improve the enabling environment for sustainable agricultural productivity growth in Southeast Asia. A country's enabling environment is defined as the multifaceted settings within which the agricultural sector and economy more broadly operates, comprising non-distorting and stable policies, adequate provision of public goods, good governance through laws and regulations that are conducive to private-sector economic activity while addressing market failures, and strong and effective institutions through which government measures and actions are operationalised (Diaz-Bonilla, Orden and Kwieciński, 2014). Collectively, these factors play an important role in ensuring a favourable environment for innovation at the farm level and by other businesses in agricultural value chains, by shaping incentives for investment and sustainable use of natural resources, and by building economic capacities.

This chapter assesses the enabling environment for sustainable agricultural productivity growth in the countries that form the Association of Southeast Asian Nations (ASEAN),[3] using the Agricultural Growth Enabling Index (AGEI) recently developed by the OECD (Diaz-Bonilla, Orden and Kwieciński, 2014). The

AGEI summarises a wide array of available information on the components of the enabling environment for agricultural growth. As such, it can be used to identify relative strengths and weaknesses in countries' enabling environments through cross-country comparisons of performances on the index and its sub-components.

The chapter is organised as follows. The following section briefly outlines the key challenges and opportunities facing the agricultural sectors of ASEAN members that motivate the need for sustainable agricultural productivity growth. For the purposes of this chapter, the AGEI has been updated and revised to better reflect the sustainability dimensions of the enabling environment for agricultural productivity growth. The revised AGEI is presented in Section 5.3. In Section 5.4, the AGEI is applied to the ASEAN member states and a wider set of relevant comparators. This section compares countries' performances on the AGEI, as well as its component "blocks" and the indicators that comprise these, in order to provide an insight into the strengths and weaknesses of countries' enabling environments for sustainable agricultural productivity growth, including those relative to other countries. Section 5.5 provides country-specific profiles of the performance of each of the ASEAN member states across all components of the AGEI. Finally, trends for the ASEAN region as a whole are outlined in the concluding section.

5.2 Challenges and opportunities for the food and agricultural sectors of ASEAN member states

Agriculture is a strategic sector for the majority of ASEAN member states. The population is predominantly rural in a number of countries, including Cambodia, Lao PDR, Myanmar, the Philippines and Viet Nam, and the majority of the region's poor live in rural areas. As such, a dynamic agricultural sector is vital for rural development and reducing poverty, and for economic growth and development more broadly.

At the same time, incomes are growing across the region, and represent a significant opportunity for the agricultural sectors of ASEAN member states. As discussed in Chapter 1, strong economic growth has resulted in incomes more than doubling in most countries in the region since the early 1990s. Together with population growth and urbanisation, this is increasing the overall level of demand, as well as demand for higher quality and more diversified food products, including meat, dairy products, fruits and vegetables (ADB, 2013).

Moreover, closer regional integration offers an opportunity for countries to access and benefit from this growth in demand – the less developed ASEAN member states in particular. The establishment of the ASEAN Economic Community (AEC) in 2015 lays the foundations for realising the goal of ASEAN as an integrated economic region. For agriculture, the AEC aims to establish a fully-integrated ASEAN market for agro-based products, characterised by a single production and market base, to transform food, agriculture and forestry into modern, competitive, sustainable industries; and to ensure that gains from deeper regional and global market integration and modernisation are equitably distributed and shared with the small producers (farmers and fishers) and small and medium-sized enterprises that dominate the sector (AMAF, 2015).

Agricultural productivity growth will be key to realising these opportunities and, ultimately, eliminating food insecurity. As described in Chapter 1, in recent decades, agricultural productivity growth has been strong for the region as a whole – above both the world average and growth rates in other regions. Moreover, improvements in agricultural productivity have played a key role in driving agricultural output growth in Southeast Asia in recent decades.

Yet several factors suggest that past trends may not be sustainable. As mentioned in Chapter 1, average farm size is small and declining in some countries, which may constrain the capacity of farmers to invest in productive assets that support future agricultural productivity growth. Agricultural census data from the early 2000s indicate that average farm size ranges from 0.7 ha in Viet Nam to 3.2 ha in Thailand. Moreover, holdings tend to be fragmented in a number of countries (including Indonesia, the Philippines and Viet Nam), further compounding scale constraints (OECD, 2012, 2015a, 2017). Second, in a number of countries, agricultural development – and economic growth more broadly – has relied on natural resource exploitation, leading to environmental degradation (OECD, 2012, 2014a, 2015a). As discussed in Chapter 1, factors such as deforestation, excessive use of agro-chemical inputs, intensive cultivation, water pollution and erosion threaten the natural resource base for agricultural production. The sector also faces increasing competition from other domestic and industrial users for water and land resources. Finally, climate change poses a significant risk to agricultural production, with the region likely to experience temperature increases, greater climate variability, and more frequent and intense extreme weather events. The current and projected impacts of climate change on agriculture in the region are explored in further detail in Chapter 2.

In view of both the importance of sustainable agricultural productivity growth for the ASEAN region and the challenges that lie ahead, the AGEI has been updated and revised to better reflect conditions within ASEAN member states and to better capture the sustainability dimensions of the enabling environment for agricultural productivity growth. The revised AGEI is presented in the following section.

5.3 Structure of the revised Agricultural Growth Enabling Index

The preliminary AGEI was constructed based on existing indicators and indices for measuring the determinants of agricultural growth and competitiveness. First, the key determinants of economic growth and development, agricultural growth and competitiveness, and associated indicators were identified through an extensive literature review. The authors then developed a typology to structure the components of the enabling environment for agricultural growth and competitiveness, which informed the choice of indicators for inclusion in the preliminary AGEI and its structure (see Diaz-Bonilla, Orden and Kwieciński [2014] for a detailed description of the construction of the AGEI).

The revised AGEI largely reflects the preliminary AGEI developed by Diaz-Bonilla et al. (2014). The first three blocks of the revised AGEI reflect economy-wide policy settings (governance quality, capital availability and market operations), each of which receive a weight of 20% in the index. The fourth block reflects the agriculture-specific and environmental sustainability aspects of the enabling environment. This block is composed of two pillars, each of which receives an equal weight of 20%. By placing 40% of the weight on the fourth block, an explicit agricultural and environmental sustainability orientation is given to the AGEI. All indicators within a block are weighted equally (Figure 5.1).

The first block, which reflects **governance quality**, is unchanged from the preliminary version of the AGEI. This block includes the macroeconomic stability and institutions indicators of the World Economic Forum's Global Competitiveness Index (GCI), and an indicator of political stability risk affecting the availability of food, which is drawn from the Economist Intelligence Unit Global Food Security Index (GFSI).

The second block reflects **investments in and availability of capital**. This block includes an indicator related to human capital (the health and primary education pillar of the GCI) and an indicator of physical capital (the infrastructure pillar of the GCI). The preliminary AGEI included an indicator of the presence of food safety net programmes, sourced from the GFSI, which has since been excluded from the revised AGEI. This indicator reflects policy settings that affect consumers' access to food, whereas the AGEI is concerned with policy settings that influence the production of food.

The third block reflects the **effectiveness of market operations** in the economy as a whole, and as such includes the CGI pillars on labour, goods and financial market operations. The revised AGEI also includes an indicator of trade facilitation sourced from the OECD Trade Facilitation Indicators database (OECD, 2015b), covering the full spectrum of border procedures (Moïsé, Orliac and Minor, 2011).

The fourth block of the preliminary AGEI focused on the agricultural and rural economy aspects of the enabling environment. Pillar A was composed of indicators that reflect institutional conditions amenable to policy decisions in the short to medium term. Pillar B was composed of indicators that reflect sector capacities and natural resources amenable to policy influence only in the medium to long term.

The fourth block (**Agriculture and Sustainability**) has been revised for the purposes of this report, in order to better reflect the sustainability dimensions of the enabling environment for agricultural productivity growth. Pillar A (**Agriculture**) of the revised AGEI combines the indicators of agricultural policies and investments in public goods from the preliminary AGEI. Indicators of access to finance for farmers, sourced from the GFSI, and land market rights and access, sourced from the Millennium Corporation Challenge, reflect the regulatory and institutional frameworks that govern financial and land market operations in the rural economy. Public agricultural research and development (R&D) expenditures as a percentage of agricultural gross domestic product (GDP) (sourced from the International Food Policy Research Institute [IFPRI] and the OECD General Services Support Estimate [GSSE]), and an indicator of agricultural infrastructure (sourced from the GFSI) reflect the extent to which agricultural policies and public investments support agricultural growth by building the capacities of the sector.

Pillar A also includes an indicator of capital intensification in agriculture. In the preliminary AGEI, this indicator was constructed as a combined indicator of fertiliser use and tractors per hectare. Fertiliser use per hectare was removed from the indicator because the increased use of fertilisers may not be consistent with sustainable improvements in productivity. Moreover, fertiliser use is already high in ASEAN, and is a significant source of water pollution (OECD, 2014a). In the revised AGEI, this indicator is constructed as an index of the gross capital stock per person employed in agriculture, and includes investments in land development, livestock, machinery and equipment, plantation crops and structures for livestock, based on data from FAOSTAT and the World Development Indicators (WDI).

Figure 5.1. Structure of the revised Agricultural Growth Enabling Index (AGEI)

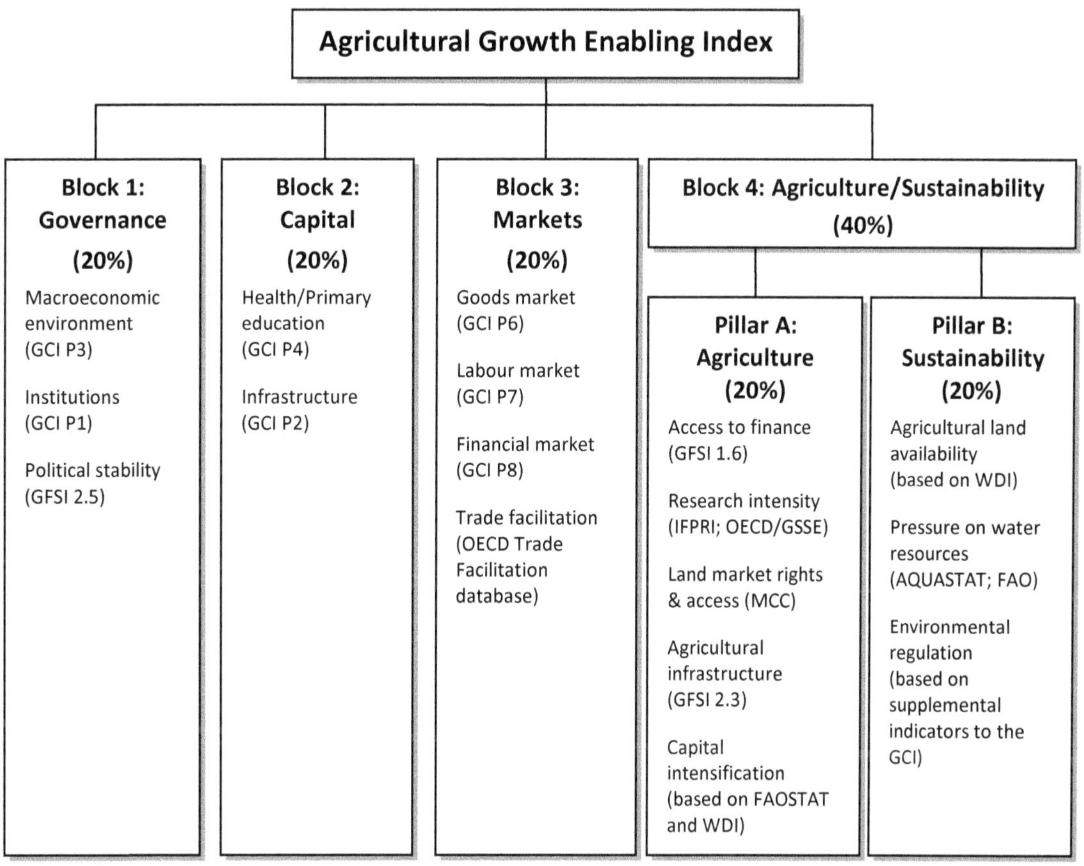

Sources: OECD estimates based on EIU (2015); FAO (2015b), *AQUASTAT*, www.fao.org/nr/aquastat/; FAO (2015c), *FAOSTAT*, http://faostat.fao.org/; IFPRI (2015); MCC (2014); OECD (2015b); World Bank (2015), *World Development Indicators*, http://data.worldbank.org; WEF (2014).

Table 5.1. Country coverage and classification

	WEF Global Competitiveness Index Classification	World Bank Classification
Chile	Transitioning from efficiency-driven to innovation-driven	High-income
Brazil	Transitioning from efficiency-driven to innovation-driven	Upper-middle-income
Costa Rica	Transitioning from efficiency-driven to innovation-driven	Upper-middle-income
Kazakhstan	Transitioning from efficiency-driven to innovation-driven	Upper-middle-income
Malaysia	**Transitioning from efficiency-driven to innovation-driven**	**Upper-middle-income**
Mexico	Transitioning from efficiency-driven to innovation-driven	Upper-middle-income
Russian Federation	Transitioning from efficiency-driven to innovation-driven	Upper-middle-income
Turkey	Transitioning from efficiency-driven to innovation-driven	Upper-middle-income
China	Efficiency-driven	Upper-middle-income
Colombia	Efficiency-driven	Upper-middle-income
South Africa	Efficiency-driven	Upper-middle-income
Thailand	**Efficiency-driven**	**Upper-middle-income**
Egypt	Efficiency-driven	Lower-middle-income
Indonesia	**Efficiency-driven**	**Lower-middle-income**
Morocco	Efficiency-driven	Lower-middle-income
Sri Lanka	Efficiency-driven	Lower-middle-income
Tunisia	Efficiency-driven	Upper-middle-income
Ukraine	Efficiency-driven	Lower-middle-income
Philippines	**Transitioning from factor-driven to efficiency-driven**	**Lower-middle-income**
Bangladesh	Factor-driven	Lower-middle-income
Cambodia	**Factor-driven**	**Lower-middle-income**
Ghana	Factor-driven	Lower-middle-income
India	Factor-driven	Lower-middle-income
Kenya	Factor-driven	Lower-middle-income
Lao PDR	**Factor-driven**	**Lower-middle-income**
Myanmar	**Factor-driven**	**Lower-middle-income**
Nigeria	Factor-driven	Lower-middle-income
Pakistan	Factor-driven	Lower-middle-income
Viet Nam	**Factor-driven**	**Lower-middle-income**
Ethiopia	Factor-driven	Low-income
Senegal	Factor-driven	Low-income
Tanzania	Factor-driven	Low-income

Note: ASEAN member countries in bold.
Sources: WEF (2014) and World Bank (2016b).

Pillar B (**Sustainability**) of the revised AGEI reflects countries' natural resource endowments, and the extent to which government policies support long-term agricultural growth by protecting the natural resource base upon which agricultural production relies. The preliminary AGEI included an indicator of combined land and water availability, which has been replaced by two separate indicators: one of agricultural land availability per person employed in agriculture (based on data from WDI), and one of pressure on water resources, defined as freshwater withdrawal as a percentage of total renewable water resources[4] (based on data from FAO AQUASTAT). Beyond the natural resource endowment, efficient management of natural resources is necessary to ensure the long-term sustainability of agricultural production. While specific indicators of agri-environmental governance are not available, Pillar B includes a general indicator of the stringency and enforcement of environmental regulations, sourced from the supplemental indicators to the GCI.

In the following section, the revised AGEI is applied to a subset of the ASEAN member states (Cambodia, Indonesia, Lao PDR, Malaysia, Myanmar, the Philippines, Thailand and Viet Nam) and a wider set of relevant non-ASEAN comparators. These include countries that are classified as factor-driven and efficiency-driven economies, as defined by the Global Competitiveness Index[5], as well as successful agricultural exporters such as Chile, Costa Rica and Brazil, which have also achieved significant reductions in the number of food-insecure people over the past 25 years. The 32 countries covered by the index are presented in Table 5.1.

5.4 Performance of ASEAN countries with respect to the AGEI blocks and indicators

Comparing how countries perform with respect to the four main blocks of the AGEI – and the indicators that comprise the blocks – provides an insight into the strengths and weakness of countries' enabling environments for sustainable agricultural productivity growth, including relative to other countries. Following a brief comparison of the performance of ASEAN and non-ASEAN countries in relation to the AGEI as a whole, this section assesses and compares the performance of ASEAN member states in aggregate and individually across all blocks of the AGEI and their indicators.

The performances of all 32 countries are shown in Figure 5.2, which ranks countries by their normalised AGEI score.[6] Figure 5.2 and all subsequent figures present the normalised scores of each country with respect to the AGEI and its component blocks and indicators, to account for differences in averages of scores of the countries and the variances of these scores across the index and its blocks (see notes to Figure 5.2). Countries with above-average scores on the AGEI have positive values, and countries with below-average scores have negative values. For example, Malaysia's normalised AGEI score is 1.96, which means that Malaysia's score is 1.96 standard deviations above the average (which is zero) for the 32 countries.

Of the 32 countries to which the AGEI was applied, the enabling environments in Chile, Malaysia and South Africa provide conditions that are most conducive to sustainable agricultural productivity growth, followed by Costa Rica and Mexico. The enabling environments in Nigeria, Myanmar and Pakistan are the least conducive to sustainable agricultural productivity growth, followed by Egypt and Ethiopia. An overview of the performance of all 32 countries can be found in Box 5.1.

A focus on the ASEAN region reveals a similarly mixed performance across the countries analysed. While the enabling environments in Malaysia and Thailand are ranked among the top ten countries, and the enabling environments in Indonesia and the Philippines receive above-average scores, the enabling environments in Viet Nam, Cambodia and Myanmar perform below the average for the 32 countries. The enabling environment in Lao PDR receives an average score. However, this result is driven by missing data and should be viewed with caution. Because Lao PDR's scores are not comparable with other countries, they are not discussed further in this chapter.

Figure 5.2. Country ranking by performance on the AGEI (normalised)

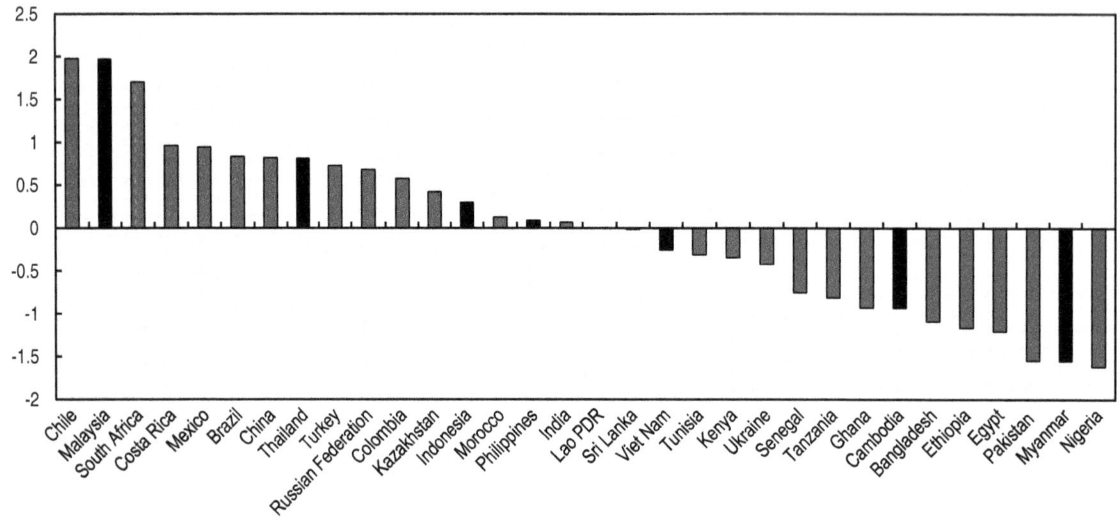

Notes: AGEI: Agricultural Growth Enabling Index. Normalised scores for each country on the AGEI index and on each component block and indicator are calculated by subtracting the average for the 32 countries from each country value, and then dividing the resulting country value by the standard deviation for the series. This creates a series with zero mean and unit standard error. ASEAN member states in black.
Source: OECD estimates.

With the exception of Malaysia, which performs above average for all four of the main blocks of the AGEI, ASEAN countries' performances, like those of the other AGEI countries, vary across blocks (Figure 5.3). In Thailand – one of the top ten performing countries in the AGEI – for example, the enabling environment for sustainable agricultural productivity growth is weakened by a below-average performance on the Governance block. In contrast, Indonesia, Myanmar and the Philippines receive their highest score on the Governance block relative to their performance on the other blocks. Viet Nam and Cambodia score relatively higher on the Capital and Markets blocks than on others, while Myanmar scores relatively lowest on both of these. Meanwhile, the majority of ASEAN member states included within the AGEI[7] – Malaysia, Indonesia, Viet Nam and Cambodia – score relatively lowest on the Agriculture/Sustainability block compared with their performance on other blocks. The following sub-section explores the performance of the ASEAN countries on each block and the indicators which comprise these.

Figure 5.3. The AGEI and its component blocks (normalised)

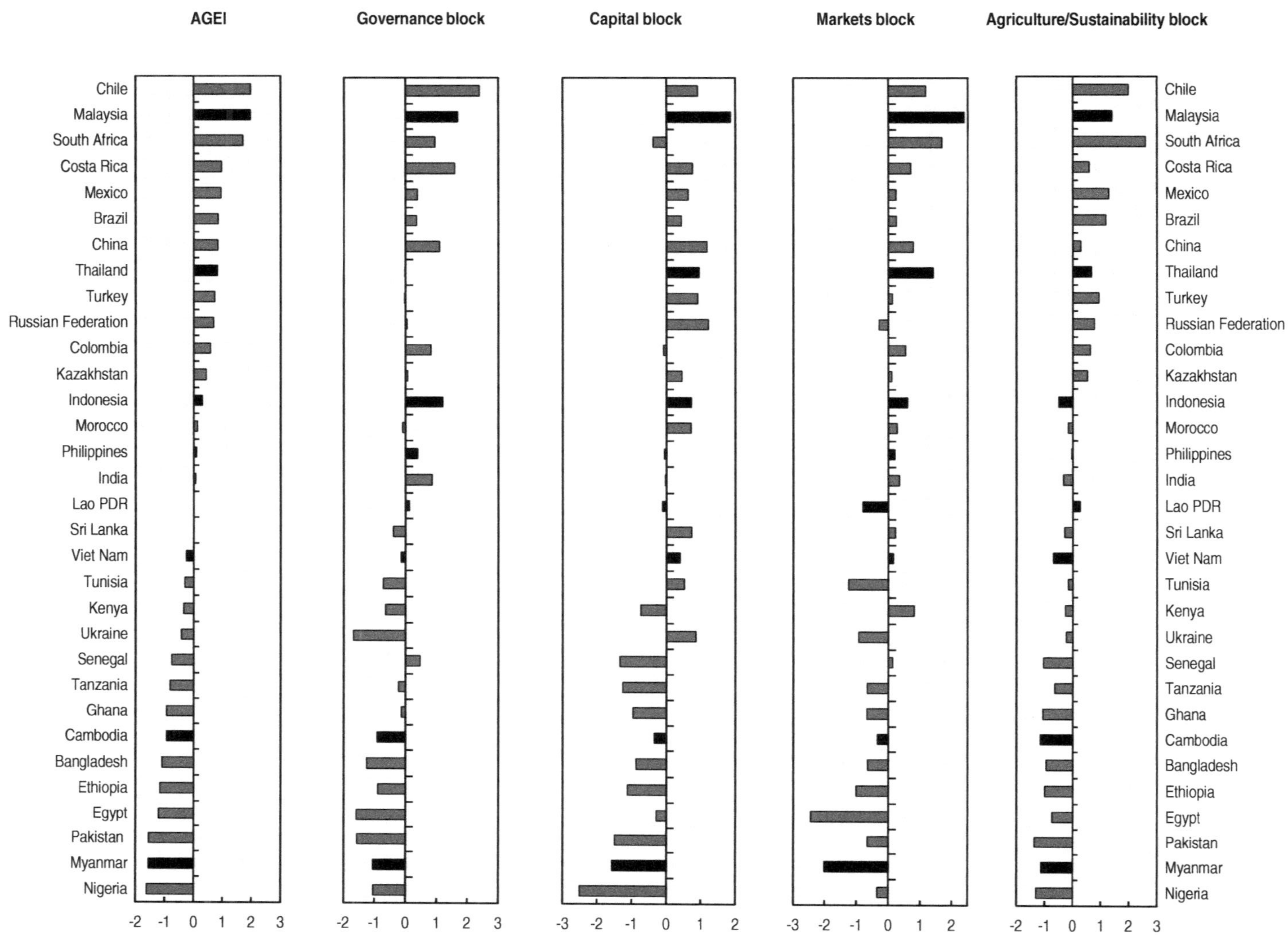

Notes: AGEI: Agricultural Growth Enabling Index. Figure shows the normalised scores for each country on the AGEI and on each component block (see notes to Figure 5.2). ASEAN member states in black.

Source: OECD estimates.

Box 5.1. Overview of the performance of all AGEI countries

Of the 32 countries to which the AGEI was applied, performances vary across the main blocks. While Chile, Malaysia, Costa Rica, Mexico, Brazil and China score above average for all four of the main blocks, in South Africa, Thailand, Turkey and the Russian Federation (the remaining top ten countries by performance in the AGEI), the enabling environment for sustainable agricultural productivity growth is weakened by a below-average performance on one component block: the Capital block in the case of South Africa, Governance in the case of Thailand and Turkey, and Markets for the Russian Federation, for example. The following is an overview of the performance of all AGEI countries across the four main blocks (Figure 5.3).

Block I: Institutions and governance (Figure 5.4): Chile, Malaysia and Costa Rica score highest on this block – largely as a result of relatively high scores on the institutions and political stability indicators – while the Ukraine, Egypt and Pakistan score lowest. Nevertheless, countries' normalised scores across the three indicators of this block display considerable variability, with the majority of countries exhibiting relative strengths and weaknesses across the different indicators.

Block II: Investment in physical and human capital (Figure 5.5): Malaysia, the Russian Federation and China score highest in the Capital block, driven by strong scores in both the health/education and infrastructure indicators. Nigeria is the weakest performer with respect to the Capital block – driven by a very low score in the health/education indicator – followed by Myanmar, Pakistan and Senegal.

Block III: Effectiveness of goods, labour and financial market operations (Figure 5.6): Malaysia has the highest score overall in the Markets block, driven by very high relative scores on the indicators of goods and labour market efficiency and financial market development. South Africa ranks second as a result of a strong performance on the trade facilitation and financial market development indicators, despite a below-average score for labour market efficiency. Egypt and Myanmar are weakest by a significant margin. In general, countries' scores on the indicators of goods and labour market efficiency and financial market development are mixed.

Block IV: Resource endowment, land institutions and investment in agriculture (Figure 5.7): South Africa scores the highest of all countries in the Agriculture/Sustainability block, driven by very strong performances in both Pillar A (Agriculture) and Pillar B (Sustainability). Chile, Malaysia, Mexico and Brazil also perform strongly, although Mexico is relatively weaker on Pillar B (Sustainability). Pakistan, Nigeria, Cambodia, Myanmar, Ghana and Senegal are the weakest in this block, largely as a result of low scores for Pillar A (Pakistan also receives a low score for Pillar B).

Pillar A (Agriculture) (Figure 5.8): South Africa, Chile and Mexico score highest on Pillar A. In the case of South Africa, this is despite relatively low scores for the indicators of access to finance for farmers and land market rights and access. Cambodia, Nigeria and Myanmar are the weakest performers on Pillar A, largely as a result of low scores for the indicators of access to finance for farmers, agricultural infrastructure (Cambodia) and land market rights and access (Nigeria and Myanmar).

Pillar B (Sustainability) (Figure 5.9): South Africa, Kazakhstan, Brazil and Malaysia scores highest on Pillar A – as a result of abundant land in South Africa and Kazakhstan, and for the stringency and enforcement of environmental regulations in Brazil and Malaysia. Egypt, Tunisia and Pakistan are weakest, reflecting significant pressure on water resources in those countries. Egypt and Pakistan also receive low scores for the stringency and enforcement of environmental regulations.

Block I: Institutions and governance

The Governance block of the AGEI is composed of indicators of the stability of the macroeconomic environment, the quality of institutions and political stability. These factors influence the overall performance of the economy, which is important for sustained growth in domestic demand – a key enabler of growth in the agricultural sector (Diaz-Bonilla, Orden and Kwieciński, 2014). Stable and sound macroeconomic policies, high-quality institutions and political stability also play an important role in ensuring a favourable environment for investment on farm and by other businesses in agricultural value chains, and enhance the effectiveness of government policies and investments aimed at supporting agricultural development. In addition, macroeconomic settings that create implicit – and possibly unintended – biases for or against the agricultural sector can distort resource allocation, constraining productivity growth (OECD, 2015c).

The Governance block is disaggregated into its component indicators in Figure 5.4. ASEAN countries' normalised scores across the three indicators display considerable variability, with the majority of countries exhibiting relative strengths and weaknesses. Only Malaysia and Indonesia have above-average scores for all three indicators, while Myanmar and Cambodia have below-average scores for all.

In general, the ASEAN member states perform relatively well with respect to the **macroeconomic stability** indicator, which the World Economic Forum (WEF, 2013) attributes to more prudent and sustainable macroeconomic management following the 1997 Asian financial crisis. Thailand, the Philippines, Indonesia and Malaysia are ranked among the top ten of all countries – non-ASEAN members included – on the indicator of macroeconomic stability. Viet Nam and Cambodia are close to the average for all 32 countries, and perform strongest on macroeconomic stability compared with other components of the governance block.

With respect to the **institutions** indicator, Malaysia scores the highest of all 32 countries, Indonesia ranks among the top ten, and the Philippines scores above average. Thailand, Viet Nam and Cambodia are below average, and Myanmar is the weakest of all countries by a significant margin. In Thailand, corruption, government instability, inefficient government bureaucracy and policy instability are identified as the most problematic factors for doing business; corruption is also an issue in Viet Nam, Cambodia and Myanmar (WEF, 2014).

ASEAN country rankings on the **political stability** indicator show some similarities to those for other governance indicators, namely that Malaysia and Indonesia once again rank among the top ten countries of the AGEI, while Viet Nam is close to average. The other ASEAN member states are below average – including the Philippines, which does not perform as strongly as it does in the other governance indicators – and Thailand and Cambodia are significantly so.

Block II: Investment in physical and human capital

The Capital block is composed of indicators of physical capital (infrastructure) and human capital (health and primary education) (Figure 5.5). Improvements in health and education play an important role in rural development by building the skills needed to innovate and improve resource use efficiency, and contribute to economic growth more broadly for the same reasons. Improving education and skills in rural areas can also facilitate normal structural adjustment in agriculture, by creating opportunities for non-farm employment in rural areas. Efficient and well-developed infrastructure, including transport and communications infrastructure, plays an important role in connecting businesses (including farms) to market opportunities and knowledge, and in providing access to economic services. In contrast, inefficient and underdeveloped infrastructure can significantly increase businesses' costs.

The World Economic Forum (2013) highlights physical and human capital factors as significant constraints to competitiveness in most ASEAN countries, in particular, poor transport, energy and communication infrastructures, and enrolment rates and/or poor quality education. Malaysia scores the highest of all countries – non-ASEAN included – on the Capital block, driven by strong scores in both the health/education and infrastructure indicators. Of the ASEAN countries in particular, Thailand and Indonesia rank second and third, while Myanmar is the weakest performer, as a result of low scores on both indicators, infrastructure in particular. The OECD (2014b) has found that roads, electricity and telecommunications infrastructures are particularly poor in rural areas of Myanmar, constraining the modernisation of the agricultural sector (among other factors).

In general, the performance of the ASEAN member states also varies across the indicators that comprise the Capital block. With the exception of Myanmar, the ASEAN member states have above-average scores on the **health/education** indicator. Viet Nam, Cambodia, the Philippines and Myanmar are relatively stronger with respect to the health/education indicator than on the **infrastructure** indicator, whereas Thailand, Malaysia and Indonesia are relatively stronger on the infrastructure indicator. Malaysia, Thailand and Indonesia all receive above-average scores on the infrastructure indicator, and Viet Nam receives an average score. The remaining ASEAN member states score below average, Myanmar by a significant margin.

Figure 5.4. Disaggregation of Block I: Governance (normalised)

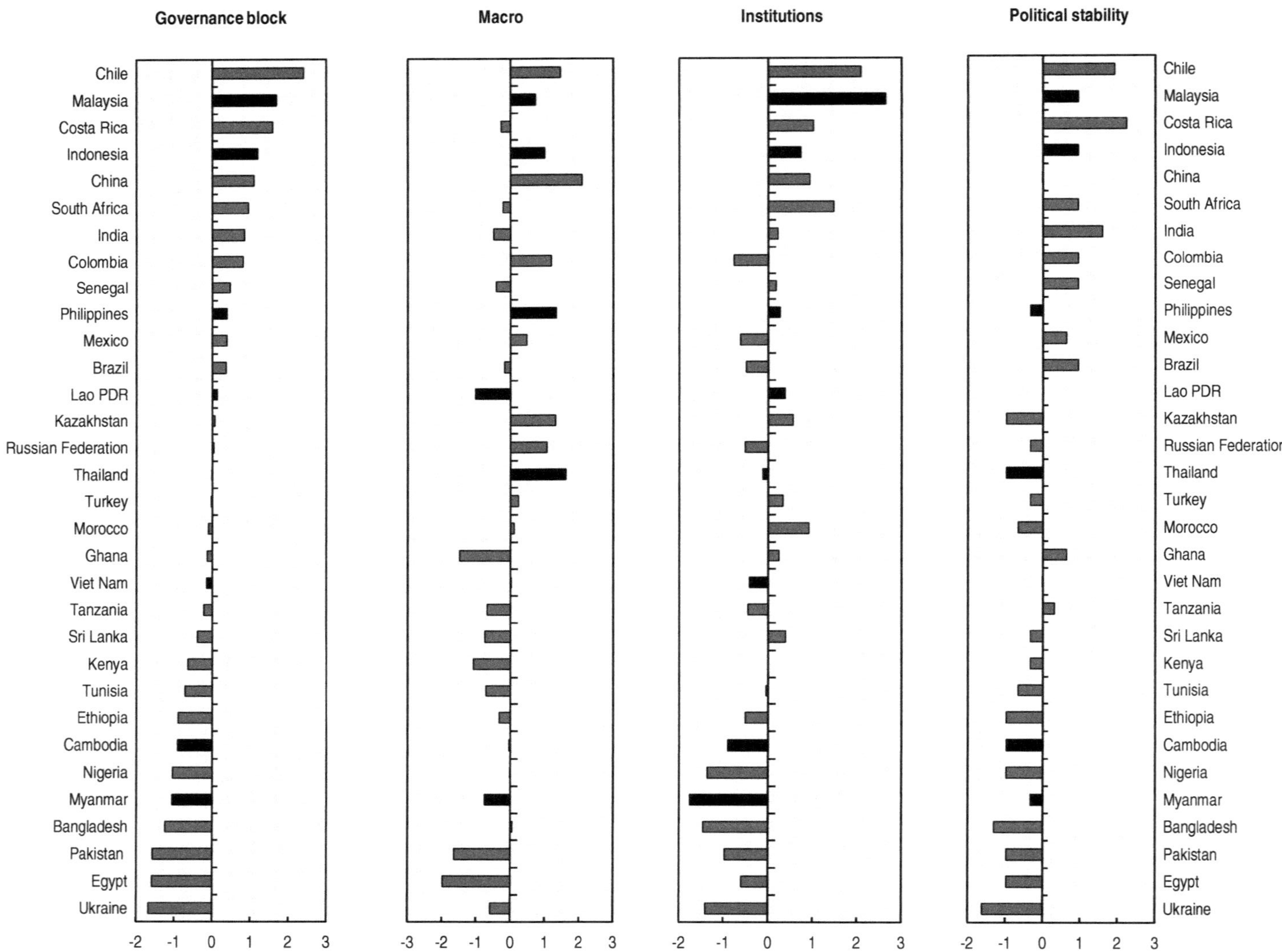

Notes: Figure shows the normalised scores for each country on the Governance block and on each indicator (see notes to Figure 5.2). ASEAN member states in black.
Source: OECD estimates.

Figure 5.5. Disaggregation of Block II: Capital (normalised)

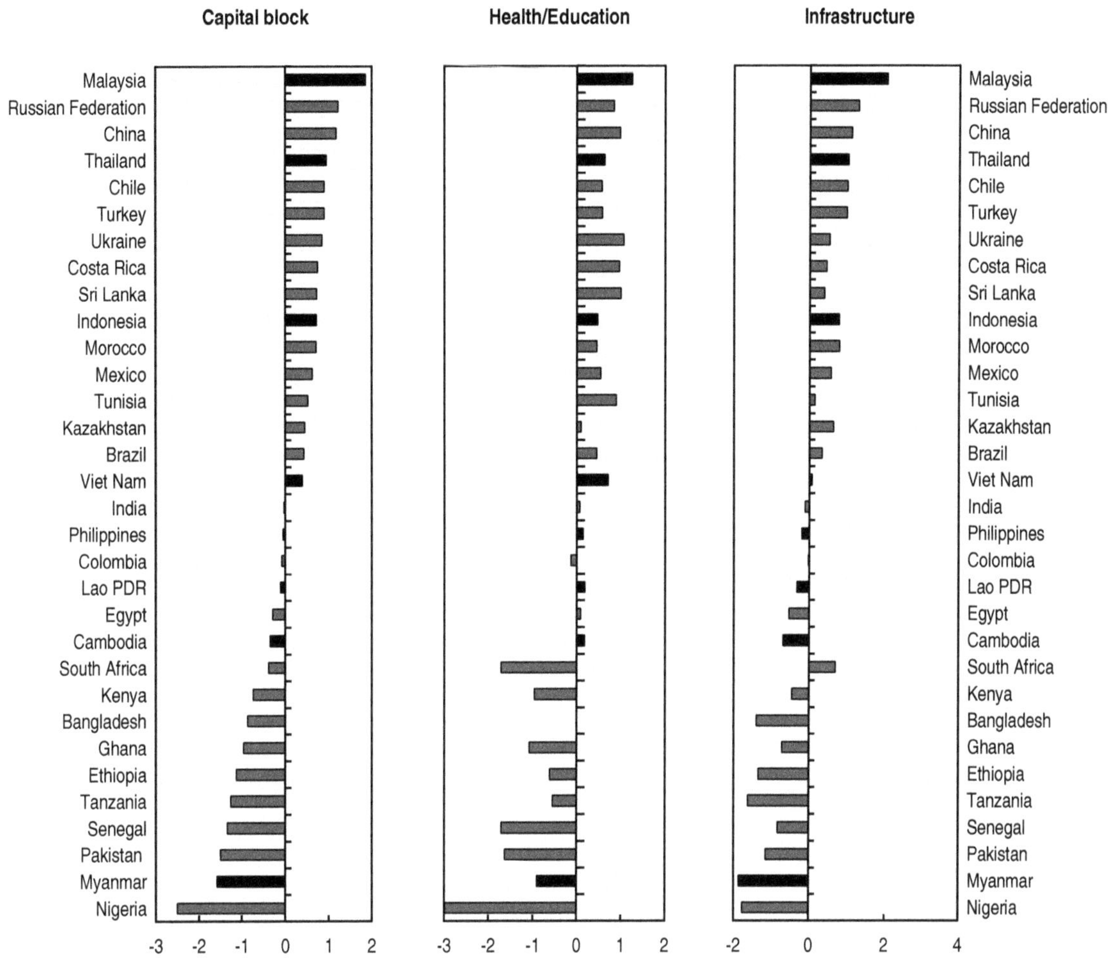

Notes: Figure shows the normalised scores for each country on the Capital block and on each indicator (see notes to Figure 5.2). ASEAN member states in black.
Source: OECD estimates.

Block III: Effectiveness of goods, labour and financial market operations

The Markets block is composed of indicators of goods and labour market efficiency, financial market development, and trade facilitation performance (Figure 5.6). Efficient markets facilitate the flow of factors of production – such as labour and capital – to their highest valued uses, including to uses within the agricultural sector, and provide undistorted incentives for investment. They also allow farms and other businesses in agricultural value chains to be flexible and responsive to changes in market conditions. Reducing agricultural trade costs via improvements in the trade facilitation environment can increase export competitiveness, and facilitates the flow of goods and capital across borders (Moïsé, Orliac and Minor, 2011). This can also lead to lower prices for farm inputs, and enhance the transfer of knowledge and technology between countries.

While scores for ASEAN member states are mixed, they generally perform well across the indicators that comprise the Markets block. Malaysia has the highest score of all ASEAN and non-ASEAN countries for this block as a whole, driven by very high relative scores on the indicators of goods and labour market efficiency, and financial market development. Myanmar is the weakest of the ASEAN countries by a significant margin, despite an above-average score on the labour market efficiency indicator.

ASEAN member states receive mixed scores with respect to **goods market efficiency**. Malaysia, Thailand and Indonesia all rank among the top ten countries of the AGEI, and the Philippines scores slightly above average, while Viet Nam, Cambodia and Myanmar score below, Myanmar significantly so.

With the exception of the Philippines and Indonesia, the ASEAN member states score above average on the **labour market** efficiency indicator. In the case of Indonesia and the Philippines, the World Economic Forum (WEF, 2014) notes that labour markets in these countries suffer from rigidities and inefficiencies, weakening competitiveness. In contrast, Malaysia and Cambodia are ranked among the top five countries – ASEAN and non-ASEAN included – for labour market efficiency, while Viet Nam is ranked tenth.

Malaysia, Thailand, Indonesia and the Philippines are ranked among the top ten countries for **financial market** development. Nevertheless, access to finance has been identified as one of the most problematic factors for doing business in Malaysia and Indonesia, as well as in Viet Nam, Cambodia and Myanmar (WEF, 2014), suggesting that there are opportunities for improvement.

Thailand is ranked second of all countries included within the AGEI with respect to **trade facilitation** performance. Malaysia, Indonesia, Viet Nam and the Philippines are also above average, while Cambodia is below average, and Myanmar receives the lowest score of all ASEAN countries.

Block IV: Resource endowment, land institutions and investment in agriculture

Two pillars comprise the Agriculture/Sustainability block. Pillar A (Agriculture) includes indicators of agricultural policies and investments in public goods that support agricultural growth by building the sector's capacities, and the regulatory and institutional frameworks that govern land and financial market operations in the rural economy. Pillar B (Sustainability) includes indicators of countries' natural resource endowments, and the extent to which government policies support long-term agricultural growth by protecting the natural resource base upon which agricultural production relies.

The Agriculture/Sustainability block is disaggregated into Pillar A and Pillar B in Figure 5.7. With the exception of Malaysia and Thailand, the ASEAN member states receive relatively low scores in this block, mainly as a result of relatively low scores in Pillar A. Malaysia scores the highest of the ASEAN countries, and the third-highest of all countries considered by the AGEI, driven by a strong performance on Pillar A in particular. Thailand takes second place among the ASEAN countries. Cambodia and Myanmar, by contrast, are the weakest of the ASEAN members with respect to the Agriculture/Sustainability block, largely as a result of very low scores for Pillar A.

Pillar A of the Agriculture/Sustainability block is disaggregated into its component indicators in Figure 5.8. Indicators for public expenditure on agricultural R&D as a percentage of agricultural GDP (hereafter, agricultural research intensity), and the quality and presence of agricultural infrastructure, capture the extent to which governments invest in building agricultural sector capacities that support higher agricultural productivity growth. With the exception of Malaysia, the ASEAN countries receive relatively low scores for **agricultural research intensity**. Of the ASEAN countries for which data are available, agricultural research intensity is lowest in Viet Nam and Cambodia. Viet Nam scores the lowest of all countries, non-ASEAN included, for which data on this component is available.

Malaysia scores the highest of the ASEAN countries for the quality and presence of **agricultural infrastructure**, ranking equally with the top three non-ASEAN countries. Cambodia scores the lowest of all countries, non-ASEAN included, for the quality and presence of agricultural infrastructure. Apart from Malaysia and Thailand, the ASEAN member states have relatively low scores on the agricultural infrastructure indicator. However, the indicator only measures the ability to store and transport crops to market. In Viet Nam and Indonesia, for example, governments have invested in irrigation infrastructure, to the extent that in Viet Nam in particular, the development and maintenance of infrastructure – namely irrigation infrastructure – is the most important component of the General Services Support Estimate (OECD, 2016a).

Figure 5.6. Disaggregation of Block III: Markets

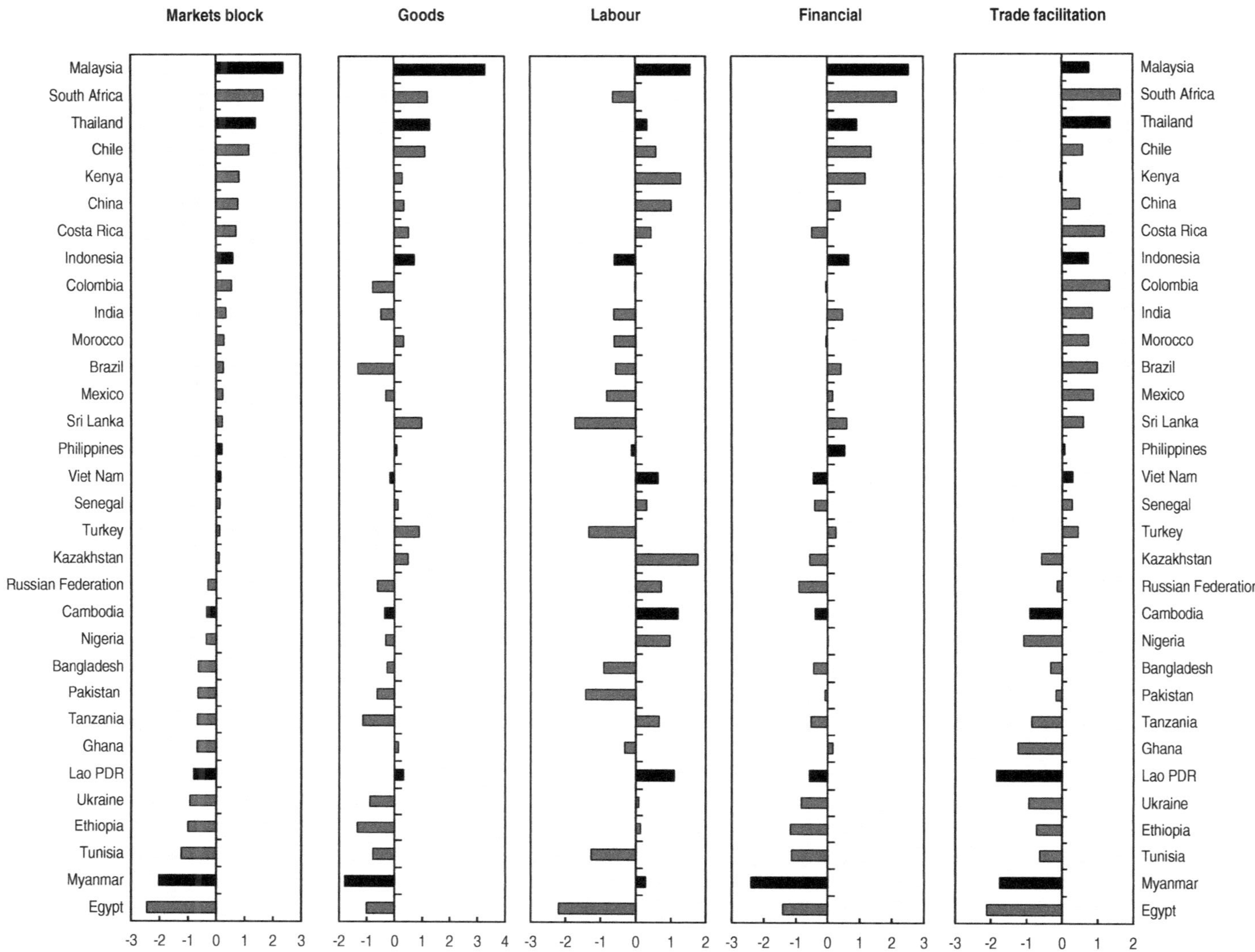

Notes: Figure shows the normalised scores for each country on the Markets block and on each indicator (see notes to Figure 5.2). ASEAN member states in black.
Source: OECD estimates.

Access to a range of **finance** options allows farmers – and smallholders in particular – to manage price shocks, and to invest in productivity- and sustainability-enhancing innovations (EIU, 2015). Malaysia and Thailand receive high scores, featuring among the top performers of all countries considered by the AGEI. The Philippines, Indonesia and Viet Nam receive slightly below-average scores. In the case of Indonesia and Viet Nam, concentrated financial markets in rural areas and a lack of sufficient collateral means that small farmers in particular have limited access to credit (OECD, 2012, 2015a). Cambodia and Myanmar receive very low scores, suggesting that farmer access to finance is minimal.

Figure 5.7. Disaggregation of Block IV: Agriculture/Sustainability (normalised)

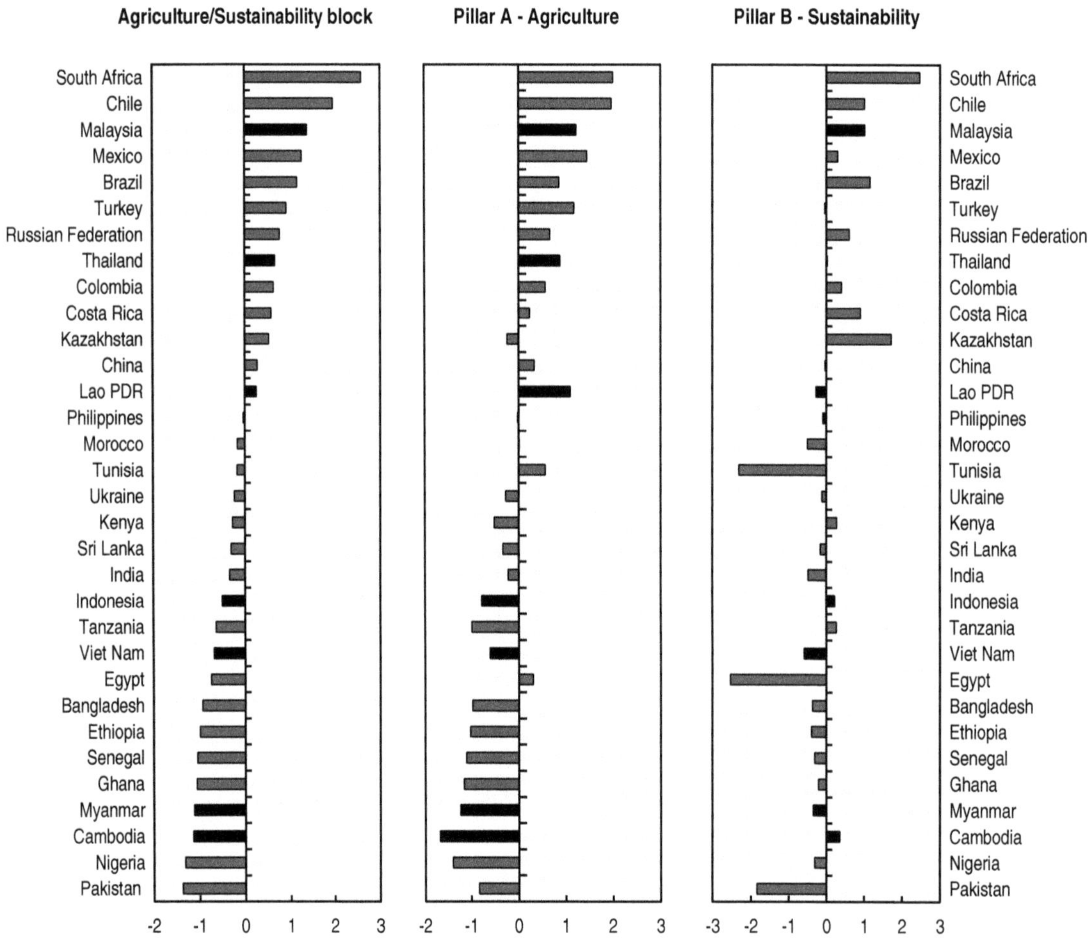

Notes: Figure shows the normalised scores for each country on the Agriculture/Sustainability block and on each pillar (see notes to Figure 5.2). ASEAN member states in black.
Source: OECD estimates.

Figure 5.8. Disaggregation of Pillar A of Block IV: Agriculture (normalised)

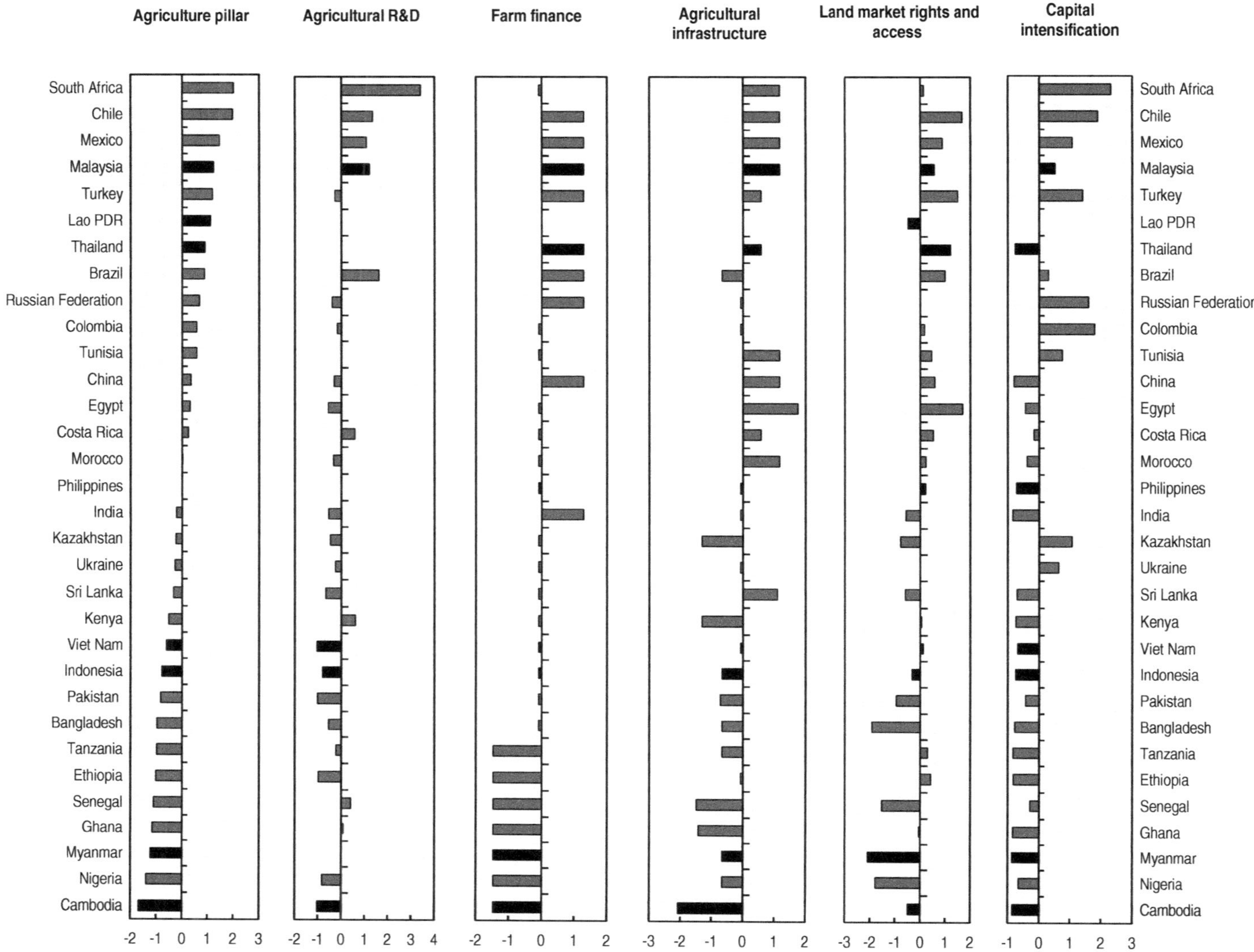

Notes: Figure shows the normalised scores for each country on the Agriculture pillar and on each indicator (see notes to Figure 5.2). ASEAN member states in black.
Source: OECD estimates.

Secure **land market rights and access** support sustainable agricultural productivity growth by enhancing landholders' incentives to make long-term investments and ensuring that users of land and water resources have a long-term interest in the integrity of the resource base to ensure future production (FAO, 2011b). Moreover, secure land market rights facilitate the use of land as collateral for loans, which can be important for funding productivity and sustainability-enhancing innovations (MCC, 2014). Thailand scores the highest of all ASEAN countries – and fourth-highest of all countries – with regard to this indicator. Myanmar meanwhile scores the lowest of all countries. Malaysia, the Philippines and Viet Nam perform above average – however, the security of land market rights and access is found to be below average in Cambodia and Indonesia, and significantly below average in the case of Myanmar. In Myanmar, state ownership of land and the underdeveloped rule of law have left land rights unclear in many cases, and some laws require permission to change land-use designations (OECD, 2016b). In Indonesia, the OECD has identified slow progress in the registration of land rights as a significant constraint, particularly for smallholders (OECD, 2012).

One outcome of a positive enabling environment is that farmers are able to invest in productive assets that support future agricultural productivity growth. This is captured by an indicator of **capital intensification**, measured as the gross capital stock per person employed in agriculture. Levels of capital intensification by farmers are relatively low in ASEAN countries – only Malaysia scores above average with respect to this indicator. This may reflect the dominance of plantation crops (palm oil and rubber) in Malaysia's agricultural sector, as well as other aspects of the enabling environment, such as farmer access to finance, the security of land market rights and access, and economy-wide influences. For example, the OECD (2014b) notes that in Myanmar, difficulty in accessing agricultural credit and poorly-defined land rights are important barriers preventing farmers from investing in fixed capital such as tractors and post-harvest technologies. Farm characteristics may also be a factor. In the ASEAN countries for which data are available, average farm size appears to be small, with estimates ranging between 0.8 ha per agricultural holding in Indonesia to 3.2 ha in Thailand (Lowder, Skoet and Singh, 2014).

Pillar B of the Agriculture/Sustainability block is disaggregated into its component indicators in Figure 5.9. This pillar includes two indicators of countries' natural resource endowments – one of pressure on water resources, and one of agricultural land availability per person employed in agriculture – and an indicator of the stringency and enforcement of environmental regulations. While ASEAN countries receive mixed scores with regard to this pillar, they remain relatively close to the average for all AGEI countries.

The ASEAN member states score relatively highly with respect to the indicator of pressure on **water resources** compared with the performance of many non-ASEAN countries in this component, suggesting that water availability is not a significant constraint in these countries. Cambodia (jointly with Colombia) receives the highest score allocated to any country, non-ASEAN included, while the Philippines scores the lowest of the ASEAN countries. In contrast, the availability of **agricultural land** is a significant constraint for the majority of ASEAN countries, as is the case with most non-ASEAN countries analysed. In part, this reflects the high endowment of agricultural land in South Africa, Kazakhstan and, to a lesser extent, the Russian Federation. The ASEAN member states all score low (below average) on this indicator and have low land availability.

Environmental governance and regulations on natural resources largely determine conditions for access to land, water and biodiversity resources, and are central to ensuring the sustainable use of natural resources in the long term (OECD, 2015c). ASEAN members receive relatively low scores for the stringency and enforcement of environmental regulations, with the exception of Malaysia, which scores the highest of all countries considered by the AGEI. While data are not available for Myanmar and Viet Nam, the remaining ASEAN member states score below average. This is consistent with more in-depth reviews, which have found that natural resources and the environment are under considerable pressure in some ASEAN member states. This is partly due to the rapid pace of industrialisation in these countries, and to the expansion of agricultural land, in addition to other practices such as high levels of fertiliser use (OECD, 2014a).

Figure 5.9. Disaggregation of Pillar B of Block IV: Sustainability

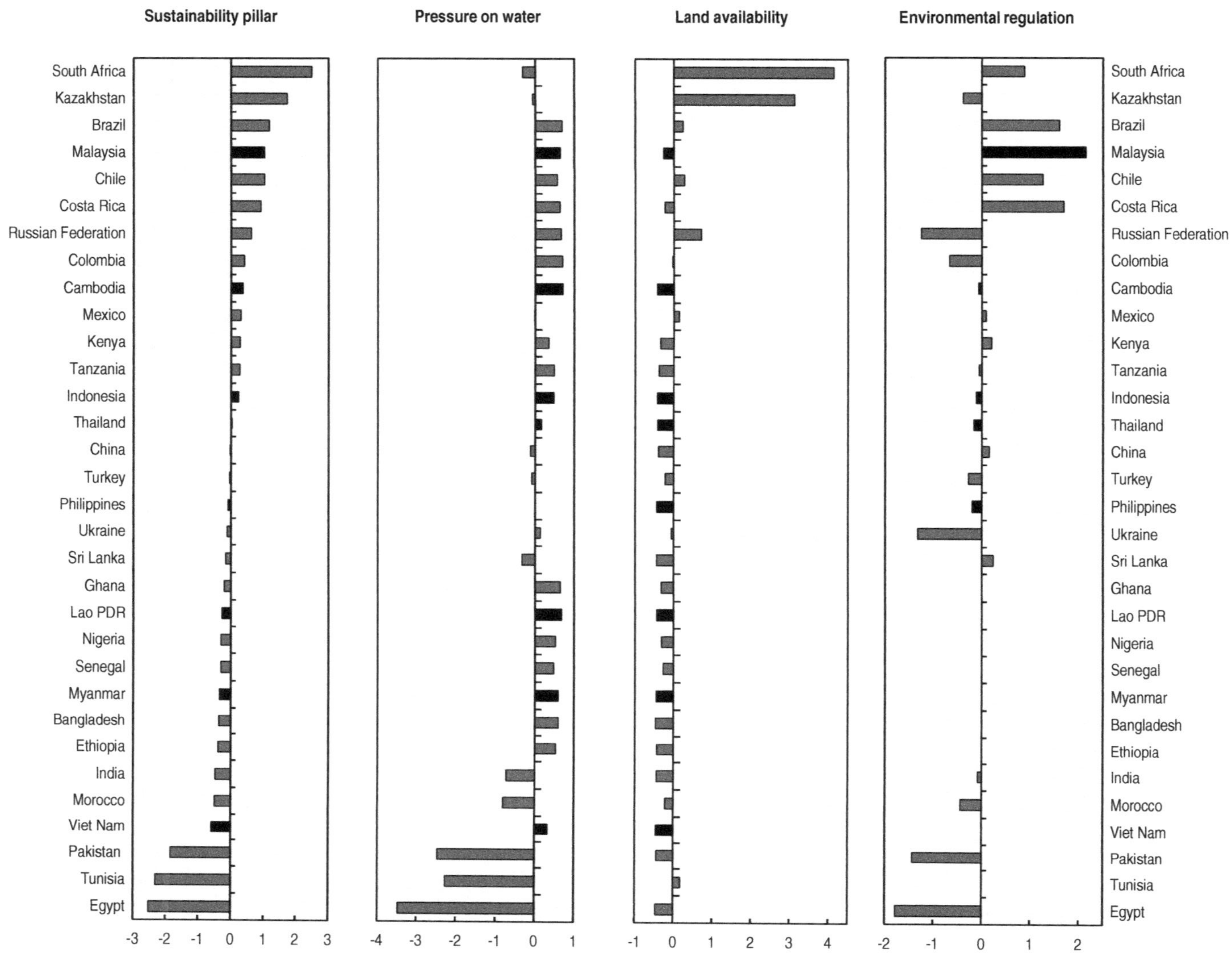

Notes: Figure shows the normalised scores for each country on the Sustainability pillar and on each indicator (see notes to Figure 5.2). ASEAN member states in black.
Source: OECD estimates.

In some countries, poor environmental outcomes are a result of weak enforcement of existing regulations, rather than inadequate legislation on environmental protection. In the Philippines, for example, there are numerous laws and regulations in force to halt unsustainable farming and fishing practices. However, their implementation is weak because of a lack of resources and political will for enforcement (NEDA, 2014). Compliance, monitoring and enforcement of environmental regulations is also weak in Indonesia and Viet Nam (OECD, 2015a; WEF, 2014).

5.5 Assessment of country-specific performances in the ASEAN region

This section provides country-specific profiles of the performance of each of the seven ASEAN countries considered by the AGEI (Malaysia, Thailand, Indonesia, the Philippines, Viet Nam, Cambodia and Myanmar) across all components of the index.

Malaysia's (normalised) scores across all components of the AGEI are presented in Figure 5.10. Malaysia is ranked among the top three countries in the AGEI and its sub-component blocks. Key areas of relative strength include the institutional environment, infrastructure, goods and financial market operations, and the stringency and enforcement of environmental regulation.

Overall, Malaysia is weakest in the Agriculture and Sustainability pillars of the enabling environment, while nevertheless ranking above average for the countries in the AGEI. In particular, strengthening land market rights and access may enhance the capacity of Malaysia's farmers to increase agricultural productivity. Within the economy-wide blocks, Malaysia scores relatively lower on macroeconomic stability and trade facilitation, suggesting that improvements in these areas would further enhance Malaysia's already strong enabling environment for sustainable agricultural productivity growth.

Figure 5.10. Enabling environment in Malaysia

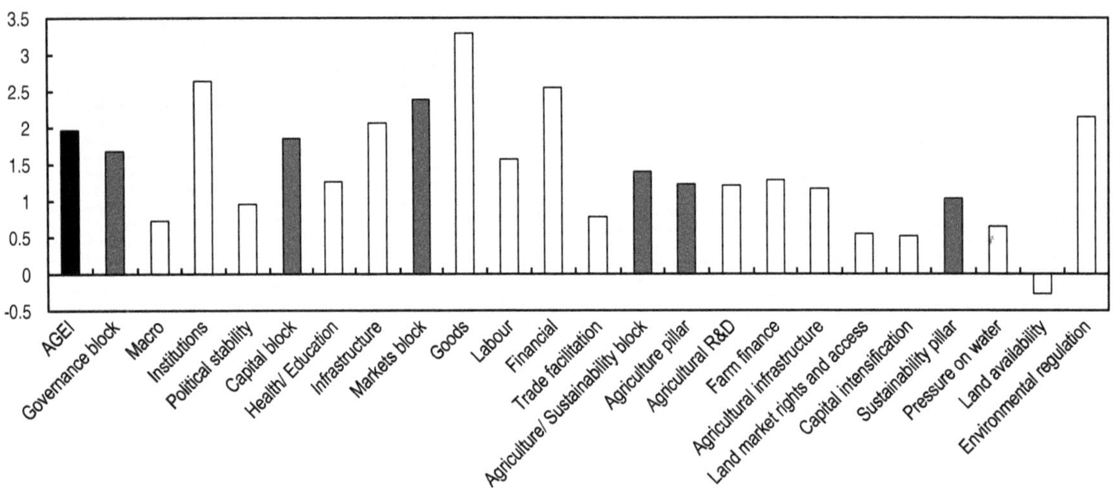

Notes: AGEI: Agricultural Growth Enabling Index. Figure shows the normalised scores for Malaysia on the AGEI and its component indicators (see notes to Figure 5.2).
Source: OECD estimates.

Figure 5.11 presents **Thailand**'s (normalised) scores across all components of the AGEI. Thailand's areas of relative strength include macroeconomic stability,[8] its investments in infrastructure, and well-functioning markets, although there is potential for the improvement of labour market operations. The institutional environment – political stability in particular – is a key weakness.

Although levels of capital intensification are low, Thailand is relatively strong in the agricultural aspects of the enabling environment, and scores highly for land market rights and farmers' access to finance. However, its performance on the sustainability aspects of the enabling environment is relatively weak, which may constrain agricultural productivity growth in the long term. This reflects the relative scarcity of agricultural land and, to a slightly lesser extent, the country's low score for the stringency and enforcement of environmental regulation.

The (normalised) scores for **Indonesia** across all AGEI components are presented in Figure 5.12. Indonesia performs consistently across the indicators that comprise the economy-wide blocks, and is relatively strong on macroeconomic and political stability and infrastructure. Labour market operations are a key exception.

Figure 5.11. Enabling environment in Thailand

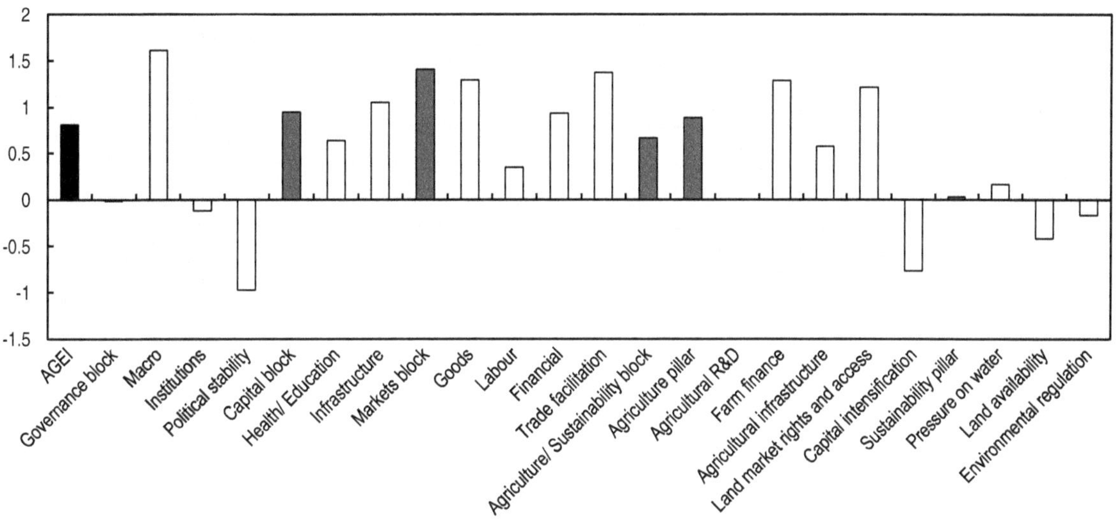

Notes: AGEI: Agricultural Growth Enabling Index. Figure shows the normalised scores for Thailand on the AGEI and its component indicators (see notes to Figure 5.2).
Source: OECD estimates.

Figure 5.12. Enabling environment in Indonesia

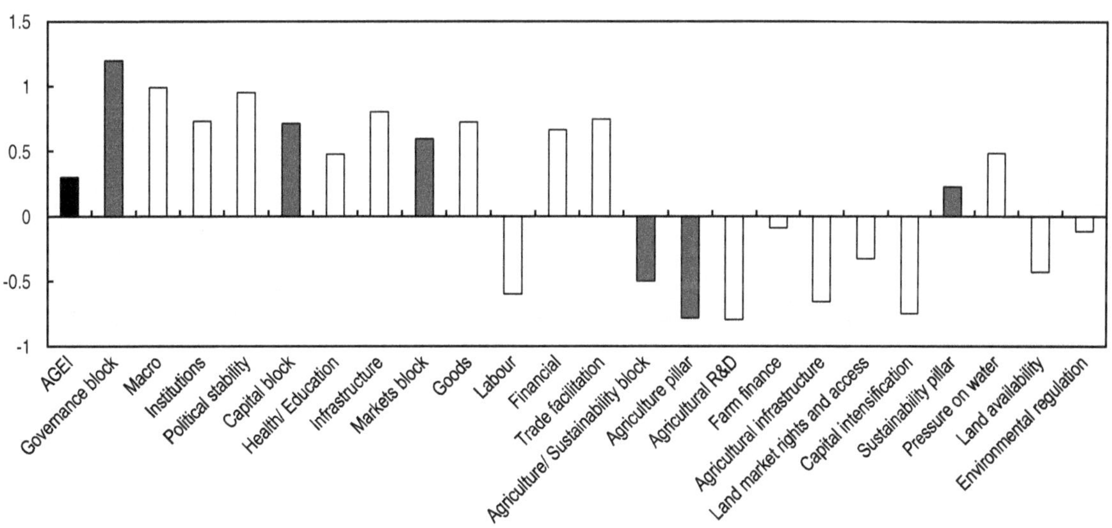

Notes: AGEI: Agricultural Growth Enabling Index. Figure shows the normalised scores for Indonesia on the AGEI and its component indicators (see notes to Figure 5.2).
Source: OECD estimates.

Compared with its performance on the economy-wide blocks, Indonesia scores relatively low on the dimensions of the Agriculture/Sustainability block. In particular, the country's level of investment in agricultural R&D and low levels of capital intensification by farmers may constrain future productivity growth. Indonesia also scores relatively low on land availability and land market rights. Environmental regulations and their enforcement are also weak relative to other countries on the AGEI.

Indonesia also receives below-average scores for farmer access to finance and, moreover, for the quality and availability of agricultural infrastructure – although, as noted earlier, significant investments have been made in irrigation infrastructure. Given its strong scores for infrastructure and financial market development (in the economy-wide blocks), this suggests that the rural economy may not be provided with public goods and economic services to the same extent as other sectors in the economy.

Figure 5.13 presents the **Philippines**' (normalised) scores across all components of the AGEI. The Philippines' areas of relative strength include macroeconomic stability – where it is among the top five countries in the AGEI – and financial market operations. However, the Philippines receives a below-average score for access to finance for farmers. Lower scores for infrastructure, including agricultural infrastructure, and relatively low levels of capital intensification by farmers suggest potential constraints to agricultural productivity growth, a fact which is recognised in the *Philippines Development Plan 2011-2016.* The plan identifies low levels of mechanisation and access to formal credit and financing among the constraints to growth in agricultural production and productivity (NEDA, 2011).

Figure 5.13. Enabling environment in the Philippines

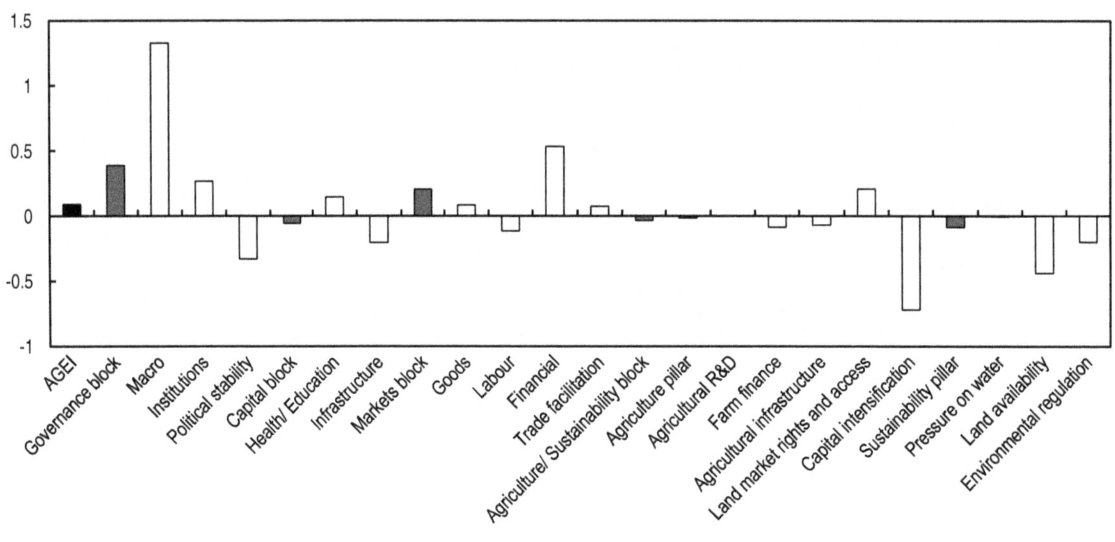

Notes: AGEI: Agricultural Growth Enabling Index. Figure shows the normalised scores for the Philippines on the AGEI and its component indicators (see notes to Figure 5.2).
Source: OECD estimates.

The Philippines is also relatively weaker on the sustainability dimensions of the enabling environment. In contrast with other ASEAN member states, the Philippines receives an average score for pressure on water resources. Agricultural land is relatively scarce, and environmental regulations and their enforcement in the Philippines are also relatively weak.

Viet Nam's (normalised) scores across all components of the AGEI are presented in Figure 5.14. While Viet Nam scores below average with respect to the AGEI as a whole, it receives above-average scores in the Capital and Market blocks, reflecting relatively high scores for its investments in developing human capital (health and primary education) and labour market operations. In contrast, the country receives relatively low scores for goods and financial market operations, and the quality of institutions.

Of the main blocks of the AGEI, Viet Nam is weakest on the Agriculture and Sustainability pillars. The country scores very low for public investments in agricultural R&D and the level of capital intensification by farmers, which may constrain productivity growth on farms. Land availability is also low – at 0.12 ha per capita, agricultural land availability in Viet Nam is the lowest of the ASEAN member states (World Bank, 2016a).

Figure 5.14. Enabling environment in Viet Nam

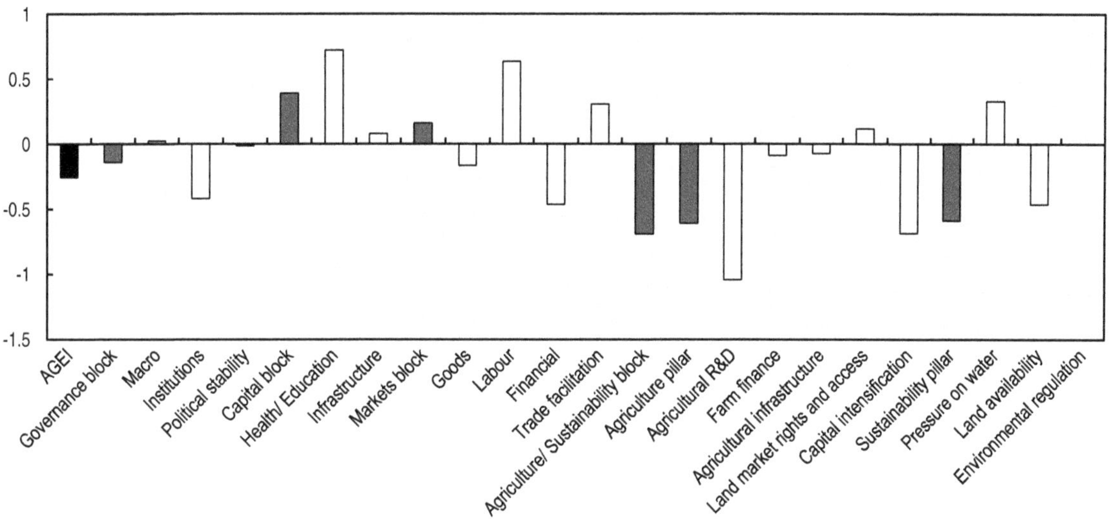

Notes: AGEI: Agricultural Growth Enabling Index. Figure shows the normalised scores for Viet Nam on the AGEI and its component indicators (see notes to Figure 5.2).
Source: OECD estimates.

Figure 5.15 presents **Cambodia**'s (normalised) scores across all AGEI components. Cambodia is among the weaker countries – non-ASEAN included – in the AGEI, scoring below average on all of the sub-component blocks. Within the blocks, Cambodia receives a relatively high score for labour market operations, and scores above average for its investments in developing human capital. Across the economy-wide blocks of the AGEI, Cambodia scores lowest for political stability, trade facilitation, the institutional environment and the availability of infrastructure.

Cambodia is weakest overall in the agricultural aspects of the enabling environment, particularly the availability of agricultural infrastructure, farmers' access to finance and investments in agricultural R&D. Cambodia scores lowest of all countries for the quality and availability of agricultural infrastructure, suggesting that the rural sector may be disadvantaged in the provision of public goods.

Myanmar's (normalised) scores across all components of the AGEI are presented in Figure 5.16. Myanmar is one of the weakest countries in the AGEI. Of the component blocks, Myanmar scores lowest in market operations – with the exception of labour market operations, which represent an important strength. The quality of institutions and infrastructure are other significant weaknesses – agricultural infrastructure to a lesser extent.

Myanmar receives low scores for all agriculture-related aspects of the enabling environment, particularly land market rights and farmers' access to finance, which suggests important constraints to agricultural productivity growth.

Figure 5.15. Enabling environment in Cambodia

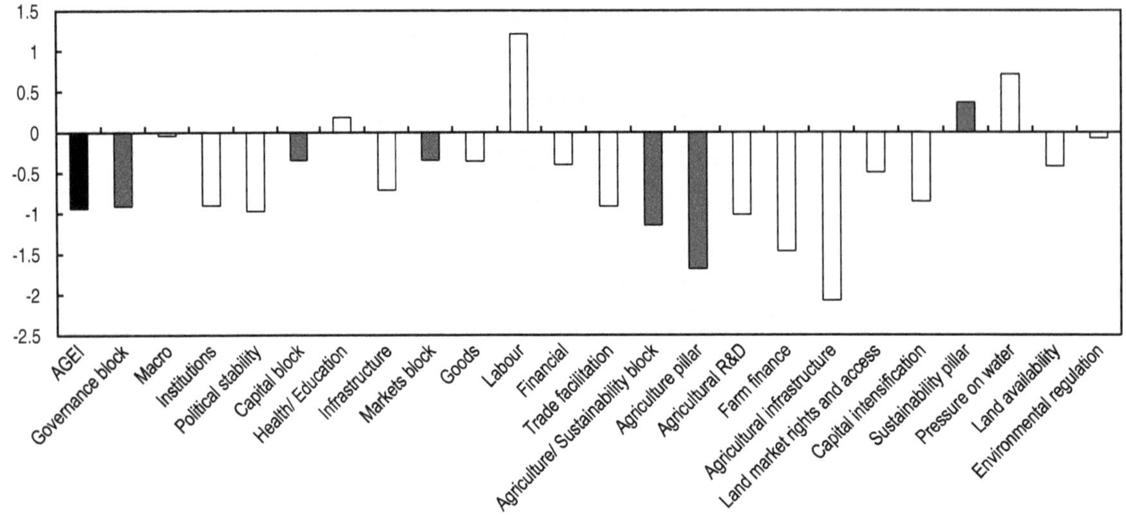

Notes: AGEI: Agricultural Growth Enabling Index. Figure shows the normalised scores for Cambodia on the AGEI and its component indicators (see notes to Figure 5.2).
Source: OECD estimates.

Figure 5.16. Enabling environment in Myanmar

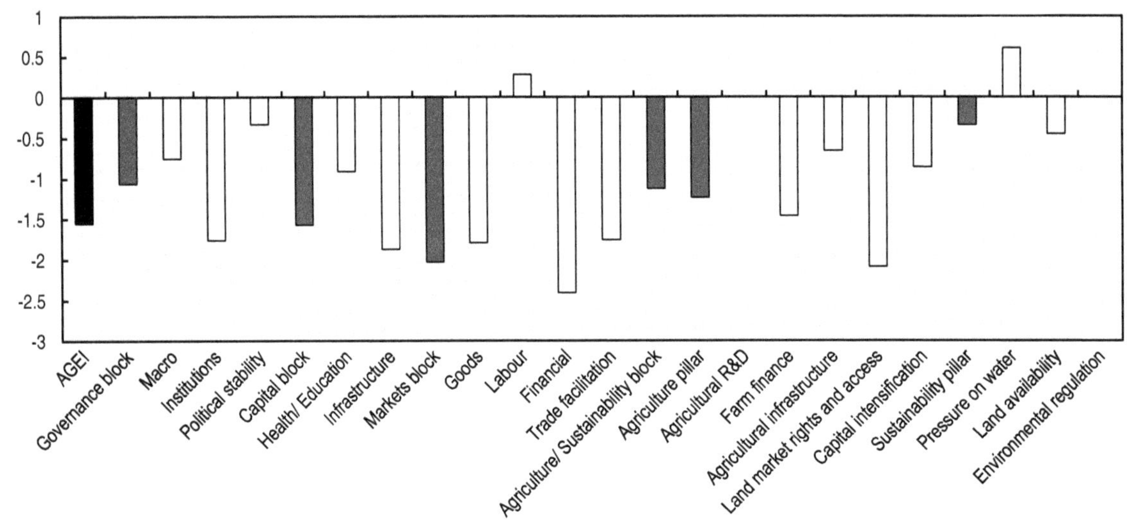

Notes: AGEI: Agricultural Growth Enabling Index. Figure shows the normalised scores for Myanmar on the AGEI and its component indicators (see notes to Figure 5.2).
Source: OECD estimates.

5.6 Concluding comments

The enabling environment plays an important role in determining food security outcomes in the longer term by creating conditions that promote sustainable agricultural productivity growth. This chapter assesses the enabling environment for sustainable agricultural productivity growth in a subset of ASEAN member states and a wider set of relevant comparators, using a revised and updated Agricultural Growth Enabling Index (AGEI).

The diversity of the ASEAN region means that countries vary significantly in their performance across the AGEI and its sub-components. Nevertheless, the analysis points to some areas of common (relative) strengths and weaknesses.

Looking first at the economy-wide blocks, the macroeconomic environment is relatively strong in the countries analysed. The ASEAN member states also generally perform well across the indicators that comprise the Markets block. For example, with the exception of Indonesia and the Philippines, the ASEAN member states receive above-average scores with respect to the labour market operations indicator. Moreover, labour market operations is an area of relative strength for Viet Nam, Cambodia and Myanmar – countries which score below average on the AGEI overall. With the exception of Myanmar, the ASEAN member states also receive above-average scores in the health/education indicator, suggesting that ASEAN member state investments in developing human capital are an area of strength relative to the other countries in the analysis. This is despite suggestions by some commentators that physical and human capital factors are significant constraints to competitiveness in most ASEAN countries (for example, WEF, 2013).

The analysis also points to common areas of relative weakness in ASEAN, most significantly, the agriculture-related aspects of the enabling environment (Pillar A of the Agriculture/Sustainability block). With the exception of Malaysia, the ASEAN member states have relatively low scores for public investments in agricultural R&D and the quality and presence of agricultural infrastructure (Thailand also scores above average for the quality and presence of agricultural infrastructure). These findings are consistent with OECD reviews that find that agricultural support is focused on irrigation infrastructure and input subsidies, leaving other agricultural infrastructure and agricultural R&D underfunded (OECD, 2012, 2014b, 2015a). Going forward, future productivity growth in ASEAN may be constrained by low levels of investment in public goods that support agricultural growth by building the sector's capacities.

The analysis also suggests that agriculture – and the rural sector more broadly – may be underprovided with public goods and other economic services relative to other sectors. Some countries receive below-average scores for agricultural infrastructure and/or farmer access to finance, despite scoring highly for infrastructure and financial market development (in the economy-wide blocks). A perceived bias against agriculture in the provision of public goods and economic services may be a disincentive to investment and constrain agricultural development.

The ASEAN member states are also found to be weaker in some sustainability aspects of the enabling environment. In particular, agricultural land is relatively scarce. On the other hand, water resources are relatively abundant. The majority of ASEAN member states also receive relatively low scores for the stringency and enforcement of environmental regulations, with the exception of Malaysia, which scores highest of all countries – non-ASEAN included – on this indicator. Again, this finding is borne out by recent reviews of agricultural policies in Indonesia, the Philippines and Viet Nam, which find that natural resources and the environment are under strong pressure in these countries, including from agriculture (OECD, 2012, 2015a, 2017).

Going forward, the diversity of countries' performances across the AGEI and its sub-components suggests that there is a significant opportunity for ASEAN member states to learn from their neighbours' experiences with policy reform. While Malaysia stands out for its strong overall performance on the AGEI and its subcomponents, other ASEAN member states also exhibit strengths relative to their peers, with Indonesia, the Philippines and Thailand ranked among the top ten countries for certain sub-components. Importantly, ASEAN offers a forum for this to take place.

This assessment provides a starting point for in-depth assessments of how ASEAN member states' policy settings influence productivity and the sustainable use of natural resources in the long run. Specifically, individual countries' policy settings could be assessed using the OECD's framework, *Analysing Policies to Improve Agricultural Productivity Growth, Sustainably* (OECD, 2015c), which considers the full range of policy incentives and disincentives that might impact agricultural productivity growth and the sustainable use of natural resources. Such a review would help the identification of opportunities for alternative policies to improve the enabling environment for sustainable agricultural productivity growth and, in turn, help to promote domestic and regional food security objectives in Southeast Asia in the long term.

Notes

1. This chapter benefited from comments provided by colleagues from the OECD Secretariat: Frank van Tongeren, Jared Greenville, Clara Thompson-Lipponen, Shingo Kimura, Silvia Sorescu. The chapter also benefited from comments provided by participants at the OECD-FAO-ASEAN Regional Conference on Policies to Enable Food Security, Agricultural Productivity and Improved Nutrition, which was supported by the FAO and the Governments of Myanmar and Australia, and which took place on 14-15 June 2016 in Nay Pyi Taw, Myanmar. The chapter also draws on a consultancy report prepared by Winona Bolislis.

2. FAO (2015a) estimates cover Brunei Darussalam, Cambodia, Indonesia, Lao PDR, Malaysia, Myanmar, the Philippines, Thailand, Timor-Leste and Viet Nam.

3. ASEAN comprises Brunei Darussalam, Cambodia, Indonesia, Lao PDR, Malaysia, Myanmar, the Philippines, Singapore, Thailand and Viet Nam. Brunei Darussalam and Singapore are not considered in the analysis.

4. A positive effect was assigned to this indicator by taking the residual part (100% – freshwater withdrawals).

5. Two criteria are used to categorise countries according to stages of development. The first is the level of GDP per capita at market exchange rates. A second criterion is used to adjust for countries that, based on income, would have moved beyond the first stage of development, but where prosperity is based on the extraction of resources (WEF, 2014).

6. Normalised scores for each country on the AGEI index and on each component block and indicator are calculated by subtracting the average for the 32 countries from each country value, and then dividing the resulting country value by the standard deviation for the series. This creates a series with zero mean and unit standard error.

7. Lao PDR excluded.

8. The World Economic Forum (2014) notes that the data used in assessing the macroeconomic environment were collected prior to the military coup of May 2014.

References

ADB (2013), *Food Security in Asia and the Pacific*, Asian Development Bank, Mandaluyong City, Philippines.

ADB (2009), *The Economics of Climate Change in Southeast Asia: A Regional Review*, Asian Development Bank, Mandaluyong City, Philippines.

AMAF (2015), *Vision and Strategic Plan for ASEAN Cooperation in Food, Agriculture and Forestry (2016-2025)*, ASEAN Ministers of Agriculture and Forestry, Makati City, Philippines.

Diaz-Bonilla, E., D. Orden and A. Kwieciński (2014), "Enabling environment for agricultural growth and competitiveness: Evaluation, indicators and indices", *OECD Food, Agriculture and Fisheries Papers*, No. 67, OECD Publishing, Paris, http://dx.doi.org/10.1787/5jz48305h4vd-en.

EIU (2015), *Global Food Security Index 2015: An Annual Measure of the State of Global Food Security*, Economist Intelligence Unit, London, http://foodsecurityindex.eiu.com/ (accessed on 7 May 2015).

FAO (2015a), *Regional Overview of Food Insecurity Asia and the Pacific: Towards a Food Secure Asia and the Pacific*, Food and Agriculture Organization, Bangkok.

FAO (2015b), *AQUASTAT* (database), Food and Agriculture Organization, Rome, www.fao.org/nr/aquastat/ (accessed on 29 June 2015).

FAO (2015c), *FAOSTAT* (database), Food and Agriculture Organization, Rome, http://faostat.fao.org/ (accessed on 05 May 2015).

FAO (2011a), *Southeast Asian Forests and Forestry to 2020*, Food and Agriculture Organization, Bangkok.

FAO (2011b), *The State of the World's Land and Water Resources for Food and Agriculture (SOLAW): Managing Systems at Risk*, Food and Agriculture Organization, Rome, and Earthscan, London.

IFPRI (2015), "Food policy indicators: Tracking change", in *2014–2015 Global Food Policy Report*, International Food Policy Research Institute, Washington, DC.

Lowder, S.K., J. Skoet and S. Singh (2014), "What do we really know about the number and distribution of farms and family farms worldwide?", Background paper for *The State of Food and Agriculture 2014*, *ESA Working Paper* No. 14-02, Food and Agriculture Organization, Rome.

MCC (2014), *Guide to the Indicators and the Selection Process, FY 2015* (webpage), Millennium Challenge Corporation, www.mcc.gov/resources/doc/report-guide-to-the-indicators-and-the-selection-process-fy-2015 (accessed on 30 April 2015).

Moïsé, E., T. Orliac and P. Minor (2011), "Trade facilitation indicators: The impact on trade costs", *OECD Trade Policy Working Papers*, No. 118, OECD Publishing, Paris, http://dx.doi.org/10.1787/5kg6nk654hmr-en.

NEDA (2014), *Philippine Development Plan 2011-2016: Midterm Update with Revalidated Results Matrices*, National Economic Development Authority, Philippines.

NEDA (2011), *Philippines Development Plan 2011–2016*, National Economic and Development Authority, Philippines.

OECD (2017), *Agricultural Policies in the Philippines*, OECD Publishing, Paris, http://dx.doi.org/10.1787/9789264269088-en.

OECD (2016a), *Agricultural Policy Monitoring and Evaluation 2016*, OECD Publishing, Paris, http://dx.doi.org/10.1787/agr_pol-2016-en.

OECD (2016b), *Multi-Dimensional Review of Myanmar: Volume 3. From Analysis to Action*, OECD Development Pathways, OECD Publishing, Paris, http://dx.doi.org/10.1787/9789264256545-en.

OECD (2015a), *Agricultural Policies in Viet Nam 2015*, OECD Publishing, Paris, http://dx.doi.org/10.1787/9789264235151-en.

OECD (2015b), *Trade Facilitation Indicators* (webpage), www.oecd.org/trade/facilitation/indicators.htm#About-TFI (accessed on 23 June 2015).

OECD (2015c), *Analysing Policies to Improve Agricultural Productivity Growth, Sustainably: Revised Framework*, Paris, www.oecd.org/tad/agricultural-policies/innovation-food-agriculture.htm.

OECD (2014a), *Towards Green Growth in Southeast Asia*, OECD Green Growth Studies, OECD Publishing, Paris, http://dx.doi.org/10.1787/9789264224100-en.

OECD (2014b), *Multi-Dimensional Review of Myanmar: Volume 2. In-Depth Analysis and Recommendations*, OECD Development Pathways, OECD Publishing, Paris, http://dx.doi.org/10.1787/9789264220577-en.

OECD (2012), *OECD Review of Agricultural Policies: Indonesia 2012*, OECD Publishing, Paris, http://dx.doi.org/10.1787/9789264179011-en.

USDA (2015), *International Agricultural Productivity* (database), United States Department of Agriculture Economic Research Service, www.ers.usda.gov/data-products/international-agricultural-productivity.aspx, (accessed on 17 August 2016).

World Bank (2016a), *World Development Indicators* (database), http://databank.worldbank.org/data/ (accessed on 08 July 2016).

World Bank (2016b), *Country and Lending Groups* (webpage), https://datahelpdesk.worldbank.org/knowledgebase/articles/906519-world-bank-country-and-lending-groups (accessed on 28 October 2016).

World Bank (2015), *World Development Indicators* (database), http://databank.worldbank.org/data/ (accessed on 6 May 2015).

WEF (2014), *The Global Competitiveness Report 2014–2015: Full Data Edition*, World Economic Forum, Geneva.

WEF (2013), *The Global Competitiveness Report 2013–2014: Full Data Edition*, World Economic Forum, Geneva.

Chapter 6

Enhancing food security by improving agricultural innovation systems in ASEAN

This chapter takes stock of agricultural innovation systems (AIS) in Association of Southeast Asian Nations (ASEAN) member states, using the OECD framework for Analysing Policies to Improve Agricultural Productivity Growth, Sustainably. It assesses governance of the AIS in ASEAN, investments in innovation, institutions for facilitating knowledge flows in the AIS, and systems for cross-country co-operation on innovation within the region. The chapter concludes with recommendations for promoting sustainable agricultural productivity growth across the region by improving the enabling environment and innovation systems in ASEAN.

Key points

- Association of Southeast Asian Nations (ASEAN) countries recognise that innovation is important for developing their agricultural sectors. But there are gaps in the agricultural innovation systems of most ASEAN countries.
- Most governments in ASEAN underinvest in agricultural research and extension. As a result, some countries do not have the necessary domestic research capacity to respond to the challenges facing their agricultural sectors, and not all farmers are reached by extension services.
- On the other hand, research agencies in ASEAN countries have strong links with a large number of international research institutes, development agencies and research networks.
- Sustained increases in funding for agricultural research are needed in most countries. Stronger governance is also needed, including mechanisms to increase farmer engagement in priority setting. Governments also need to improve the co-ordination of extension systems to ensure that all farmers are reached.
- To maximise the payoff to agricultural research, governments in ASEAN should also strengthen the agricultural enabling environment, including by improving environmental governance, investing in infrastructure, and improving land market rights and access, in addition to farmers' access to finance.

6.1 Introduction[1]

The assessment in the previous chapter found that the agriculture-related aspects of the enabling environment (Pillar A of the Agriculture/Sustainability block of the Agricultural Growth Enabling Index [AGEI]) are an area of relative weakness in countries that comprise the Association of Southeast Asian Nations (ASEAN).[2] In particular, with the exception of Malaysia, ASEAN member states receive relatively low scores on the AGEI for public investments in agricultural research and development (R&D). Investments in agricultural research are important for ensuring that farmers have access to a supply of innovations, and are a key driver of agricultural productivity growth over time. As such, current low funding for research may constrain sustainable agricultural productivity growth in future and, in turn, the capacity of countries to eliminate food insecurity.

A number of factors determine the capacity of agricultural innovation systems (AIS) to provide farmers with innovations and the capacity of farmers to adopt and implement these. Beyond funding for agricultural research, development and extension, the OECD (2013a) has stressed the importance of agricultural innovation policy[3] – and governments' role in the AIS more broadly. The following specific aspects of the AIS are likely to be particularly important for emerging and developing economies:

- governance frameworks that support an efficient and effective AIS that is demand-driven and responsive to the needs of farmers, and encourages collaboration between AIS actors to avoid duplication of research efforts
- sufficient research capacity (funding, researchers and infrastructure) to develop and/or adapt innovations to address the challenges facing their agricultural sectors
- effective training, extension and advisory services to facilitate farmers' access to and adoption of technology and knowledge, particularly in countries characterised by a large number of very small-scale farmers
- networks and capacity for cross-country research collaborations to facilitate the sharing of resources and outcomes, particularly for research addressing shared challenges and for small countries with limited domestic research capacity.

Box 6.1. Analysing Policies to Improve Agricultural Productivity Growth, Sustainably: Revised Framework

The food and agricultural sector is subject to a wide range of economy-wide policies as well as measures specific to the sector itself. The OECD has developed a framework to systematically analyse the full range of policy incentives and disincentives that might impact agricultural productivity growth, sustainable use of natural assets such as land, water and biodiversity resources, and climate change.

Figure 6.1 provides a schematic overview of the relationship between policies and productivity and sustainability outcomes. Innovation, structural change, and access to and impact on natural resources and climate change are all key drivers of productivity growth and sustainability. Policies affect these drivers through four main channels or incentive areas:

- economic stability and trust in institutions (justice, security, property rights), which are essential to attract long-term investment in the economy
- private investment, through a regulatory environment that enables competition, ensures sustainable use of resources, and facilitates the adoption of new technologies; trade that facilitate flows of goods, capital and knowledge; access to finance and tax provisions
- capacity building, including the provision of essential public services, which facilitates access to markets and knowledge, and improves skills needed to innovate and improve resource use efficiency
- sector-specific incentives for innovation, structural change and sustainable resource use in the food and agricultural system from:
  - agricultural policy: domestic measures, including price and income support, investment support, input subsidies, risk management, adjustment and agri-environmental measures, and agriculture-specific trade measures
  - agricultural innovation policy: the role of government in agricultural innovation systems, namely in providing governance, funds for innovation activities, and incentives for private investment in and the adoption of innovation.

For each policy area, the framework considers the likely impacts of the country's policy measures on productivity growth and sustainability, through the incentives and intended or unintended disincentives that they create for the three drivers of sustainable productivity growth: innovation, structural change, and sustainable use of resources and climate change.

Figure 6.1. Policy drivers of innovation, productivity and sustainability in the agricultural sector

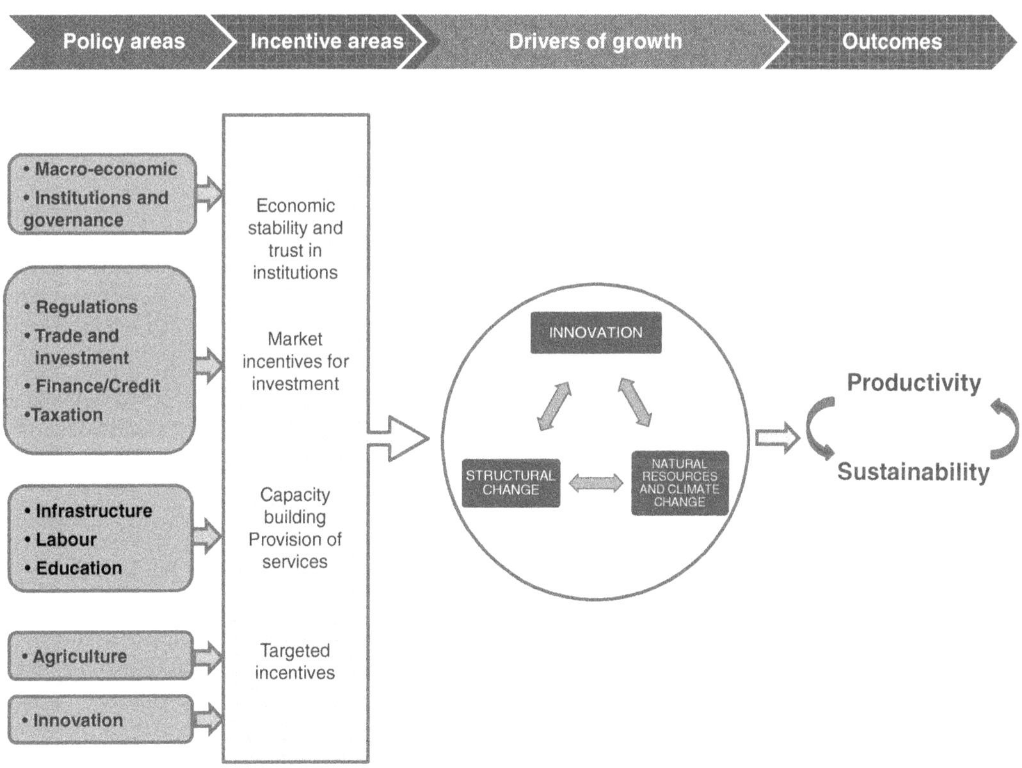

Source: OECD (2015a).

This chapter takes stock of agricultural innovation systems in ASEAN member states. To this end, it uses a framework developed by the OECD to analyse policies for innovation, productivity and sustainability in the food and agricultural sectors (Box 6.1 and OECD, 2015a). The stocktake assesses: (i) governance of the AIS, (ii) investments in innovation, (iii) institutions for facilitating knowledge flows in the AIS, and (iv) systems for cross-country co-operation on innovation within the region. The stocktake will provide a basis to identify gaps in the AIS of the region and therefore pinpoint areas for additional policy action that will contribute to improving the region's food security.

The availability of data and information, and hence the level of detail, varies between ASEAN member states. The assessment draws on a consultancy report prepared by SEARCA, as well as data from IFPRI's Agricultural Science and Technology Indicators (IFPRI, 2016). As with the previous chapter, Brunei Darussalam and Singapore are excluded from the analysis.

6.2 Governance of agricultural innovation systems in ASEAN

Effective governance of AIS is necessary to maximise the payoffs to investments in agricultural research, development and extension (RD&E). An institutional infrastructure that co-ordinates and encourages collaboration between AIS actors – including government, research institutions and the private sector – can help to ensure policy coherence, create synergies and avoid unnecessary duplication of efforts (OECD, 2013a). This includes establishing clear and coherent research priorities and having systems in place for monitoring and evaluating AIS performance.

Unsurprisingly, governance arrangements for the AIS vary across ASEAN. In all ASEAN member states, the line ministry for the agricultural sector – or a specialised institution or agency under it – performs oversight functions for the AIS. In Cambodia and Lao PDR, the ministries of agriculture head the AIS, in addition to being the main performers of agricultural research through the research institutes that operate under their responsibility. In Malaysia and Myanmar, there are several ministries with responsibilities for agricultural sector policy and research. For example, in Malaysia, the Ministry of Agriculture and Agro-based Industries (MOA) and the Ministry of Plantation Industries and Commodities (MPIC) each have responsibility for large national research institutes. In addition, two of Malaysia's states, Sabah and Sarawak, operate their own agencies for agriculture, fisheries and forestry research (Flaherty and Dardak, 2013).

The national innovation system is more developed in Indonesia, Malaysia, the Philippines, Thailand and Viet Nam, such that agriculture is, to a greater or lesser extent, integrated into national innovation frameworks. In these countries, other ministries and agencies in the science and technology policy space also have a role in the governance of AIS, through their broader policy formulating, co-ordinating and implementing roles. In the Philippines, for example, the Department of Science and Technology (DOST) has an important role in AIS policy and governance through the Philippine Council for Agriculture, Aquatic and Natural Resources Research and Development (PCAARRD) (Remøe, 2014). PCAARRD is the central co-ordinating body for agricultural research activities in the Philippines. The Department of Environment and Natural Resources also has a role in technology creation in agriculture (OECD, 2017). In Indonesia, Malaysia and Thailand, additional oversight of the AIS is provided by ministries responsible for science and technology policy, as well as various councils with a role in innovation policy advice, priority setting and co-ordination.

Main actors in the AIS

The public sector has a dominant role in the AIS in ASEAN, both as a funder and performer of agricultural R&D. Most agricultural research takes place in government research institutes (including in other ministries), in the research departments of the line ministries for agriculture, and in public universities. Box 6.2 provides an overview of some of the main public sector actors for agricultural research in ASEAN.

Box 6.2. The main public sector actors in agricultural innovation systems in ASEAN

In **Cambodia**, the main research organisations operate under the responsibility of the Ministry of Agriculture, Forestry and Fisheries (MAFF). There are five semi-autonomous research institutes, which are organised according to commodity lines. The largest of these are the Cambodian Agricultural Research and Development Institute (CARDI) for crop research, and the Cambodian Rubber Research Institute (CRRI) for research into rubber. The higher education sector is increasingly important for agricultural research in Cambodia, accounting for over a quarter of total public agricultural research capacity in 2010, measured in full-time equivalents (IFPRI, 2016). In particular, the Royal University of Agriculture (RUA) is a significant research performer, including in collaboration with CARDI, CRRI and MAFF's other research institutes for livestock, forestry and fisheries research. At the national level, the Department of Agricultural Extension (DAE) leads and co-ordinates extension and technology transfer activities in Cambodia (MAFF, 2005).

In **Indonesia**, the Indonesian Agency for Agricultural Research and Development (IAARD) is the research arm of the Ministry of Agriculture (MoA). IAARD is responsible for formulating and implementing research and agricultural development on priority commodities and cross-commodity issues, and for disseminating agricultural innovations. There are 11 research and development (R&D) centres in IAARD, which manage R&D on food crops, horticulture, estate crops, livestock (veterinary), land resources (soil and agro-climate), agro-socio economics, machinery development, post-harvest, biotechnology and agricultural technology assessment. Within these centres, IAARD manages 15 research institutions, 3 research stations and 31 Assessment Institutes for Agricultural Technology (AIAT). The AIATs are responsible for testing research findings in the provinces so that technologies can be adapted to suit each location. As such, they link research at the central level and extension agents in the districts.

Outside of IAARD, the largest agricultural research centre in Indonesia is the Indonesian Research Institute for Estate Crops (IRIEC) – IRIEC is linked with IAARD, but not formally part of it. IRIEC manages five commodity research centres and a biotechnology unit for R&D on estate crops (OECD, 2012). Forestry R&D is conducted by the Forestry Research and Development Agency in the Ministry of Forestry, and R&D on fisheries by the Agency of Marine and Fisheries Research in the Ministry of Marine Affairs and Fisheries.

The higher education sector plays an important role in agricultural research, accounting for 24% of the total public agricultural research capacity in 2003 (Stads, Haryono and Nurjayanti, 2007). The higher education sector is dominated by the Bogor Agricultural University, which is recognised as a pocket of excellence in science and technology and R&D expertise in Indonesia (Trienes et al., 2014).

In **Lao PDR**, the National Agriculture and Forestry Research Institute (NAFRI), under the Ministry of Agriculture and Forestry (MAF) is the only government agency involved in agricultural research. NAFRI manages eight research centres on: agriculture, coffee, forestry, horticulture, livestock, living aquatic resources, northern agriculture and forestry, and soil survey and land classification. The higher education sector has a more limited role in the Lao PDR AIS, accounting for around 16% of total public agricultural research capacity in 2010. The National University of Laos (NUOL) is the main university that performs agricultural research in Lao PDR (Stads and Manivong, 2006). The National Agriculture and Forestry Extension Service (NAFES) is the lead extension institution.

In **Malaysia**, the largest actors in the Malaysian AIS are national agricultural research institutes operating under the Ministries of Agriculture and Agro-based Industries (MOA) and Plantation Industries & Commodities (MPIC). The Malaysian Agricultural Research and Development Institute (MARDI), a research institute under the jurisdiction of MOA, is Malaysia's main agricultural R&D agency. MARDI has three branches (Research, Technology Transfer and Commercialisation, and Operations) and oversees 29 regional research stations. Three commodity boards that report to MPIC – the Malaysian Palm Oil Board (MPOB), the Malaysian Cocoa Board (MCB), and the Malaysian Rubber Board (MRB) – play an important role in agricultural research as well (Flaherty and Dardak, 2013), and are recognised as an area of strength in Malaysia's national innovation system (OECD, 2016a).

Two of Malaysia's states, Sabah and Sarawak, exercise a greater degree of autonomy and, as such, operate their own agencies for agriculture, fisheries and forestry research.

The higher education sector is also a significant performer of research, accounting for almost 15% of total agricultural research capacity in Malaysia in 2010. Universiti Putra Malaysia accounts for more than half of that capacity (Flaherty and Dardak, 2013).

In **Myanmar**, there are three agencies with responsibilities for the agricultural sector: (i) the Ministry of Agriculture and Irrigation (MOAI), which is responsible for crops, irrigation, agricultural credit and mechanisation; (ii) the Ministry of Livestock, Fisheries and Rural Development (MLFRD); and (iii) the Ministry of Environmental Conservation and Forestry (MOECAF). The Department of Agricultural Research (DAR) and the Department of Industrial Crop Development Enterprise in MOAI, and the Livestock Breeding and Veterinary Department (LBVD) in MOLFRD, are the principal government agencies involved in agricultural R&D. MOECAF also performs some research, development and extension (RD&E) (Stads and Kam, 2007).

The Yezin Agricultural University (YAU) under MOAI operates seven regional research stations where it deploys students to conduct research.

In the **Philippines**, the Philippine Council for Agriculture, Aquatic and Natural Resources Research and Development (PCAARRD), a sectoral council under the Department of Science and Technology (DOST), is the central co-ordinating body for agricultural research activities in the Philippines (Remøe, 2014). PCAARRD provides support to a network of 132 agricultural research centres, state universities and colleges, and government and private agencies (the National Agriculture and Resources Research and Development Network), in addition to 14 region-based consortia (Stads, Faylon and

Buendia, 2007). PCAARRD formulates policies and research programmes for science and technology-based R&D in the agricultural, aquatic and natural resources sectors, and also has a role in funding, monitoring and evaluating R&D.

The Philippines Department of Agriculture (DA) is also responsible for agricultural research. Within DA, the Bureau of Agricultural Research (BAR) is the central co-ordinating agency for agricultural and fisheries research, but does not perform research itself. BAR co-ordinates and funds agricultural R&D activities, develops partnerships with local and international research organisations, strengthens institutional capabilities, manages knowledge, and advocates policies to improve the governance of agricultural and fisheries research (Stads, Faylon, and Buendia, 2007).

There are also a number of agencies attached to DA that undertake agricultural R&D as part of their mandate to develop specific commodities. This includes the Fibre Industry Development Authority and the Philippine Carabao Centre. In particular, the Philippine Rice Research Institute, PhilRice, originally attached to the DA but now under the Office of the President, is in charge of developing high-yielding and cost-reducing technologies for rice production.

In **Thailand**, the bulk of public agricultural research falls under the Ministry of Agriculture and Cooperatives (MOAC). MOAC has specialised research Departments of Agriculture (DOA), Rice, Fisheries and Livestock, which consist of national and regional research centres that are responsible for specific product groups as well as thematic research centres. For example, DOA manages research institutes for rubber, horticulture, agricultural engineering, and field and renewable energy crops; R&D offices for plant protection, agricultural production sciences, biotechnology, and post-harvest and products processing; and regional R&D offices (DOA, 2016).

The National Science and Technology Development Agency (NSTDA), which is under the responsibility of the Ministry of Science and Technology (MOST), is an important funder and performer of research, including in agriculture, and undertakes research with emphasis on biotechnology (Degelsegger and Sukprasertchai, 2014). NSTDA's programme on Agriculture and Food includes seven programmes on rice, rubber, tapioca, seeds, plants for the future, animal production and health, and food innovation.

Another important organisation is the Agricultural Research Development Agency (ARDA). ARDA is the public organisation that manages funding for agricultural research in Thailand, under the guidance of the Minister of Agriculture and Cooperatives.

Thailand's public universities are also significant performers of agricultural research, Kasetsart University in particular. Each agricultural region has at least one national research university that undertakes agricultural R&D with a more region-specific focus (UNCTAD, 2015).

In **Viet Nam**, agricultural sector science and technology policy is the responsibility of the Ministry of Agriculture and Rural Development (MARD). Within MARD, there are 16 research organisations (12 research institutes and four universities). The largest of MARD's research institutes is the Vietnam Academy of Agricultural Sciences (VAAS), which oversees a further 18 institutes and research centres. VAAS' purpose is provide a comprehensive vision, strategic direction and oversight of agricultural R&D programmes in Viet Nam; to conduct basic and applied research and foster the transfer of new technologies; and to provide post-graduate and professional training (OECD, 2015b). The National Agriculture Extension Center (NAEC) within MARD is responsible for extension.

The higher education sector accounts for around 25% of total public research capacity in Viet Nam (IFPRI, 2016). The Hanoi University of Agriculture and Can Tho University are significant in agricultural research. Most agricultural research by universities has focused on lowland farming systems, particularly the development, adaptation and dissemination of new rice varieties (OECD, 2013b).

In general, the majority of agricultural researchers are employed in government research agencies (Figure 6.2), sometimes in a small number of research agencies. In Malaysia, around 36% of researchers are employed by the Malaysian Agricultural Research and Development Institute (MARDI), a research institute under the Ministry of Agriculture and Agro-based Industries (MOA). A further 19% of researchers are employed by three commodity boards under the Ministry of Plantation Industries and Commodities (MPIC) (Flaherty and Dardak, 2013). In Indonesia in 2003, the research centres of the Indonesian Agency for Agricultural Research and Development (IAARD) employed 45% of agricultural researchers, while Bogor Agricultural University accounted for almost a quarter.

In other ASEAN member states, the structure of public agricultural research systems is more complex, with multiple government and higher education agencies engaged in agricultural research. In Viet Nam, the Ministry of Agriculture and Rural Development (MARD) supervises 16 research organisations (12 research institutes and four universities). The largest of these, the Vietnam Academy of Agricultural Sciences (VAAS), oversees a further 18 institutes and research centres (OECD, 2015b; Box 6.2). MARD accounted for more than 70% of Viet Nam's agricultural researchers in 2002-03. In the Philippines, the AIS has a multi-level institutional structure in which three departments have roles in technology creation. The central co-ordinating body, PCAARRD, provides support to an extensive network of 132 research centres, state universities and colleges, and government and private agencies (including research centres and agencies under the Department of Agriculture), as well as 14 regional R&D consortia charged with developing regional capacity for managing research (Stads, Faylon, and Buendia, 2007; Box 6.2).

The non-profit sector – including international research and development agencies and donors – plays an important role in the AIS in Cambodia and Lao PDR, and, to a lesser extent, Viet Nam. Donors are an important source of funding for research in Cambodia and Lao PDR, including for research organisations in the agricultural ministries. Between 2007 and 2010, donors provided almost half of total funding for the Cambodian Agricultural Research and Development Institute (CARDI), the main research institute in Cambodia's Ministry of Agriculture, Forestry and Fisheries (MAFF) (IFPRI, 2016). In 2010, donors provided 70% of funding to Lao PDR's National Agriculture and Forestry Research Institute (NAFRI), in the Ministry of Agriculture and Forestry (MAF), and around 55% of funding over the period 2010-14 (Stads, 2015). Important actors in the non-profit sector include the Food and Agriculture Organization (FAO), the International Fund for Agricultural Development (IFAD), the Asian Development Bank (ADB), and the World Bank – all of which have funded rural development programmes – as well as a number of governments.

The role of the private sector in agricultural research is increasing in ASEAN, although this varies across countries. Estimates of the share of private sector research in total investments in agricultural research suggest that it could be as high as 30% in Thailand (Meerod et al., 2011). For the Philippines, Stads, Faylon and Buendia (2007) estimated that the private sector accounted for 16% of agricultural researchers and 18% of total agricultural R&D spending in 2002. For Indonesia, Stads, Haryono and Nurjayanti (2007) estimated that the share of the private sector in total agricultural R&D spending was 19%.

Figure 6.2. Composition of the agricultural research system, by type of agency (share of total FTEs, %)

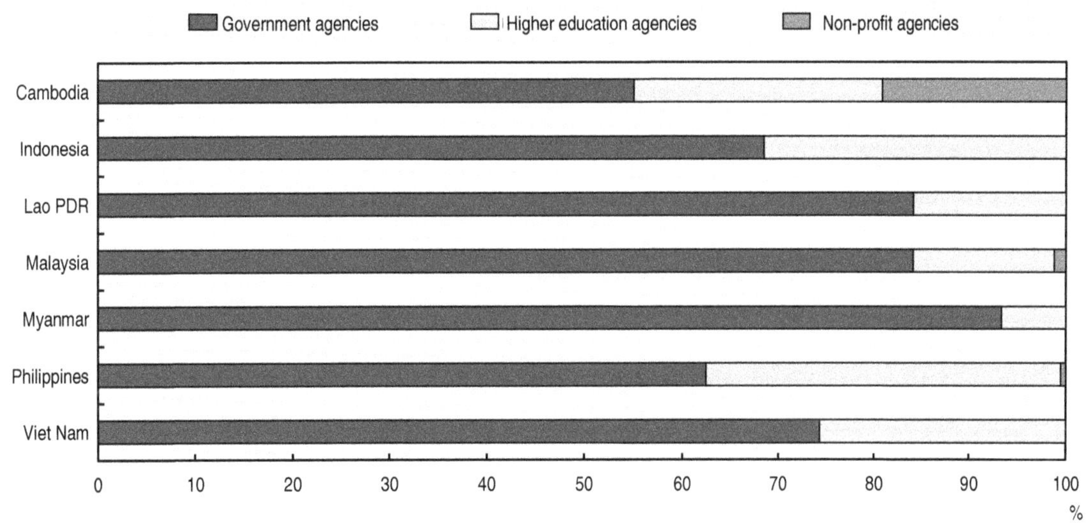

Notes: 2010, 2003 for Indonesia and Myanmar; 2002 for the Philippines. Data for Thailand are not available. Share of total researchers (excluding the private sector), measured in full time equivalents (FTEs).
Sources: IFPRI (2016), *Agricultural Science and Technology Indicators*, www.asti.cgiar.org/data, for Cambodia, Lao PDR, Malaysia and Viet Nam; Stads, Haryono and Nurjayanti (2007) for Indonesia; Stads and Kam (2007) for Myanmar; Stads, Faylon and Buendia (2007) for the Philippines.

Private sector investments appear to be focused on higher-valued, market-oriented commodities such as plantation and industrial and horticulture, and on agricultural inputs. In Malaysia, private agricultural research has focused on plantation crops such as oil palm and sago palm (OECD, 2013b). In contrast, private sector research in the Philippines has focused on plantation crops such as bananas and pineapples, as well as agricultural chemicals and pest management, breeding of plant varieties and livestock (poultry) and, to some extent, agricultural machinery (Stads, Faylon and Buendia, 2007). In Thailand, private sector research has focused on hybrid seeds (vegetables, maize and rice), genetic improvement of rubber trees, cultural practices, and post-harvest technology (Poapsongsakorn, 2011).

In contrast, the private sector has a smaller role in the AIS of Cambodia, Lao PDR and Viet Nam. In Cambodia, some companies finance and implement research programmes on their own or with MAFF research institutes and the universities, on industrial crops in particular. The private sector also collaborates in non-profit research (MAFF, 2005). Reflecting the fact that Viet Nam's companies were government-owned until recently, the private sector has traditionally had a small role in agricultural research (Stads and Nguyen, 2006). More recently, private sector research has increased, including in improved plant varieties, agricultural chemicals, and production technologies (OECD, 2015b).

Research priorities for AIS in ASEAN

Across ASEAN, agricultural priorities are guided by national development strategies and strategic plans for the agricultural sector. This reflects the fact that in all countries, agriculture is viewed as a strategic sector, and that agricultural innovation – and the AIS more broadly – is seen as having a vital role in developing the sector and ensuring that agriculture contributes to economic growth and development. In the more developed ASEAN member states, research priorities are also informed by national innovation system plans. Some of the main frameworks for agricultural research priorities are outlined in Box 6.3.

Box 6.3. Frameworks guiding agricultural research priorities in ASEAN

In **Cambodia**, research priorities for agriculture are established in national development frameworks, including: the Rectangular Strategy for Growth, Employment, Equity and Efficiency Phase III (2014-18), which provides the development framework for Cambodia; and the National Strategic Development Plan 2014-2018 (NSDP), which sets out the planned actions to implement the policy agenda of the Rectangular Strategy Phase III (RGC, 2014). The Rectangular Strategy Phase III and the NSDP promote the enhancement of agricultural productivity, diversification and commercialisation, promotion of livestock farming and aquaculture, land reform and sustainable management of natural resources. Both the Rectangular Strategy Phase III and the NSDP specify research and development (R&D) activities and actions to strengthen the agricultural research, development and extension (RD&E) system (RGC, 2014). MAFF implements these priorities through programmes outlined in the Agricultural Sector Strategic Development Plan 2014-2018, which sets out planned RD&E activities to meet the goals for the sector (MAFF, 2015a).

MAFF also developed a Master Plan for National Agricultural Research for the 2006-15 period (MAFF, 2005).

In **Indonesia**, agricultural research priorities are informed by Indonesia's national development plans and the research priorities established by other councils in the national innovation system. The National Research Council (*Dewan Riset Nasional*) prepares National Research Agendas that reflect the national priorities outlined in Indonesia's long- and medium-term national development plans.[1] The current National Research Agenda 2015-2019 identifies priority areas for public research in agriculture, including: 1) agriculture in sub-optimal land; 2) food diversification, quality of nutrition, and food safety; 3) smallholder farmer empowerment; and 4) resource management, infrastructure and climate change (Prasvita, 2014).

Indonesia has also implemented a Master Plan for the Acceleration and Expansion of Indonesian Economic Development 2011-2025 (MP3EI) to transform Indonesia into an innovation-driven economy.[2] The MP3EI identifies agriculture as a strategic sector, and five of its 22 activities relate to agricultural production: palm oil, rubber, cocoa, food crops and animal husbandry. A common theme across the agricultural activities is to increase yields and stimulate further processing, such as by increasing planting with high-quality seeds, developing port capacity and establishing research centres (OECD, 2012).

In **Lao PDR**, research priorities were established as part of the goals and programmes for the agricultural sector outlined in the Agricultural Development Strategy 2011-2020, the long-term framework for the development of the agricultural sector; and the Agricultural Master Plan 2011-2015, the '"roadmap" for implementing the Strategy over the 2011-15 period (MAF, 2010). The Master Plan prioritises applied research for development in the agricultural, livestock, aquaculture and fisheries, and forestry sectors, and in climate change scenarios and climate change impacts. These goals and programmes feed into the government's framework for growth and development.[3]

In **Malaysia**, agriculture is embedded in national policy plans that emphasise science, technology and innovation (STI), such as the 10th Malaysia Plan (2011-15) (the Tenth Plan) and the Economic Transformation Plan. These plans identify palm oil and related products, and high-value agriculture as key economic areas for advancing national economic growth.

For agriculture, the Tenth Plan includes: strategies to promote innovation-based growth and production processes that utilise both modern farm technology and information and communications technology (ICT), including an ICT-based Agriculture Flagship Project; and strategies to intensify collaborative R&D with established agricultural research institutes to leapfrog innovation in production processes, disease control, safety and quality control, including the development of new high-value added products (such as aquaculture, seaweed farming, swiftlet nests, herbal products, fruits and vegetables, and food processing) (Arshad, Noh and Zainalabidin, 2015). Related R&D activities at MARDI and other agencies focus on the development and improvement of new varieties, biodiversity, and downstream activities (Flaherty and Dardak, 2013).

In **Myanmar**, priorities for the crop, livestock and fisheries sub-sectors have been established in various national policies and priority frameworks, such as the National Economic and Social Development Plan (2011-12 to 2015-16).[4] MOAI and MLFRD have adopted a number of measures to achieve these priorities, which include RD&E activities. DAR's research focuses on increasing crop production through improved seed, crop management, and crop protection techniques; and cropping systems tailored to suit the country's various agro-ecological zones (Stads and Kam, 2007).

In the **Philippines**, there are several major policy frameworks that guide research priorities. The Philippine Development Plan 2011–2016 defines the overarching development strategy for the Philippines. The plan identifies agriculture as an economic sector for prioritisation, and outlines strategies for using RD&E to boost competitiveness and sustainability in agriculture, including increased investments in R&D and extension (NEDA, 2011). The medium-term update of the plan sets out three priorities to achieve competitiveness and sustainability in the agricultural sector: increase productivity; increase forward linkages with the industry and services sectors; and improve sector resilience to risks, including climate change (NEDA, 2014).

The National Science and Technology Plan 2002–2020 (NSTP), developed by DOST, sets the agenda for the country's medium-term strategies for science and technology development. The NSTP identifies twelve priority areas, including various agricultural and biotechnology-related fields. DOST also prepared the Harmonized National Research and Development Agenda, which further develops the NSTP and is aligned with the Philippine Development Plan.

PCAARRD and the Department of Agriculture also have a role in research priority setting. PCAARRD has developed ten Industry Strategic Plans (ISPs), which set the agenda for R&D for the crop, livestock and forestry sectors. The ISPs support the Harmonized National Research and Development Agenda (PCAARRD, n.d.).

Thailand's national development and STI policies address the agricultural sector. The 11th National Economic and Social Development Plan (2012-2016) (NESDP) includes objectives for agriculture as well guidelines for the achievement of the objectives. The National STI Policy and Plan 2012-2021 identifies a number of agricultural products as priorities for future STI efforts, including capacity building. The National Research Policy and Strategy Plan (2012-2016) identifies rice and rice-based products, bio-based energy, rubber and rubber products, and processed food as four of the 12 target economic sectors.

Since 2012, **Viet Nam** has introduced policies to develop R&D activities in agriculture that are consistent with the goal of modernising the sector. In the first instance, the National Assembly identified a prioritised agricultural investment portfolio, with the aim of raising the effectiveness and efficiency of public investment for agriculture, farmers and rural areas.[5] Priorities include science and technology in biotechnology, post-harvest processing, crop seeds, livestock and fishery breeds.

MARD has set in place a strategy for the development of science and technology for agriculture and rural development over the period 2013-20,[6] which is based on the overall strategy of science and technology in Viet Nam for the period 2011-20.[7] It sets specific targets for science and technology to become a key driving force for the industrialisation and modernisation of agriculture and rural development. The programme of activities to support these strategies includes R&D on staple crops, livestock husbandry and animal health, agricultural engineering and post-harvest technology, irrigation technology, and research on policies for agriculture and rural development (OECD, 2015b).[8]

Notes

1. Indonesia's Long-Term National Development Plan (*Rencana Pembangunan Jangka Panjang Nasional* [RPJPN]) sets out the vision for the nation for the period 2005 to 2025. Medium-Term National Development Plans (*Rencana Pembangunan Jangka Menengah Nasional* [RPJMN]) describe the government's development strategy, outline national priorities, and serve as a basis for setting annual budgets over a five-year period.
2. *Masterplan Percepatan dan Perluasan Pembangunan Ekonomi Indonesia.*
3. The National Growth and Poverty Eradication Strategy.
4. Other key documents include the National Comprehensive Development Plan (2011-31); the Framework for Economic and Social Reforms, which acts as a bridge between the National Economic and Social Development Plan (2011-12 to 2015-16) and the National Comprehensive Development Plan; the National Strategy on Rural Development and Poverty Alleviation; the 2004 Agriculture Sector Review; and the National Millennium Development Goals (FAO and GoRUM, 2012; OECD, 2015c).
5. Resolution No. 26/2012/QH13.
6. Decision No. 3246/QD-BNN-KHCN, dated 27 December 2012.
7. Decision No. 418/QD-TTg, dated 4 November 2012.
8. Decision No. 1259/QD-BNN-KHCN, dated 4 June 2013.

Figure 6.3. Research focus of scientists in 2010 (share of total FTEs)

Research focus of scientists, by major commodity area (share of total FTEs)

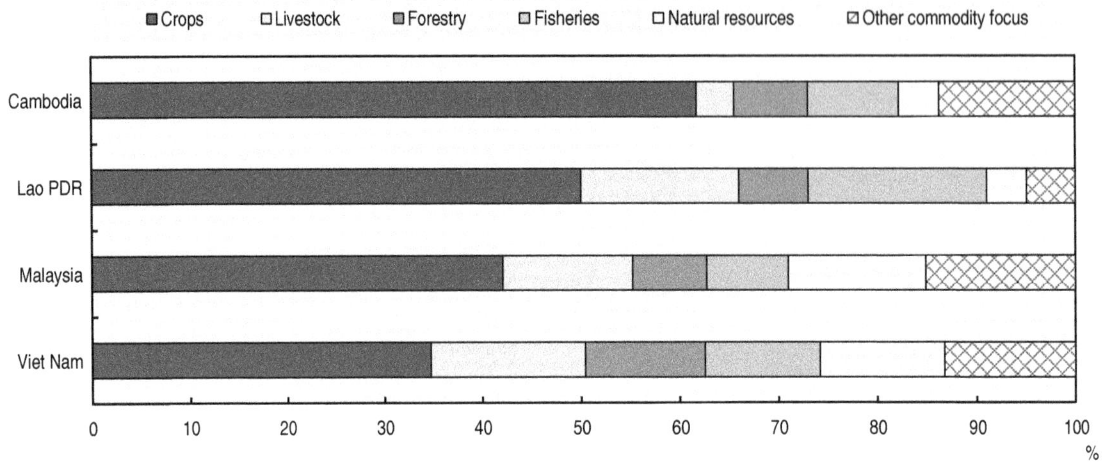

Research focus of crop scientists, by major crop commodity (share of total FTEs)

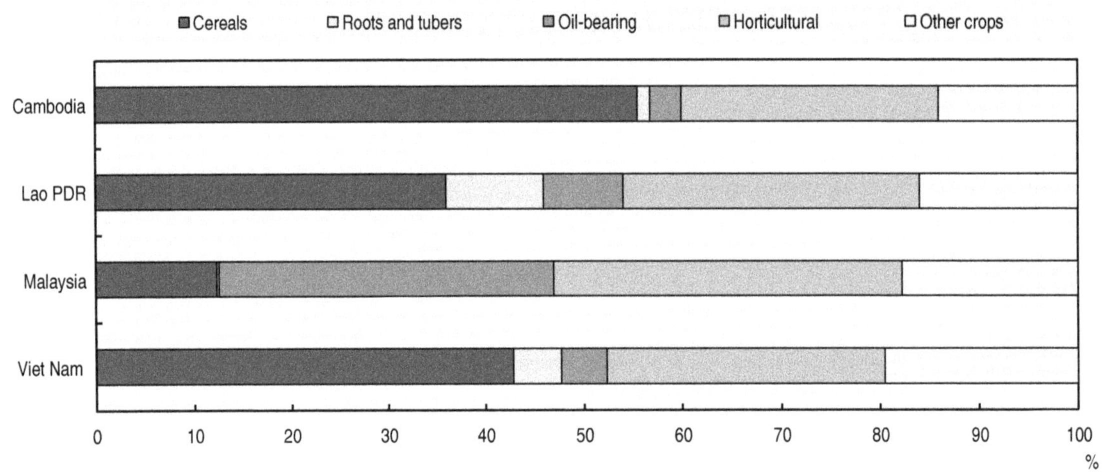

Notes: FTEs: Full time equivalents. Other commodity focus includes pastures and forages, post-harvest issues, agricultural engineering, socioeconomic research and other commodities.
Source: IFPRI (2016), *Agricultural Science and Technology Indicators*, www.asti.cgiar.org/data.

While this can help to ensure that agricultural research is consistent with national priorities, some stakeholders have suggested that national priorities may not reflect farmers' needs at a local level. IFAD (2014) has suggested that in Myanmar, DAR's research is driven by political priorities rather than by the needs of farmers. Similarly, in its review of agricultural policies in Indonesia, the OECD (2012) found that while efforts are made to adapt research to local production conditions, the initial research is driven by national priorities rather than by consideration of local comparative advantage. Further issues may arise if national priorities are not developed into clear priorities and plans for the AIS.

In practice, agriculture ministries and research agencies also have a role in determining research priorities. In the case of Thailand, UNCTAD (2015) has suggested that within the parameters of the NESDP and other STI and research policy plans, agricultural research activities are determined according to priorities set by MOAC, with government and higher education research agencies generally acting independently. In Indonesia, IAARD is responsible for the allocation of public sector funding for agricultural research (OECD, 2012), which suggests that it has a role in determining research priorities.

Some ASEAN member states have mechanisms in place to ensure greater stakeholder engagement in priority setting. For Cambodia's Master Plan for National Agricultural Research, 2006-2015, research priorities were established through stakeholder consultation with staff from MAFF research institutes and departments; representatives from universities and agricultural schools; the private sector, including agro-industry and farmers; and non-governmental organisations (NGOs) and donors. Stakeholders identified the strengths, weaknesses, opportunities and threats of Cambodia's agricultural sector and its main research institutes. Working groups consisting of representatives from each agricultural sub-sector (for crops, livestock, fisheries, forestry, agro-industry and rubber) then prepared projects to address the identified constraints (MAFF, 2005). In Malaysia, the private sector is directly involved in the research programmes of the palm oil and rubber commodity boards – a factor which contributes to those boards' success in using commodity levies as a source of research funding (Stads, 2015). PCAARRD's Industry Strategic Plan (ISP)-derived R&D agenda is informed by stakeholder consultations (PCAARRD, n.d.).

As defined in the frameworks listed in Box 6.3, agricultural R&D is expected to contribute to high-level objectives for the agricultural sector, such as enhancing food security, rural development and the modernisation of the sector, sustainable management of natural resources, adaptation and responsiveness to climate change, and the development of high-value agriculture.

Further insight into countries' research priorities can be derived by assessing how research resources are allocated across commodities and thematic areas. ASTI data for Cambodia, Lao PDR, Malaysia and Viet Nam indicate that in 2010, countries prioritised research on crops, including cereals (particularly rice), horticulture (fruits and vegetables), and oil palm (in Malaysia) (Figure 6.3). Livestock, forestry and fisheries research received significantly smaller shares of resources. Thematic and cross-cutting research also received relatively small shares of resources. 14% of researchers in Malaysia and 13% of researchers in Viet Nam focused on natural resources research, compared with 4% of researchers in Cambodia and Lao PDR. In Cambodia, Malaysia and Viet Nam, 13-15% of researchers focused on pastures and forages, post-harvest issues, agricultural engineering and socioeconomic research.

6.3 Investing in innovation

Public investments in innovation, including R&D, extension and research infrastructure, are an important source of innovations, technologies and practices that drive sustainable productivity growth. In developing countries, public investments may be particularly important, to the extent that private sector research is underdeveloped or does not meet the innovation needs of some farmers – for example small-scale, poorer farmers – or address cross-cutting issues. Governments can also support private-sector research through a range of instruments.

Public investments in research

Public agricultural research expenditure varies across the ASEAN member states (Table 6.1). In absolute terms, Indonesia invests the most in agricultural research, at USD 1.07 billion in 2010 (2011 PPP). In contrast, in 2010, Lao PDR invested USD 16 million and Cambodia invested USD 22 million in agricultural research. Comparable ASTI data were not available for Myanmar and the Philippines. However, Flaherty and Dardak (2013) report that Myanmar invested USD 6 million in agricultural research in 2002 and the Philippines invested USD 133 million in 2008 (2005 PPP). Between 2000 and 2010, research expenditure increased in Cambodia, Indonesia, Malaysia and Thailand, and more than doubled in Viet Nam. In Lao PDR, recent increases in government funding were offset by high inflation levels and reduced donor support, prompting an overall decline in agricultural research expenditure (in real terms) (Stads, 2015). It is worth noting that annual levels of funding for agricultural research tend to be more volatile in Cambodia and Lao PDR, a consequence of those countries' dependence on donor funding.

Table 6.1. Public investments in agricultural research (in million 2011 PPP dollars), 2000-10

	2000	2005	2010
Cambodia	18	20	22
Indonesia	580	915	1 068
Lao PDR	37	21	16
Malaysia	495	557	592
Thailand	327	278	440
Viet Nam	62	109	136

Notes: PPP: Purchasing power parity. Data for 2000-10 for Cambodia, Malaysia and Viet Nam from IFPRI (2016). Data for Indonesia and Lao PDR from ASTI survey data in Stads (2015). Data for Thailand from the National Research Council of Thailand (various years), in Stads (2015).
Sources: IFPRI (2016), *Agricultural Science and Technology Indicators*, www.asti.cgiar.org/data; and Stads (2015).

Figure 6.4. Funding for agricultural research as a percentage of agricultural gross value added (GVA)

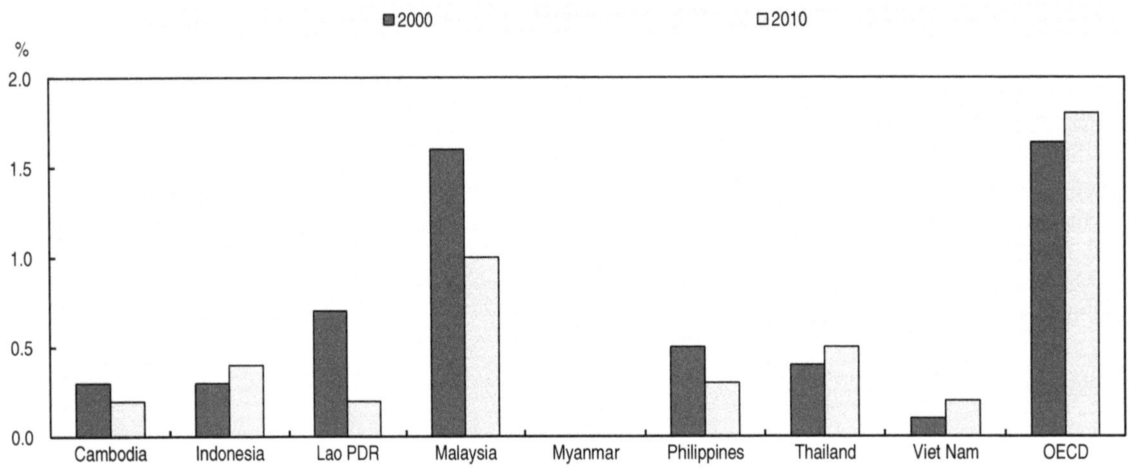

Notes: Data for Myanmar and Philippines are for 2002 and 2008. Data for Indonesia is for 2009. In 2002, agricultural research intensity was 0.03% in Myanmar.
Sources: IFPRI (2016), *Agricultural Science and Technology Indicators*, www.asti.cgiar.org/data; IFPRI (2015); and Stads (2015). OECD (2016b), *OECD Research and Development Statistics,* http://stats.oecd.org/Index.aspx?DataSetCode=GERD_SCIENCE, and OECD (2016c), *System of National Accounts,* https://stats.oecd.org for the OECD average.

Given differences in data sets, agricultural research intensity ratios (expressing agricultural research as a percentage of agricultural gross value added [GVA]) are a more useful indicator for comparing R&D spending levels across countries (Figure 6.4). With the exception of Malaysia, the ASEAN member states are investing in agricultural research at levels that are considerably below 1% of agricultural GVA.[4] While research intensity increased in Indonesia, Thailand and Viet Nam in the 2000-10 period, it declined in the other ASEAN member states, suggesting that public investments in agricultural research are not keeping pace with growth in the sector (although in the Philippines, at least, government expenditure on agricultural research has increased substantially since 2010 [OECD, 2017]). While agricultural research intensities have some limitations as an indicator of R&D expenditure,[5] Flaherty and Dardak (2013) argue that countries with R&D intensity ratios of less than 0.3% – levels that are seen in Cambodia, Lao PDR, Myanmar and Viet Nam – are significantly underinvesting in agricultural research.

A consequence of low levels of funding for agricultural research is that some countries lack the research capacities to develop and adapt innovations to address the challenges facing their agricultural sectors. Beyond programme funding, this includes research infrastructure – such as equipment and laboratory facilities – and sufficiently experienced and skilled research staff. For example, Cambodia's NSDP 2014-2018 identifies a lack of human resources, infrastructure and budget for research as a significant constraint for farmers, and further notes that key research agencies – CARDI, for example – lack the resources to retain skilled researchers (RGC, 2014).

Sources of public funding for agricultural research

Governments provide the majority of funding for agricultural research in most ASEAN member states. The exceptions are Cambodia and Lao PDR, which receive a substantial share of funding from donors, including funding for government research agencies. Further information on funding sources for agricultural research is provided in Box 6.4.

Some government research agencies have the capacity to generate revenue. For example, Viet Nam's VAAS and CARDI in Cambodia – and, to a lesser extent, NAFRI in Lao PDR – generate a small share of funds from the sale of goods and services, such as seed and other products from research stations. In contrast, a significant proportion of funding for Indonesia's IRIEC comes from the sale of plantation crops and technology inputs, such as seed stock, as well as funding from public and private enterprises for contract research (Beintema and Stads, 2008). In terms of expenditure, IRIEC is Indonesia's largest research agency.

Commodity levies are also a source of research funding in some countries. In Malaysia, the palm oil and rubber commodity boards, MPOB and MRB, receive a substantial share of funding from commodity levies on rubber and oil palm. In 2010, these levies accounted for 78% of MPOB's funding and 27% of MRB's funding. One reason for the success of these commodity levies is that the private sector is directly involved in the research programmes of the commodity boards (Flaherty and Dardak, 2013). In Cambodia's Master Plan for National Agricultural Research, MAFF raised the prospect of a levy on the sale of rubber products to increase the level of funding for CRRI (MAFF, 2005).

Box 6.4. Sources of funding for agricultural research

In **Cambodia**, donors provided almost half of total funding for CARDI between 2007 and 2010 (IFPRI, 2016). Typically, donor funding has prioritised agricultural services, particularly extension. In 2007, 39% of total donor funding to MAFF was for extension, while agricultural research received only 3% of donor funds (ADB, 2012). CARDI also generates funds through the sale of goods and services (4.6% of funding in 2010) (IFPRI, 2016). The remainder of agricultural research funding is provided from the national government budget.

In **Indonesia**, the government is the main source of funding for both IAARD and agricultural research carried out by universities. In contrast, IRIEC receives a significant share of revenue from the sale of plantation crops and technology inputs (e.g. seed stock), as well as funding from public and private enterprises for contract research. Beintema and Stads (2008) note that this is because unlike IAARD agencies, IRIEC has semi-autonomous status and can keep any revenues that it generates from product sales.

The MoA also offers research grants to researchers in all working units in the Ministry, covering a range of topics, and to fund the activities of researchers and field instructors in empowering local farmers and local markets. A further funding scheme, the National Research and Development Partnership on Agriculture, is available to support collaborative research between the working units in the research and development (R&D) centre of the Ministry of Agriculture and universities and/or other national research agencies (Trienes et al., 2014).

In **Lao PDR,** donors provided 70% of NAFRI's funding in 2010, and around 55% of NAFRI's funding over the period 2010-14 (Stads, 2015). NAFRI also generates funds through the sale of goods and services (1% of funding in 2010) (IFPRI, 2016). The remainder of agricultural research funding is provided from the national government budget.

In **Malaysia**, the government provides the majority of funding for agricultural research. Nearly all of MARDI's funding is provided by the Malaysian government, either through direct allocations or through a competitive grant system (Beintema and Stads, 2008). The palm oil and rubber commodity boards, MPOB and MRB, also receive funding from commodity levies on rubber and oil palm. In 2010, these levies accounted for 78% of MPOB's funding and 27% of MRB's funding. Given the high value of export crops and related commodity-based resources (such as the levies), MPOB and MRB are better funded than MARDI, despite being smaller in terms of researcher numbers (Flaherty and Dardak, 2013).

Historically, agricultural research in **Myanmar** has been almost entirely financed by the national government (Stads and Kam, 2007).

In the **Philippines**, public agricultural research agencies are mainly funded by the national government. The government provides direct support (institutional funding) to each public agricultural research organisation, and channels project funding for strategic research via specialised government agencies (Beintema and Stads, 2008). In the case of PCAARRD, the council receives the majority of its funding from the government. The budget is long term, but the additional funding has to be defended in Congress every year. Externally-generated funds from donor agencies and local partners comprise around 5% of PCAARRD's overall budget (Remøe, 2014).

In **Thailand**, MOAC's own budget funds most of the R&D undertaken by its departments. However, each research programme and project has to be approved by the Bureau of the Budget, and funding is approved on an annual basis (UNCTAD, 2015). MOAC also receives funding from ARDA to enhance its research capacity. There are also five public funding agencies which provide funds for agricultural research. These are the National Research Council of Thailand, NSTDA, the Thai Research Fund, ARDA, and the Thailand Tapioca Development Institute (Poapongsakorn, 2011).

The national government is the main funder of **Viet Nam**'s public agricultural research institutes, although donor funding, internally generated resources, and funding from private and state enterprises comprise a smaller but important share in some of MARD's agencies (Beintema and Stads, 2008). For example, two-thirds (66.6%) of VAAS' funding in 2010 came from the government, sales of goods and services accounted for 9.9%, donor and development bank funding accounted for 9.3%, commodity levies and producer organisations accounted for 7.4%, and other sources for 6.8% (IFPRI, 2016).

Support for private investment in innovation

The share of private sector R&D has increased in ASEAN member states, particularly in higher-valued, market-oriented commodities such as plantation and industrial crops and horticulture, and in agricultural inputs. This stocktake did not identify a large number of measures to support private investment in innovation. However, Malaysia, the Philippines, Thailand and Viet Nam offer tax incentives for firms that undertake R&D and provide R&D services, which may also apply to the agro-food sector (Box 6.5). Indonesia does not have an R&D-based tax incentive scheme. Nevertheless, there is a range of tax incentives available to Indonesian companies seeking new investments in Indonesia (Ernst & Young, 2013).

Box 6.5. R&D-based tax incentives for innovation in ASEAN

In **Malaysia**, companies that provide research and development (R&D) services are eligible for Pioneer Status (income tax exemption) or an Investment Tax Allowance for qualifying R&D capital expenditure. A double tax deduction is available for R&D revenue expenditure incurred by companies carrying out in-house R&D or expenditure for the services of approved R&D service providers. There are also a variety of financial assistance schemes.

In the **Philippines**, R&D expenditure may be treated as a current expense deductible at 100%, or a deferred expense ratably distributed over a period of not less than 60 months, as chosen by the taxpayer. Moreover, the 2012 Investments Priorities Plan identified R&D activities as investment priorities which promote the economic development of the Philippines. Enterprises engaged in R&D activities (as a R&D service provider) that qualify for registration with the Board of Investments may be entitled to a four-year income tax holiday and other incentives.

Thailand provides a 200% deduction for the cost of engaging approved Thai R&D Service Providers with no requirement for foreign-majority owned companies to own the results of the R&D activities. Companies providing eligible R&D services may also be entitled to other incentives.

In **Viet Nam**, newly-established companies in high technology, scientific, research and technology domains are entitled to a reduced tax rate for 15 years, which can be extended to 30 years, subject to approval. There is also a one-year exemption for income derived from performing R&D, the sale of products during test production, and products made from new technology applied for the first time in Viet Nam.

Source : Ernst & Young (2013).

Strong and effective intellectual property rights (IPR) regimes are another important incentive for private sector investment in agricultural R&D, including in improved plant varieties, agricultural chemicals, and production technologies. All ASEAN member states have joined the World Intellectual Property Organization (WIPO) and the WTO Agreement on Trade-Related Aspects of Intellectual Property Rights (TRIPS). All countries also have their own patent laws in place. However, Myanmar's legislation may not be compliant with the minimum provisions of the TRIPS Agreement. Cambodia, Lao PDR and Myanmar do not have laws governing the protection of plant varieties (Table 6.2). Viet Nam is the only country to have joined the International Convention for the Protection of New Varieties of Plants (UPOV) (the 1991 Act). However, Malaysia and the Philippines have initiated the procedure for acceding to the UPOV Convention, and Cambodia, Indonesia, Lao PDR, Myanmar and Thailand have been in contact with the Office of the Union for assistance in developing laws based on the UPOV Convention (UPOV, 2016).

Nevertheless, while IPR legislation is generally adequate, recent OECD reviews have identified areas of weakness in IPR regimes of some countries, which likely affect the functioning of innovation systems. For example: (i) the less-developed economies in the region often lack the institutional capacity to manage and provide legal support to IPR cases (OECD, 2013b), (ii) Viet Nam could better protect patents and plant varieties to improve the performance of its AIS (OECD, 2015b), and (iii) in Indonesia, the enforcement of intellectual property rights, such as patents and licenses, could induce the private sector to invest more in agricultural R&D (OECD, 2012).

Table 6.2. Laws governing intellectual property rights (IPR) in ASEAN

	Year joined WIPO	Year joined WTO-TRIPs	Latest versions of IPR Laws			
			Patent	Copyright	Trademarks	Plant variety protection
Cambodia	1995	2004	2003	2003	2002	-
Indonesia	1979	1995	2001	2014	2001	2000
Lao PDR	1995	2013	2011	2011	2011	-
Malaysia	1989	1995	2006	2006	2002	2004
Myanmar	2001	1995	1946	1911	1889	-
Philippines	1980	1995	1998	2013	1998	2002
Thailand	1989	1995	1999	2015	2000	1999
Viet Nam	1976	2007	2009	2009	2009	2004

Source: WIPO (2016).

6.4 Facilitating knowledge flows in the AIS

Agricultural education and extension and advisory services play an important role in ensuring both the timely adoption of innovations by farmers and a supply of skilled labour for the sector. Producers with a good general, technical and business education will generally be more willing and better able to adopt innovations. Effective AIS also require well-educated agricultural researchers, policy makers and extension officers. Moreover, to the extent that extension activities provide an opportunity for a "two-way" exchange of information, they can contribute to ensuring that agricultural innovation systems are demand-driven.

Agricultural education

The capacity of agricultural education varies across the ASEAN member states (Box 6.6). All countries have higher education institutes (HEIs) that offer courses and degree programmes in agricultural disciplines. In Cambodia, Lao PDR and Viet Nam, this includes universities and colleges under the responsibility of the ministries of agriculture. The number of HEIs ranges between countries. In Lao PDR, the National University of Laos (NUOL) is the only HEI to offer post-graduate programmes in agriculture and food and forestry (there are also five agricultural colleges). In contrast, there is a large number of HEIs in the Philippines – in 2011, there were 97 state colleges and universities (SCUs) offering degree programmes in forestry and other related courses, and at least 106 offering degree programmes in agriculture (Cruz et al., 2013).

Box 6.6. Capacity of agricultural education in ASEAN

In **Cambodia**, MAFF supervises three HEIs: the Royal University of Agriculture (RUA) and two national colleges, the Prek Leap National College of Agriculture (PNSA) and the Kampong Cham National School of Agriculture (KCNSA). The RUA is the leading agricultural university in Cambodia, with faculties of Agronomy, Animal Science, Veterinary Medicine, Forestry Science, Fisheries Science, Agricultural Engineering, Agricultural Economics and Rural Development, Agro-Industry, Land Management and Land Administration, and Rubber Sciences. RUA offers an Associate programme, a Bachelor's programme, a Master's programme, and a Doctoral programme. RUA had almost 6 000 enrolled students in 2013.

PNSA and KCNSA offer Short Course training programmes, Associate programmes, and Bachelor programmes in a range of agricultural subjects. PNSA and KCNSA each had around 1 560 students enrolled in 2013 (MAFF, 2015a).

Other universities offering agriculture courses are the Moharussey Vedic University and the University of Battambang. The University of Battambang's Faculty of Agriculture and Food Processing offers majors in Horticulture, Animal Science and Veterinary Medicine, Fisheries/Aquaculture, and Food Processing (James, Gill and Bates, 2013).

In **Lao PDR**, the National University of Laos (NUOL) is the only HEI to offer post-graduate programmes in agriculture and food and forestry.

There are a further five agricultural colleges under the MAF: the Northern Agriculture and Forestry College, the Savannakhet Agriculture and Forestry College, and the Champasak Agriculture and Forestry College (specialising in agriculture, livestock and forestry); the Thangone Irrigation College (specialising in irrigation techniques); and the Bolikhamxay Agriculture and Forestry College (specialising in lowland agriculture).

In **Indonesia**, more than 50 universities and polytechnics offer various agriculture-related courses. The higher education sector is dominated by Bogor Agricultural University.

In **Malaysia**, there are 19 universities offering programmes related to food and agriculture. Of these, six are major universities, including the University of Putra Malaysia, University of Malaysia Sabah, University of Malaysia Terengganu, University of Sultan Zainal Abidin, University of Teknologi Mara, and University of Malaysia Kelantan. There are also colleges and training institutes in Malaysia that offer non-degree programmes related to agriculture and food production (GFRAS, 2016).

The agricultural education system in **Myanmar** includes three universities focused on different segments of the agricultural sector. The Yezin Agricultural University (YAU) under MOAI is the main agricultural university in Myanmar, and most agricultural scientists in government agencies have graduated from this university. YAU covers crop sciences and also offers some courses in animal sciences and fisheries. YAU offers Bachelor's, Master's and Doctoral programmes. The University of Veterinary Science (under MLFRD) offers degrees in veterinary sciences and fisheries. The University of Forestry (under MOECAF) specialises in land management, environment and forestry.

There are also seven state agricultural institutes that offer diplomas in agriculture.

The **Philippines** has a large number of higher education institutions offering agricultural programmes. In 2011, there were 97 state colleges and universities (SCUs) offering forestry and other related courses, and at least 106 SCUs offering degree programmes in agriculture. The more prominent universities offering agricultural programmes include: the University of the Philippines Los Baños; Benguet State University; Visayas State University; University of Southern Mindanao; Central Luzon State University; Central Bicol State University of Agriculture; University of Southeastern Philippines; Cavite State University; Bicol University College of Agriculture and Forestry; West Visayas State University-College of Agriculture and Forestry; Xavier University, Ateneo de Cagayan; University of Southern Mindanao; and Central Mindanao University.

In 2011, the Commission on Higher Education identified the priority courses for the next five years, which included agroforestry, veterinary medicine, agricultural engineering, agribusiness/management, agricultural entrepreneurship, agri-tech, agriculture and fisheries. Despite the proliferation of HEIs and programmes, agriculture, forestry and fishery programmes are experiencing declining enrolments (Cruz et al., 2013).

In **Thailand**, the major research universities, such as Kasetsart University, provide high-quality agricultural education, as do more teaching-focused universities. There are four agricultural education institutions that provide formal technical and vocational education training (UNCTAD, 2015).

In **Viet Nam**, MARD supervises four universities: the Viet Nam National University of Agriculture, the Viet Nam Forestry University, Water Resource University, and Bac Giang Agriculture and Forestry University.

Other universities that provide degree programmes in agriculture include: the Hanoi University of Agriculture, the Ho Chi Minh City University of Agriculture and Silviculture, the Can Tho University, the Hue University College of Agriculture and Forestry, Thai Nguyen University of Agriculture and Forestry, Phu Yen University (Faculty of Agriculture), Bac Giang University of Agriculture and Forestry, Vinh University (Faculty of Agriculture, Forestry and Fish), and the Nghe An Economic University (Faculty of Agriculture, Forestry and Fish).

Source: SEARCA (2016).

Enrolments in agricultural programmes are relatively low or declining in some countries. In Cambodia in 2009-10, students enrolled in agricultural programmes accounted for 2.3% of total bachelor students, while students enrolled in animal science and veterinary medicine programmes accounted for a further 0.5% (James, Gill and Bates, 2013). In the Philippines, enrolments in agriculture, forestry and fisheries-related programmes have declined, even as the number of SCUs has increased (Cruz et al., 2013), while UNCTAD (2015) suggests that in Thailand, university departments specialising in agricultural subjects are finding it increasingly difficult to recruit students (and lecturers).

A lack of resources potentially affects the quality of agricultural education in some ASEAN member states. HEIs in Cambodia and Viet Nam have only a small proportion of university lecturers and researchers qualified at PhD level – around 16% in Viet Nam (OECD, 2015b) – while around 8% of full time faculty staff at Cambodia's RUA have PhDs (De la Pena and Taruno, 2012). RUA is further constrained by a lack of proper equipment and laboratory facilities (James, Gill and Bates, 2013).

Agricultural extension

Across ASEAN, the public sector – and central governments in particular – has a key role in agricultural extension. In most countries, extension programme planning and priority-setting occurs at the national level under the responsibility of an agency in the ministry of agriculture (local governments have primary responsibility for this in the Philippines and Viet Nam). Offices and departments at the provincial level are responsible for managing the delivery of extension services and the training of staff. Extension services are then provided at a district and/or village level (Box 6.7).

Box 6.7. The organisation of public agricultural extension systems in ASEAN

Cambodia has a pluralistic agricultural extension system, wherein extension services are provided by the public and private sectors, NGOs and as part of donor programmes.

The public agricultural extension system is hierarchical in structure. At the national level, the Department of Agricultural Extension (DAE) in MAFF is responsible for developing extension policies and programmes, providing technical support to extension offices in provinces and districts, and for monitoring and evaluating programme implementation. Provincial Departments of Agriculture co-ordinate and manage the delivery of extension services and provide support for extension staff in the District Agricultural Offices (DAOs). The majority of public sector extension staff work at the provincial level – however, extension activities are dependent on external funding, such as from NGOs and donors. The DAOs are responsible for providing extension and technology transfer programmes to farmers, farmer groups and other actors (for example in the value chain) at the district or village level (Soeun, 2012). In practice, the DAOs have no annual budget to deliver agricultural services to farmers, but work as counterparts on NGO and donor projects.

NGOs and donors have a significant role in providing extension services (often in partnership with public extension providers), particularly for smallholder farmers.

The RUA and other agricultural colleges and universities, technical departments within MAFF, and MAFF's research institutes (such as CARDI) also provide some level of extension services.

Indonesia has a pluralistic and decentralised agricultural extension system. In 2006, the agricultural extension system was reorganised at national, provincial and district levels under Law No 16/2006 on the Agriculture, Fishery and Forestry Extension System. The goal is for every one of the almost 70 000 villages to be supported by one extension worker and one voluntary extension worker (OECD, 2012).

At the national level, the Agency for Agricultural Extension and Human Resources Development in the MoA is responsible for developing reliable systems for agricultural extension; training in management, leadership and entrepreneurship; agricultural education; and the empowerment of farmers, farmer institutions, and competitive farming. Within this agency, the National Centre for Agricultural Extension Development is responsible for extension policy. Provincial Agricultural Extension Coordination Offices are responsible for managing education and training for extension personnel at all levels, including the training of field extension agents. District Agricultural Extension Offices have a key role in extension programme planning and execution. Their responsibilities include: developing policies and programmes for managing extension at the district level; providing education and training support to farmer groups; ensuring co-ordination and co-operation among farming communities, extension staff and researchers; and providing education and training to field extension agents on a regular basis. At the village level, there is a village extension post.

Public sector extension primarily serves small or medium-scale commercial farmers and small-scale subsistence farmers, in addition to landless farmers. Large-scale commercial farmers and farmers growing cacao, rubber and horticultural products are also considered important.

Lao PDR has a pluralistic and decentralised extension system, with extension services provided by the public and private sectors, NGOs, and as part of donor programmes. The extension system is guided by the Lao Extension Approach, which is based on the following principles: decentralised, pluralistic, participatory, needs-based, integrated, gender-sensitive, group-based, self-motivated, and sustainable. Extension services cover agriculture, livestock, forestry and irrigation.

The MAF primarily provides extension services through the National Agricultural and Forestry Extension Service (NAFES), which has offices in all provinces and districts in the country. At the provincial level, Provincial Agricultural and Forestry Extension Service offices are staffed with subject-matter specialists, while District Agricultural and Forestry Extension Offices are staffed with extension generalists (Farming Systems Extension Workers). The government does not provide extension at the village level, which is instead covered by the Village Extension System (VES). The VES is an informally private extension system, supported by the government's district extension staff.

NAFRI and the public universities (such as NUOL) and colleges in Lao PDR also undertake some extension activities.

Malaysia has a public sector-driven agricultural extension system. Under the MOA, the Department of Agriculture (DOA) is the leading agency providing extension services for food production (the Department of Fisheries and Department of Veterinary Services also undertake extension-related activities). While the agricultural extension programme has traditionally targeted rural farmers, the service has expanded and also provides advisory and consultancy services for entrepreneurs and investors who are interested in entering into an agriculture-based business. Clients are categorised by annual income, from large-scale agro-entrepreneurs to micro agro-entrepreneurs. DOA's extension in food crop production consists of six different types of extension: crop production, pesticide, diagnostic, harvesting, food processing and marketing.

The DOA's Extension, Advisory and Consultancy service (*Doktor Pokok* or the "Tree Doctor" Development Clinic) provides consultancy services on Good Agricultural Practice, agricultural project feasibility studies, agricultural investment and consultation, diagnosis of plant problems, marketing of agricultural produce, and agro-based industry production technology.

The commodity boards also provide advisory services to companies and entrepreneurs in the plantation sector. For example, the MPOB has a Technical Advisory Services Unit and the MRB has a consulting arm (RRIM - Consult Corporation). The Malaysian Cocoa Board also provides advisory services.

Myanmar has a public sector-driven agricultural extension system. Agricultural extension services are the responsibility of the Department of Agriculture of the MOAI, which is tasked with technology transfer and training of farmers in modern agricultural practices. The LBVD also provides extension services (OECD, 2014). Within the Department of Agriculture, the Agricultural Extension Division undertakes a range of extension activities, including training and capacity-building of extension agents and the training of farmers (Cho, 2013).

At the village level, the Agricultural Extension Division has a Village Tract Extension Service, in which Village Managers maintain contact with farmers over a few village tracts or villages, covering an area of around 1 215 to 2 450 hectares of cropland.

The **Philippines** has a pluralistic and decentralised extension system. It consists of six types of service providers: i) national departments, led by the DA and their respective bureaus and agencies; ii) Local Government Units (LGU); iii) state colleges and universities; iv) farmer associations, such as co-operatives, irrigator associations and agrarian reform communities; v) non-government and other civil society organisations; and vi) the private sector, e.g. agribusiness and banks. The DA is responsible for agriculture and fisheries extension at the national level. However, under decentralisation, extension services were devolved to LGUs, which provide extension services at the village level (OECD, 2017).

The extension and training arm of DA is the Agricultural Training Institute (ATI). ATI co-ordinates diversified agricultural extension delivery systems for the local government sector and other stakeholders to facilitate the flow of information on technology and other services. ATI does not deliver extension services to farmers, but assists local governments in doing so. ATI also collaborates with other bureaus and agencies of DA that need extension support for their individual programmes, as well as the various autonomous agencies attached to DA.

Thailand has a pluralistic extension system, wherein the public and private sectors, NGOs and farmer organisations provide extension services. Within MOAC, the Department of Agricultural Extension (DOAE) and the Department of Livestock provide extension services. DOAE is the main institution responsible for providing extension services to farmers.

Agricultural extension is decentralised (to provinces, districts, and sub-districts), and extension activities are supposed to be supervised by provincial offices. However, extension programme planning remains central.

State universities also provide extension services and pre-service training. For example, Kasetsart University's National Agricultural Extension and Training Centre has several extension programmes for developing human resources, regional and international development, developing farmers' institutions, and producing media for developing agricultural learning aids.

Viet Nam has a pluralistic and decentralised extension system. The public extension system is organised into five levels. At the national level, the National Agriculture Extension Center (NAEC) within MARD is responsible for developing policies and mechanisms for providing agricultural, forestry, fishery and rural industry extension services.

At the provincial level, the Departments of Agriculture and Rural Development have their own extension centres, and are responsible for proposing extension projects that suit provincial conditions, providing extension guidelines at district level and co-operating with district offices to carry out extension activities, as well as implementing extension communication and training activities for district extension staff and key farmers in the provinces. District extension stations directly carry out extension activities and offer training courses for extension staff at the commune and village levels. Extension staff at the commune and village levels are responsible for mobilising farmers to participate in extension activities, communicating farmers' needs to higher levels, and directly implementing activities at the village level.

Central government funding for agricultural extension is allocated through an open bidding process. As a result, a large number of research institutions, universities, enterprises and NGOs are involved in the provision of public extension activities in addition to the government extension service. In many cases, these other providers sub-contract government extension workers, particularly at the provincial level, to implement projects awarded to them. NAEC supervises the implementation of the projects awarded through this process (OECD, 2015b).

Source: GFRAS (2016), *World Wide Extension Study*, www.g-fras.org/en/.

Nevertheless, agricultural extension systems in most ASEAN member states include a range of actors. Extension systems in Cambodia, Indonesia, Lao PDR, the Philippines, Thailand and Viet Nam are pluralistic, and include actors from the non-profit and private sectors, in addition to farmer organisations. The non-profit sector (NGOs and donors) has a significant role in providing extension services to farmers in some countries (Cambodia and Lao PDR in particular), often in partnership with public sector extension providers. Donors have also supported extension system reforms in some countries, such as Indonesia. The private sector has a more limited role, particularly in terms of providing extension services to farmers on a regular, fee-for-service basis. In countries with contract farming, private companies generally provide extension services to farmers in the form of training, demonstrations and field days. Farmer organisations and co-operatives are also active in providing extension services in a number of countries, Indonesia, the Philippines and Thailand in particular. In Indonesia, farmer organisations also have a role in specifying extension programmes and priorities (GFRAS, 2016).

While pluralistic extension systems offer a number of advantages,[6] poor co-ordination of extension providers in some countries has resulted in gaps in the coverage of farmers. In Cambodia, for example, it is common to find duplication of agricultural services provided to villages by government, NGOs and donor programmes, while some small-scale farmers are not reached by extension services (GFRAS, 2016).

Some countries are progressing towards the decentralisation of agricultural extension as a means to make extension systems more efficient, participatory and responsive to farmers' needs. Indonesia, the Philippines and Thailand have decentralised extension systems – although Indonesia's and the Philippines' experiences with decentralisation have been mixed. In these countries, decentralisation *weakened* extension service provision, largely because regional governments lacked sufficient resources for extension activities or placed a low priority on extension (or agriculture more broadly) (NEDA, 2011; OECD, 2012). Meanwhile, Cambodia intends to decentralise agricultural extension services and activities under its recently developed Agricultural Extension Policy, as part of a strategy to increase the effectiveness of agricultural extension services (MAFF, 2015b). Lao PDR's extension model, the Lao Extension Approach, is based on the principle of a decentralised system, among other principles (GFRAS, 2016).

In some ASEAN member states, shifts in the focus of agricultural extension services and the extension methods used also suggest developments towards more participatory and responsive extension systems. Agricultural extension systems in ASEAN have traditionally followed a top-down, supply-driven model, focused on agricultural production (for example, emphasising technology transfer to increase yields). While a production focus continues to exist in a number of countries, other extension initiatives across the region demonstrate a shift away from this approach (Box 6.8).

Financial and human resource constraints can also reduce the effectiveness of extension systems. While data on funding for extension are limited, government and development stakeholders have highlighted the resourcing challenges of some ASEAN member states. For example, the ADB (2012) notes that government extension services in Cambodia would find it difficult to operate without donor subsidies. IFAD (2014) notes that agriculture in Myanmar is constrained by weak extension services, in that the extension staff are inadequate in number and quality, and that insufficient budgetary resources prevent field visits and hinder knowledge-sharing. The OECD (2015b) has noted that extension services in Viet Nam face several challenges, including limited human resources (one public extension worker per 330 farming households, and expenditures on extension services amounting to about USD 3.30 per farming household). Similarly, extension in the Philippines suffers from low levels of financing – by the Local Government Units in particular – and falling numbers of extension staff (OECD, 2017).

Box 6.8. The use of innovative and participatory extension methods in ASEAN

Across ASEAN, agricultural extension systems are using more innovative and participatory extension methods. Some developments include:

- *Use of farmer-to-farmer approaches in the transfer of knowledge and skills.* Under Lao PDR's Village Extension System, village extension workers are in charge of organising farmers into "production groups" that focus on gaining knowledge through "learning projects" that could be applied to enhance production. This focus follows the idea that agricultural extension in Lao PDR is an educational process, not a production process. Indonesia's farmer organisations also have a key role in facilitating farmer-to-farmer extension activities.
- *Recognition of indigenous knowledge.* Thailand's extension service providers have historically followed a participatory extension approach that recognises the value of farmers' indigenous knowledge and capability. Cambodia's recently developed Agricultural Extension Policy requires agricultural extension service providers to take into consideration indigenous or farmer-invented knowledge and technologies (MAFF, 2015a). Similarly, Lao PDR's Agricultural Master Plan 2011 to 2015 calls for indigenous and gender knowledge to be applied in combination with modern farming technologies to enhance the resilience and sustainability of the agricultural sector (MAF, 2010).
- *Expanding the focus of extension services.* A number of countries are shifting from production-oriented extension services to include broader aspects of farm management, including more sustainable production methods and marketing skills. Most significantly, Malaysia's DOA has expanded the focus of its agricultural extension programme to also provide advisory and consultancy services for entrepreneurs and investors who are interested in entering into an agriculture-based business.
- *Greater use of information and communication technology.* The Philippines' e-Extension programme aims to provide an alternative to the traditional extension system of the agricultural, fisheries and natural resources sectors, by integrating an ICT-based extension delivery system. The e-Farming component delivers farm and business advisory services, primarily technical assistance, to farmers and agricultural extension agents via text and voice (using a pre-defined toll-free number), as well as through chat, email, and online forums. Thailand's Smart Agriculture Programme aims to develop databases and knowledge management systems, including a consolidated national agricultural information system and a system to provide useful data to farmers in specific localities and for specific agricultural products. There are also plans to set up online communities of farmers and others to exchange knowledge and information (UNCTAD, 2015). A number of countries are also using the internet, radio and mobile phones to provide technical information and training, as well as weather and market information.

Source: GFRAS (2016), *World Wide Extension Study*, www.g-fras.org/en/.

The capacity of agricultural extension systems, as indicated by estimates of human resources in extension, varies across the ASEAN member states (Table 6.3). Data from the joint IFPRI/FAO/IICA Worldwide Extension Study (GFRAS, 2016) show that for the period 2009 to 2012, Indonesia had the highest number of extension staff (53 944), followed by Viet Nam (34 747), the Philippines (32 328), Thailand (18 881) and Myanmar (10 947). Agricultural extension systems in the remaining countries had significantly fewer extension staff: 1 355 in Malaysia (for food crops); 1 244 in Cambodia; and 752 in Lao PDR (GFRAS, 2016).

Table 6.3. Capacity of public agricultural extension systems in ASEAN, 2009-13

	Total extension staff	
Cambodia	1 244	Includes 1 120 field extension staff and 66 subject matter specialists (crops, livestock, farm management, agricultural marketing, and land, soil, water and forestry management).
Indonesia	53 944	Includes 27 922 field extension staff on permanent contracts and 24 551 field extension staff on three-year contracts, and 28 subject matter specialists.
Lao PDR	752	Includes 201 agriculture (crops and livestock) extension staff, 170 forestry extension staff, 186 fisheries (marine and aquaculture) extension staff, and 195 rural development extension staff. There are a further 210 extension staff from the non-profit sector (for example, extension staff attached to major projects).
Malaysia	1 355	Includes 1 000 field extension staff and 190 subject matter specialists on food crops (excluding palm oil and rubber).
Myanmar	10 947	Includes 5 361 agriculture (crops and livestock) extension staff, 5 296 rural development extension staff, and 20 fisheries (marine and aquaculture) extension staff.
Philippines	32 328	Includes 25 000 extension staff at the Local Government Unit level, 3 390 in DA bureaus and attached agencies, 2 111 in NGOs, 1 250 in state colleges and universities, and 577 in private agribusiness.
Thailand	18 881	Includes 16 986 agriculture (crops and livestock) extension staff, 795 forestry extension staff, and 1 100 fisheries (marine and aquaculture) extension staff.
Viet Nam	34 747	Includes 17 587 staff at the village level (mainly part-time, non-professional extension staff at the village level), and 11 232 extension staff at the commune level.

Source: GFRAS (2016), *World Wide Extension Study*, www.g-fras.org/en/.

6.5 Systems for cross-country co-operation on innovation in ASEAN

Cross-country co-operation on agricultural R&D offers many benefits to ASEAN member states. Regional (and international) research collaboration allows countries to avoid costly duplication of research efforts, particularly in the face of shared challenges, such as responding to climate change, natural resource constraints, and food security. Moreover, cross-country co-operation allows countries to better leverage their domestic research resources and benefit from specialisation and international research spillovers. This may be particularly important for small countries with a limited research capacity.

The stocktake identified a large number of initiatives, suggesting that in general, research and higher-education agencies in ASEAN face few constraints to cross-country co-operation on innovation. Key mechanisms for cross-country co-operation in the region include collaborations with international research institutes and development agencies, membership of international research networks, and co-operation between higher education institutes. ASEAN country research agencies also co-operate regionally. Furthermore, ASEAN is itself a platform for regional co-operation on innovation. What follows is a (non-exhaustive) outline of the systems for cross-country co-operation on innovation that exist within the region. Bibliometrics of food, agriculture and biotechnology publications provides an additional insight into ASEAN countries' international co-operation on agricultural R&D (Box 6.9).

Research collaborations between government agencies and various research centres of the CGIAR Consortium represent an important component of ASEAN member state cross-country co-operation on innovation. The International Rice Research Institute (IRRI) has bilateral programmes with ASEAN member states and also works closely with ASEAN. This collaboration takes the form of strategic technical assistance, building the next generation of rice scientists and enhancing regional co-operation (IRRI, n.d.). Other CGIAR research centres that are active in Southeast Asia include: the International Livestock Research Institute (ILRI), the International Crops Research Institute for the Semi-Arid Tropics (ICRISAT), the International Center for Tropical Agriculture (CIAT), the International Institute of Tropical Agriculture (IITA), the International Maize and Wheat Improvement Center (CIMMYT), Bioversity International, and the World Fish Centre. In addition, several CGIAR research centres have offices in Southeast Asia: IRRI's headquarters are in the Philippines, and the two CGIAR centres with mandates for forestry and natural resources research – the World Agroforestry Centre (ICRAF) and the Center for International Forestry Research (CIFOR) – have offices in Indonesia.

A large number of international development agencies undertake projects in ASEAN member states, in co-operation with public research agencies. This includes the World Bank, FAO, IFAD and the ADB. Research institutes and development agencies from a number of countries are also active in Southeast Asia, including the Australian Centre for International Agricultural Research (ACIAR), the Deutsche Gesellschaft für Internationale Zusammenarbeit GmbH (GIZ), the French Agricultural Research Centre for International Development (CIRAD), the Japan International Cooperation Agency (JICA), the Japan International Research Center for Agricultural Sciences (JIRCAS), and USAID.

ASEAN member states, represented by the main government research agencies, are also members of international research networks, including for research on cross-cutting issues. Notable research networks and platforms include the Global Research Alliance on Agricultural Greenhouse Gases (GRA), Conservation Agriculture in Southeast Asia (CANSEA), the Asia-Pacific Association of Agricultural Research Institutions (APAARI) and SEA-EU-NET.

Indonesia, Malaysia (represented by MARDI), Philippines, Thailand and Viet Nam (represented by VAAS) are part of the GRA, and are variously members of the livestock, paddy rice, croplands and integrative research groups. Member countries of the GRA collaborate on the research, development and extension of technologies and practices to help deliver more climate-resilient food systems without increasing greenhouse gas emissions.

CANSEA is a network that promotes the development of innovative farming systems based on an agro-ecological approach that combines rural development with environmental preservation. Its membership includes Cambodia (represented by MAFF), Indonesia (represented by IAARD), Lao PDR (represented by NAFRI), Thailand (represented by Kasetsart University), and Viet Nam (represented by MARD agencies), and the network is linked with institutions from Australia, China and France.

APAARI promotes the development of national agricultural research systems in the Asia-Pacific region through inter-regional and inter-institutional co-operation. The main research agencies from Malaysia (MARDI), the Philippines (BAR and PCAARRD), Thailand (DOA) and Viet Nam (MARD) are regular members of APAARI.

Box 6.9 International co-operation on agricultural R&D

Agricultural research and development (R&D) is a significant and increasing area for international research co-operation in ASEAN member states.

The share of internationally co-authored publications in total publications provides an insight into the degree of international collaboration by ASEAN countries on agricultural R&D. Between 2004 and 2008, around 57% of publications by ASEAN countries[1] on food, agriculture and biotechnology had international co-authorship, exceeding the share of publications in Australia, Korea, Japan and China (UNU-IIST, 2011). More recent analysis suggests that the European Union, United States and Japan are significant collaborators. Between 2004 and 2013, 16% of ASEAN journal publications on agriculture, fisheries and forestry by Southeast Asia-based authors had at least one co-author from Europe, 10% of publications had at least one co-author from Japan, and 9% had at least one co-author from the United States (Lampert and Stefan, forthcoming).

The high share of publications with international co-authorship seen in ASEAN countries points to the openness of researchers from the region to collaboration, and the fact that research agencies have developed strong international research linkages with international research institutes and development agencies. However, it may also reflect the weaker domestic research capacity of some countries – the share of publications with international co-authorship ranged from 38% in Malaysia to around 90% in Cambodia, Lao PDR, Myanmar and Viet Nam (UNU-IIST, 2011). This suggests that in some ASEAN countries, international research co-operation is an important mechanism for building research capacity.

1. Including Brunei Darussalam and Singapore.

SEA-EU-NET is a network for deepening science and technology co-operation between Europe and Southeast Asia. The network involves institutions from the two regions, including Indonesia, Lao PDR, the Philippines, Thailand and Viet Nam. Food security and safety is one of four thematic areas, and priorities for research collaboration include: helping to develop ecologically intensive agriculture to feed Europe and Southeast Asia; innovating to make food accessible, varied and safe; foreseeing and managing infectious disease risks linked to wildlife and domestic animals; and supporting public policies aimed at reducing structural inequality and poverty (SEA-EU-NET, 2015).

Cross-country linkages between higher-education institutes represent another important mechanism to strengthen the AIS, through training and building the capacity of agricultural researchers. Moreover, relationships built through student and researcher exchanges can provide a foundation for future research collaborations.

Across ASEAN, there are networks for co-operation among higher education institutes. The most notable of these is the Southeast Asian Regional Center for Graduate Study and Research in Agriculture (SEARCA). SEARCA is an international organisation that works to strengthen institutional capacities toward inclusive and sustainable agricultural and rural development in Southeast Asia through graduate education, R&D, and knowledge management. SEARCA uses its Umbrella Programs to develop and implement an integrated agenda for collaborative research, capacity development, and knowledge management on strategic development concerns in the region. The Umbrella Programs outline the priority areas of convergence and collaboration at the regional and sub-regional levels, where institutions can work together, pool resources, and address gaps or move in directions where joint efforts provide a strategic advantage for food security and climate resilience in the region. Currently, SEARCA has two Umbrella Programs: (i) the Umbrella Program on Food and Nutrition Security for Southeast Asia 2014-2019, and (ii) the Umbrella Program on Climate Change Adaptation and Mitigation for Southeast Asia 2015-2020.

Other SEARCA initiatives include the Southeast Asian University Consortium for Graduate Education in Agriculture and Natural Resources, launched with the five premier agricultural universities in the region in 1989. The University Consortium is a network of universities in Southeast Asia, Canada, Germany and Japan, and pursues agricultural human resource development in Southeast Asia by linking top agricultural universities in the region to facilitate free exchange of information, facilities, and expertise. In addition, SEARCA's Institutional Development Assistance Program aims to enhance the teaching, research and extension of agricultural capacities that play strategic roles in providing the human resources necessary to support economic development in their countries – and the region more broadly. Cambodia's RUA and Myanmar's YAU are part of the Institutional Development Assistance Program.

Cambodia, Indonesia, Lao PDR, the Philippines, Thailand and Viet Nam are also partners in the Erasmus Mundus Programme of the European Commission, a co-operation and mobility programme in the field of higher education that has funded projects for human development in agriculture. For example, the EXPERTS Consortium established a collaborative framework for human resource development through training and upgrading the skills of junior faculty staff, undergraduates, postgraduates and post-doctoral researchers, including in the field of agricultural sciences. The ALFABET project aims to contribute to increasing the relevance, capacities and efficiency of the higher-education system in Asia by strengthening the existing network of co-operation among universities oriented to life sciences, food, agriculture, biology, environment and technology.

Other networks for co-operation between higher-education institutes are the Greater Mekong Subregion Academic and Research Network, which has members from academic and research institutions from Cambodia, Lao PDR, Myanmar, Thailand and Viet Nam; and the ASEAN University Network, which includes institutes from all ASEAN member states.

ASEAN is itself a forum for cross-country co-operation on innovation. ASEAN's Vision and Strategic Plan for ASEAN Cooperation in Food, Agriculture and Forestry (2016-2025) calls for the strengthening of existing regional collaboration among ASEAN member states and with international research institutes, IRRI in particular (ASEAN Secretariat, 2015).

The ASEAN Technical Working Group on Agricultural Research and Development (ATWGARD) provides a collaborative platform for improving knowledge flows and exchange on agricultural R&D between the ASEAN member states by: i) providing policy inputs for decision-making in agricultural research; ii) providing a framework for agricultural research prioritisation; and iii) facilitating co-operation with existing ASEAN bodies for inter-governmental networking, technical assistance, and knowledge-sharing in agricultural research. For example, ATWGARD, through Thailand's DOA, initiated a project, Promotion of Climate Resilience in Rice and Other Crops. This initiative laid the foundation for the formation of the ASEAN Climate Resilience Network (ASEAN-CRN), a platform for the regional exchange of information, experiences, and expertise on climate-smart agriculture. ASEAN-CRN is supported by the ASEAN German-Programme on Response to Climate Change in Agriculture and Forestry (GAP-CC).

Other notable initiatives for co-operation through ASEAN include: the ASEAN Social Forestry Network (ASFN), a platform that links government forestry policy makers directly with other network members from civil society, research organisations, academia, private sector and experts of related fields; and Grow Asia, a multi-stakeholder partnership platform that catalyses action on inclusive and sustainable agricultural development in Southeast Asia. Grow Asia has initiated country partnerships with Cambodia, Indonesia, Myanmar, the Philippines and Viet Nam.

6.6 Concluding comments and next steps

Maintaining current, strong rates of agricultural productivity growth is critical for ASEAN member states to achieve their goal of eliminating food insecurity. For the region as a whole, agricultural output has grown significantly in recent decades, driven by improvements in productivity in most countries. However, past growth may not be sustainable, since agricultural development has, in large part, relied on natural resource exploitation, leading to the degradation of land and water resources. Going forward, ASEAN countries will need to shift to an agricultural development strategy that prioritises productivity growth while protecting the natural resource base on which the sector relies.

A key policy option for governments to promote productivity growth is to invest in agricultural research and development (R&D). Funding for agricultural research is important to ensure that farmers have access to a supply of innovations that meet diverse and complex needs. There is ample evidence that the returns to public agricultural R&D expenditures are positive and substantial for agricultural productivity and growth. Moreover, public spending on agricultural R&D outperforms other public expenditures in agriculture, such as irrigation and fertiliser subsidies, in terms of raising agricultural productivity (Diaz-Bonilla, Orden and Kwieciński, 2014).

Yet across ASEAN, most governments significantly underinvest in agricultural R&D. As a result, agricultural innovation systems (AIS) in some countries lack the research capacities to develop and adapt innovations that address the challenges facing their agricultural sectors. Similarly, in the countries where donors provide a substantial share of funding for R&D, the short-term and volatile nature of funding may severely restrict the types of research undertaken. Given lags between research investments, the creation of innovations and their adoption by farmers, continued underinvestment in agricultural R&D may constrain long-term agricultural productivity growth in ASEAN, and the capacity of countries to eliminate food insecurity.

Beyond funding for agricultural R&D, the capacity of AIS in ASEAN to supply farmers with innovations will also depend on factors such as effective governance arrangements and extension systems that facilitate farmers' access to and adoption of technology and knowledge. Regional – and international – research collaboration will be important to address shared challenges, including climate change, natural resource constraints, and food security; and to avoid costly duplication of research efforts by countries. Given the low funding and limited research capacity in some ASEAN member states, cross-country collaboration will also be crucial for countries to make the most of their domestic resources.

To maximise the payoff to investments in agricultural R&D, further efforts to enhance the agricultural enabling environment are also needed. Governments in ASEAN can promote sustainable agricultural productivity growth by providing an economic and policy environment that creates conditions that support private investment in innovation, including on-farm, and provides incentives for the sustainable use of natural resources. This means providing an economic and policy environment that is characterised by good governance, non-distorting and stable policies, adequate provision of public goods, strong and effective institutions, and efficient markets that facilitate the flow of factors of production to their highest-valued uses.

While certain economy-wide aspects of the enabling environment are strong in ASEAN countries, further reforms and investments are needed to strengthen some agriculture-related aspects in order for the sector to benefit from increased investments in agricultural R&D. For example, the macroeconomic environment, market operations, and investments by countries in developing human capital are areas of strength for the region, relative to other countries. By influencing the overall performance of the economy and opportunities for trade, these aspects of the enabling environment support sustained growth in domestic and export demand for agricultural products – an important enabler of agricultural productivity growth. In contrast, areas of weakness in most ASEAN countries include agricultural infrastructure and the regulatory and institutional frameworks that govern land and financial market operations in the rural economy. Collectively, these factors are likely to constrain on-farm innovation even if AIS are able to provide a supply of appropriate innovations "on-the-shelf".

Moreover, further efforts are required to strengthen incentives for sustainable use of resources in the long term. Environmental governance and regulations on land, water and biodiversity resources are generally weak in ASEAN, which has contributed to the degradation of the natural resource base. However, other factors are also at play. For example, without secure land tenure, users of land and water resources have limited incentive to protect the natural resource base and ensure future production.

A number of recommendations follow from the stocktake of AIS in ASEAN and the assessment of the agricultural enabling environment in the preceding chapters. These are outlined below.

Recommendations

Stronger governance of AIS is needed across ASEAN to ensure that these systems are efficient and effective, responsive to the needs of farmers, and encourage collaboration between actors to avoid duplication of research efforts.

ASEAN governments should develop forward-looking agricultural innovation strategies – where this is lacking – that establish priorities for agricultural research and co-ordinate the efforts of key actors in the AIS. Strategies should take into account the future development needs of the agricultural sector and the nature of the challenges and opportunities facing it, as well as national development and innovation priorities. In addition, strategies should be informed by stakeholder consultation, including with government and higher education research agencies, the private sector (farmers and businesses in agricultural value chains), and development partners.

Sustained increases in funding for agricultural R&D are needed in most countries. To address persistent underinvestment in agricultural R&D, ASEAN governments should identify opportunities to reallocate funding from other, more-distorting forms of support to the sector – input subsidies, for example. In countries where donors account for a substantial share of funding for agricultural R&D, governments should develop long-term and stable funding strategies to reduce the dependence of AIS on donor support, while continuing to work closely with development partners.

Countries should also explore opportunities to provide incentives for increased private investment in agricultural innovation. This includes strengthening intellectual property rights regimes through the better enforcement of IPR and the protection of plant varieties. Together with industry, governments should also consider other sources of funding, such as commodity levies.

ASEAN member states should strengthen agricultural extension systems by continuing reforms that shift away from a top-down, supply-driven approach focused on agricultural production, and by engaging farmers in priority setting (for example, by holding consultations and workshops at a local level, and engaging with farmers' organisations, co-operatives and commodity boards). To support pluralistic extension systems, greater government co-ordination is needed to clarify the roles of different extension providers, to reduce duplication of activities and ensure that all farmers are reached. In ASEAN member states with decentralised extension systems or that are implementing reforms to decentralise extension, governments should invest in building the capacity of local governments and other providers.

Agriculture curricula in ASEAN member states might benefit from a governmental review to determine whether they provide graduates with the skills and knowledge they need for a modernised and market-oriented sector.

Research agencies across ASEAN are active participants in collaborative research within the ASEAN region and internationally, facilitating access to overseas knowledge and expertise. Governments and research agencies should pursue these efforts and seek to learn from successes in adapting the model, where needed.

Further reforms and investments are also needed to strengthen the enabling environment for sustainable agricultural productivity growth. In particular, investments to develop transport and communication infrastructure, particularly in rural areas, would improve farmers' connectedness to markets, knowledge and other services, including extension.

Greater security over land tenure arrangements is essential for farmers to make productivity- and sustainability-enhancing investments. Governments should continue with reforms to improve regulatory and institutional frameworks that govern land market operations in the rural economy, as recommended in recent OECD reviews (OECD, 2012, 2014, 2015b, 2017).

Governments should also consider opportunities to increase the range of financing options available to farmers, including smallholders. This includes greater involvement by the commercial banking sector or adoption of micro-financing models, as well as efforts to facilitate the process of applying for loans.

Finally, stronger environmental governance is required to protect the natural resource base and provide incentives for more environmentally-friendly production practices. Governments in ASEAN should strengthen regulations on access to and the use of natural resources, including land and water, and ensure that existing regulations are enforced.

Notes

1. The stocktake of agricultural innovation systems in ASEAN draws on a consultancy report provided by Bernie S. Justimbaste of the Southeast Asian Regional Center for Graduate Study and Research in Agriculture (SEARCA) in the Philippines. This chapter also benefited from input provided by Bessie M. Burgos from SEARCA, and comments provided by participants at the *OECD-FAO-ASEAN Regional Conference on Policies to Enable Food Security, Agricultural Productivity and Improved Nutrition,* which was supported by the FAO and the Governments of Myanmar and Australia, and which took place on 14-15 June 2016 in Nay Pyi Taw, Myanmar. Comments were also provided by colleagues from the OECD Secretariat: Frank van Tongeren, Olga Melyukhina, Catherine Moreddu, Shingo Kimura, Jared Greenville, Clara Thompson-Lipponen, Martina Abderrahmane, Lihan Wei and Andrzej Kwieciński.

2. The ten ASEAN member states are Brunei Darussalam, Cambodia, Indonesia, Lao People's Democratic Republic (PDR), Malaysia, Myanmar, Philippines, Singapore, Thailand and Viet Nam.

3. Innovation policy includes investments in public R&D institutions to fund staff and equipment, as well as projects and programmes; support to private R&D through tax rebates, competitive grants and funding of public-private partnerships (PPPs); the provision of knowledge infrastructure such as information and communications technology (ICT), life-science infrastructure (gene banks) and information systems; and regulations regarding intellectual property rights (IPR). Government policy also supports the creation and

functioning of networks in addition to centres of excellence, and provides platforms for partnerships (OECD, 2013a).

4. The United Nations has called for minimum agricultural R&D investment targets of at least 1% of agricultural gross value added (GVA) (Stads, 2015).

5. While research intensity is a good comparative indicator of R&D spending levels, it fails to take into consideration the policy context and institutional environment of a country's agricultural R&D system or the broader size and structure of a country's agricultural sector and economy (Flaherty and Dardak, 2013).

6. Diversity in farmers' characteristics is best matched by diversity in extension services, approaches and providers. This is because the specific characteristics and needs of farmers (for example, poor and small-scale farms versus larger-scale commercial enterprises; production and conservation objectives) will determine which organisations can best provide them with extension services, and by which methods (Heemskerk and Davis, 2012).

References

ADB (2012), *Rural Development for Cambodia: Key Issues and Constraints*, Asian Development Bank, Mandaluyong City, Philippines.

Arshad, F.M., K.M. Noh, and S.M. Zainalabidin (2015), *Agricultural Policy and Institutional Reforms in Malaysia: Experiences, Impacts and Lessons*, SAAD Primer Series, SEARCA, Philippines.

ASEAN Secretariat (2015), *Vision and Strategic Plan for ASEAN Cooperation in Food, Agriculture and Forestry (2016-2025)*, ASEAN Secretariat, Makati City, Philippines.

Beintema, N.M. and G.J. Stads (2008), *Diversity in Agricultural Research Resources in the Asia-Pacific Region*, International Food Policy Research Institute and Asia Pacific Association of Agricultural Research Institutions, Bangkok.

Cho, K.M. (2013), "Current situation and future opportunities in agricultural education research and extension in Burma", *Background Paper No. 5*, USAID, Michigan State University, and MIDRI-CESD.

Cruz, R.V.O., R.B. Bantayan, L.D. Landicho and N.C. Bantayan, (2013), "Reformulating agriculture and forestry education in the Philippines: Issues and concerns", *Journal of Developments in Sustainable Agriculture*, Vol. 8/1, pp. 49-62.

De la Pena, F.T. and W.P. Taruno, (2012), *Study on the State of S&T Development in ASEAN,* Committee on Science and Technology of Association for Southeast Asian Nations, Taguig City, Philippines.

Degelsegger, A. and W. Sukprasertchai (2014), "Thailand", in Degelsegger, A., F. Gruber, S.O. Remøe and R. Trienes (eds.), *Spotlight on: Stimulating Innovation in Southeast Asia*, SEA-EU-NET report, Centre for Social Innovation, Vienna, pp. 78-91.

Diaz-Bonilla, E., D. Orden and A. Kwieciński (2014), "Enabling environment for agricultural growth and competitiveness: Evaluation, indicators and indices", *OECD Food, Agriculture and Fisheries Papers*, No. 67, OECD Publishing, Paris, http://dx.doi.org/10.1787/5jz48305h4vd-en.

DOA (2016), *Organization*, Department of Agriculture Thailand (webpage), www.doa.go.th/en/ (accessed on 12 December 2016).

Ernst & Young (2013), *2013 Asia Pacific R&D Incentives*, Ernst & Young Australia.

FAO and GoRUM (2012), *Country Programming Framework 2012-2016 for the Cooperation and Partnership between the Food and Agriculture Organization and the Republic of the Union of Myanmar*, Myanmar.

Flaherty, K. and R.A. Dardak (2013), *Malaysia: Recent Developments in Agricultural Research*, ASTI Country Note, International Food Policy Research Institute, Washington, DC.

GFRAS (2016), *World Wide Extension Study* (database), Global Forum for Rural Advisory Services, www.g-fras.org/en/ (accessed on 21 September 2016).

Heemskerk, W. and K. Davis (2012), "Pluralistic Extension Systems", in World Bank (ed.), *Agricultural Innovation Systems: An Investment Sourcebook,* World Bank, Washington, DC, pp. 194-203.

IFAD (2014), *Fostering Agricultural Revitalization in Myanmar (FARM): Final Project Design Report*, International Fund for Agricultural Development.

IFPRI (2016), *ASTI: Agricultural Science and Technology Indicators* (database), International Food Policy Research Institute, www.asti.cgiar.org/data (accessed 13 July 2016).

IFPRI (2015), "Food policy indicators: Tracking change", in *2014–2015 Global Food Policy Report*, International Food Policy Research Institute, Washington, DC.

IRRI (n.d.), *Association of Southeast Asian Nations* (webpage), http://irri.org/our-work/locations/asean (accessed 24 October 2016).

James, D., T. Gill and R. Bates (2013), *innovAT – Cambodia: Desktop Study*, Pennsylvania State University, Innovation for Agricultural Training and Education, Pennsylvania, United States of America.

Lampert, D. and P. Stefan (forthcoming), *Analysis of Co-Publications and Co-Patenting among ASEAN Countries and their Collaboration with the EU*, SEA-EU-NET, Bonn and Vienna.

MAF (2010), *Agricultural Master Plan 2011 to 2015, Program Approach, Roadmap: Agriculture and Forestry for Sustainable Development, Food and Income Security*, Final Draft, Ministry of Agriculture and Forestry, Vientiane.

MAFF (2015a), *Agricultural Sector Strategic Development Plan, 2014-2018*, Ministry of Agriculture, Forestry and Fisheries, Royal Kingdom of Cambodia.

MAFF (2015b), *Agricultural Extension Policy in Cambodia*, Ministry of Agriculture, Forestry and Fisheries, Royal Kingdom of Cambodia.

MAFF (2005), *Master Plan for National Agricultural Research, 2006-2015*, Ministry of Agriculture, Forestry and Fisheries, Royal Kingdom of Cambodia.

Meerod, W., N. Chanvarasuth, S. Wattanasrirungkul, V. Mongkolchaisit, U. Grudloyma, and M. Tanticharoen, (2011), "Networking of R&D institutions for enhancing innovation: Strategies and approaches in Thailand", *Tech Monitor*, National Science and Technology Development Agency (NSTDA), Thailand.

NEDA (2014), *Philippine Development Plan 2011-2016: Midterm Update with Revalidated Results Matrices*, National Economic Development Authority, Pasig City, Philippines.

NEDA (2011), *Philippines Development Plan 2011–2016*, National Economic and Development Authority, Pasig City, Philippines.

OECD (2017), *Agricultural Policies in the Philippines*, OECD Publishing, Paris, http://dx.doi.org/10.1787/9789264269088-en.

OECD (2016a), *OECD Reviews of Innovation Policy: Malaysia 2016*, OECD Reviews of Innovation Policy, OECD Publishing, Paris, http://dx.doi.org/10.1787/9789264255340-en.

OECD (2016b), "Gross domestic expenditure on R-D by sector of performance and field of science", *OECD Research and Development Statistics* (database), http://stats.oecd.org/Index.aspx?DataSetCode=GERD_SCIENCE (accessed on 24 June 2016).

OECD (2016c), *System of National Accounts* (database), https://stats.oecd.org (accessed on 24 June 2016).

OECD (2015a), *Analysing Policies to Improve Agricultural Productivity Growth, Sustainably: Revised Framework*, Paris, www.oecd.org/tad/agricultural-policies/innovation-food-agriculture.htm.

OECD (2015b), *Agricultural Policies in Viet Nam 2015*, OECD Publishing, Paris, http://dx.doi.org/10.1787/9789264235151-en.

OECD (2015c), *Economic Outlook for Southeast Asia, China and India 2015: Strengthening Institutional Capacity*, OECD Publishing, Paris, http://dx.doi.org/10.1787/saeo-2015-en.

OECD (2014), *Multi-Dimensional Review of Myanmar: Volume 2. In-Depth Analysis and Recommendations*, OECD Development Pathways, OECD Publishing, Paris. http://dx.doi.org/10.1787/9789264220577-en.

OECD (2013a), *Agricultural Innovation Systems: A Framework for Analysing the Role of the Government*, OECD Publishing, Paris, http://dx.doi.org/10.1787/9789264200593-en.

OECD (2013b), *Innovation in Southeast Asia*, OECD Publishing, Paris, http://dx.doi.org/10.1787/9789264128712-en.

OECD (2012), *OECD Review of Agricultural Policies: Indonesia 2012*, OECD Publishing, Paris, http://dx.doi.org/10.1787/9789264179011-en.

PCAARRD (n.d.), *Banner Programs: Strategic R&D* (webpage), www.pcaarrd.dost.gov.ph/home/portal/index.php/transparency/banner-programs#strategic-r-d (accessed 3 October 2016).

Poapongsakorn, N. (2011), "R&D and performance of the Thai agriculture and food processing industry: The role of government, agribusiness firms, and farmers", in Intal, P.S. Jr., S. Oum, and M.J.O. Simorangkir (eds.), *Agricultural Development, Trade and Regional Cooperation in Developing East Asia*, Economic Research Institute for ASEAN and East Asia (ERIA), Jakarta.

Prasvita, L. (2014), *Public Private Partnerships in Agriculture R&D: An Innovative Approach to Help Indonesian Smallholder Farmers*, Discussion paper for Australia-Indonesia Research Centre Research Summit on Food and Agriculture, 22 May 2014, Jakarta.

Remøe, S.O. (2014), "Philippines", in Degelsegger, A., Gruber, F., Remøe, S.O. and Trienes, R., (eds.), *Spotlight on: Stimulating innovation in Southeast Asia*, SEA-EU-NET report, Vienna, pp. 46-62.

RGC (2014), *National Strategic Development Plan 2014-2018*, Royal Government of Cambodia.

SEA-EU-NET (2015), *EU-ASEAN S&T Cooperation to Jointly Tackle Societal Challenges: Food Security and Safety*, SEA-EU-NET brochure.

SEARCA (2016), *Overview and Assessment of Agricultural Innovation Systems in ASEAN*, consultant report prepared for the OECD (unpublished).

Soeun, M. (2012), *Overview of Agricultural Extension System in Cambodia*, Réseau FAR, Atelier Yaoundé, 20-22 November 2012.

Stads, G.-J. (2015), *A Snapshot of Agricultural Research Investment and Capacity in Asia*, Asia-Pacific Association of Agricultural Research Institutions (APAARI) and International Food Policy Research Institute (IFPRI), Washington, DC and Bangkok.

Stads, G.-J., P.S. Faylon and L.J. Buendia (2007), *Agricultural R&D in the Philippines: Policy, Investments, and Institutional Profile: ASTI Country Report*, International Food Policy Research Institute and the Philippine Council for Agriculture, Forestry and Natural Resources Research and Development.

Stads, G.-J., I. Haryono and S. Nurjayanti, (2007), *Agricultural R&D in Indonesia: Policy, Investments, and Institutional Profile: ASTI Country Report*, International Food Policy Research Institute and the Indonesian Agency for Agricultural Research and Development.

Stads, G.-J. and P.S. Kam (2007), "Myanmar", *ASTI Country Brief* No. 38, International Food Policy Research Institute and the National Agricultural and Forestry Research Institute.

Stads, G.-J. and K. Manivong (2006), "Laos", *ASTI Country Brief* No. 32, International Food Policy Research Institute and the National Agricultural and Forestry Research Institute.

Stads, G.-J. and H.V. Nguyen (2006), "Vietnam", *ASTI Country Brief* No. 33, International Food Policy Research Institute and the Ministry of Agriculture and Rural Development.

Trienes, R., T. Sundari, H. Hadisetiawati and F. Gruber (2014), "Indonesia", in Degelsegger, A., F. Gruber, S.O. Remøe, and R. Trienes, (eds.), *Spotlight on: Stimulating innovation in Southeast Asia*, SEA-EU-NET report, pp. 20-31.

UNCTAD (2015), *Science, Technology & Innovation Policy Review: Thailand*, United Nations Conference on Trade and Development, Geneva.

UNU-IIST (2011), *Analyses of Research Strengths of SEA Countries for SEA-EU-NET under Task 4.9 Bibliometric Analysis of S&T Strengths in Southeast Asia (Deliverable # 1)*, United Nations University International Institute for Software Technology.

UPOV (2016), *Status in Relation to the International Union for the Protection of New Varieties of Plants (UPOV) as of October 31, 2016* (webpage), Union for the Protection of New Varieties of Plants, Geneva, www.upov.int/portal/index.html.en (accessed on 08 November 2016).

WIPO (2016), *WIPO* (website), www.wipo.int/ (accessed on 30 November 2015).

Chapter 7

Lessons from Indonesia on fishing for food security

This chapter explores how fisheries and aquaculture policies in Indonesia can contribute to improved food security and nutrition. It takes stock of the challenges to food security in Indonesia, describes how fisheries and aquaculture contribute to nutritious food supplies and incomes, and outlines the strategy of the Government of Indonesia for the sector. Policies currently in place are then analysed, and alternative policy options are proposed, where relevant, in light of three main objectives: sustainable management of natural resources for the future, increased economic opportunities for fishers and aqua-farmers, and improved consumer access to seafood.

Key points

- Food security has improved significantly in Indonesia since the early 1990s, but nearly 20 million people remain undernourished. Capture fisheries and aquaculture are important components of the Indonesian diet, and sustain livelihoods for about 20 million people.
- Seafood production in Indonesia is achieved at high cost to the country's natural resources. These costs are compounded by harvest capacity support and trade and investment restrictions, which also place added pressure on consumer prices.
- The Indonesian government has taken significant measures to combat illegal fishing. These will improve both the resource base of fisheries and governmental controls over operators, but nevertheless need to be accompanied by increased regulation of the main domestic fisheries, notably with the implementation of long-term, measurable capture targets.
- Stricter regulation of the aquaculture sub-sector is also necessary, notably more comprehensive and regular impact assessment of large farm operations, and clearer allocation of monitoring responsibilities to improve monitoring efficiency.
- To enable efficiency improvements to the fisheries and aquaculture sector without increasing consumer prices or pressure on resources, governmental support should take the form of investment in research and development (R&D), transport and energy provision infrastructure, together with the provision of targeted social safety nets and training to enable the transition to more remunerative activities.

7.1 Contribution of fisheries and aquaculture to food security in Indonesia[1]

In Indonesia, as in most Association of Southeast Asian Nations (ASEAN) countries, food security and nutrition remain key policy concerns.[2] Despite significant economic development and reductions in poverty and undernourishment rates, the country still counts 19.4 million people as undernourished (FAO, 2015), the largest absolute number in the region. Indicators of observed malnutrition also remain at alarming levels: in 2014, more than a third of children under the age of five were stunted, while obesity is a growing concern, with around one in ten children under five considered to be overweight (WHO, 2015). The challenge of eliminating food insecurity is wide-ranging, multi-faceted and linked to other major policy agendas, including those of tackling poverty, using scarce natural resources sustainably, improving sanitation conditions, raising education levels and mitigating the effects of climate change. This chapter focuses on those aspects of the challenge that can directly be affected by fisheries and aquaculture policies.

Capture fisheries and aquaculture already play a key role for food security in Indonesia. The sector supplies an important component of Indonesian diet and nutrition, and sustains livelihoods for about 20 million people, including those in poor and remote areas of the country. At the same time, the sector has potential to further contribute to food security, as the country is the world's largest archipelago, three-quarters of its territory is marine environment, and it is home to some of the richest and most diverse marine resources. Indonesia recently became the world's largest producer of seafood after the People's Republic of China (FAO, 2016b).

This first section summarises the food security situation in Indonesia; describes how fisheries and aquaculture contribute to nutritious food supplies and to incomes, especially for the poor; and analyses the Indonesian government strategy for the sector in light of OECD policy recommendations for food security.

Food insecurity and malnutrition in Indonesia

According to the Food and Agriculture Organization (FAO) definition agreed at the 1996 World Food Summit, and expanded upon at the 2001 Summit, food security exists when all people, at all times, have physical, social and economic access to sufficient, safe and nutritious food to meet their dietary needs and food preferences for an active and healthy life.[3] This definition gives rise to four dimensions of food security. It suggests that people will only be food secure when sufficient food is available, when they have access to this supply of food, and that it is well utilised, in a stable way, over time. Food insecurity may therefore arise from disruptions in food availability that can result from pest outbreaks, climatic events, conflicts or barriers to trade; problems in accessing food as a result of price or income shocks and volatility; or from inadequate use of food, notably in terms of diet composition, which is influenced by cultural, religious and social norms, as well as anticipation of future shocks. Nutritional outcomes also depend on wider determinants of health, including maternal and child care, water and sanitation, and health services (OECD, 2013).

The multi-dimensional nature of food security means that several indicators are required to capture current and past levels of food security. The most commonly used indicators are the FAO prevalence of undernourishment, which compares estimated average calorie consumption with energy requirement norms, and observed child growth indicators, using data reported by the WHO.

In Indonesia, as in most ASEAN countries, the rate of undernourishment has improved significantly since the early 1990s, dropping from almost 20% to slightly more than 7%, although due to the 1997 Asian Financial Crisis, this decline has not been linear (FAO, 2015).[4] Indonesia has experienced steady economic growth, with overall poverty rates falling by 7% from 2002 to just above 11% in 2014. Gross national income per capita rose from USD 720 to USD 3 600 over the same time period (Indonesia Investments, 2016b). These trends are foreseen to continue. Based on FAO projections for future food availability, the OECD projects that undernourishment will decline to just over 3% in 2025 (Figure 7.1).

Figure 7.1. Selected food insecurity indicators in Indonesia, 1991-2025 (projection)

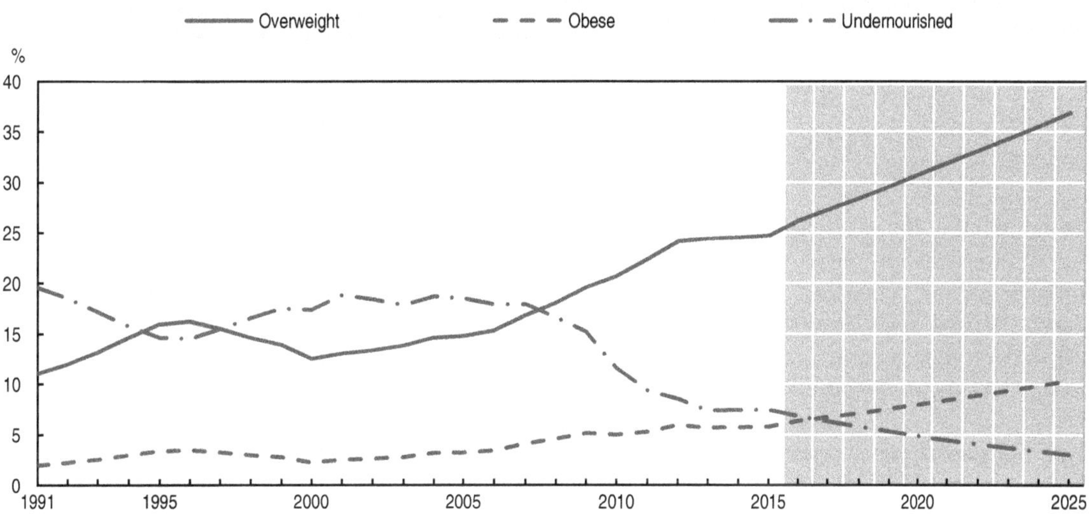

Source: OECD calculations based on FAO (2015).

In spite of this progress, challenges remain. While nearly 20 million people are still undernourished (FAO, 2015), almost twice as many people are found to be at risk of *becoming* undernourished, as specified by the FAO indicator of food inadequacy.[5] At the same time, the level of stunting (when children fail to reach their linear growth potential as a result of suboptimal health or nutritional conditions) is of serious concern, at over 36% among children under the age of five in 2014 (WHO, 2015). In addition, micronutrient deficiency – in particular, of iodine and vitamin A – is still a moderate public health concern, bordering on severe (ASEAN, 2016). This is believed to be due, in part, to inadequate diets, notably with insufficient intakes of protein and vitamins and excessive intakes of carbohydrates. Overall, about one-third of children below the age of two do not meet the minimum meal frequency; one-quarter do not achieve the minimum dietary diversity, and nearly half do not meet the recommended quality of diet (ASEAN, 2016). On the other hand, overweight is already a concern for about a quarter of the population, and obesity is foreseen to affect one in ten people by 2025 (Figure 7.1).

Fisheries and aquaculture sector contribution to food security and incomes

Seafood consumption in Indonesia has grown significantly since the early 1990s. According to the results of the Indonesian national socio-economic survey (MMAF, 2016), seafood consumption reached more than 41 kg per capita per annum in 2015, one of the world's highest rates. Its contribution to protein intake increased to over half of the animal protein supply and 15% of the total protein supply in 2013 (FAO, 2015).

The role of fisheries and aquaculture products in nutrition is particularly important for the poorest populations, for whom fish is often the cheapest and most easily accessible source of protein, available year-round, including when other sources of protein are at a seasonal low (HLPE, 2014). Fish, especially when eaten whole, is an important source of essential fatty acids and micronutrients, which are critical complements to the predominantly carbohydrate-based diets of many poor people. These micronutrients include vitamins A, B and D as well as iodine, iron, zinc and calcium (IFPRI, 2014).

At the same time, the fisheries and aquaculture sector contributes significantly to the Indonesian economy. It generated almost 4% of total gross domestic product (GDP) and earned USD 4.6 billion from seafood product exports in 2012 (MMAF, 2015). Most importantly, as 95% of fishing boats are small-scale artisanal vessels and aquaculture production is mainly artisanal (Box 7.1), the sector is an important source of income generation. It accounted for an estimated 6.4 million direct jobs for Indonesians in 2014 (MMAF, 2016), of which 3.7 million were on-farm (MMAF, 2016) and 2.7 million were for fishers (MMAF, 2015). In addition, it is estimated that as many people are indirectly employed in associated services and industries, such that about 20 million Indonesians are believed to rely on fishing for their basic livelihoods (Koeshendrajana, 2016). The number of fisheries jobs is likely to grow in the future. The number of people employed in capture fisheries increased by 84% between 2000 and 2010, while employment in fish farming increased by 156 % (FAO, 2012a). By 2030, it is estimated that almost 9 million people will be directly employed in aquaculture, with another 6 million people indirectly dependent on the sub-sector (Phillips et al., 2015).

The potential of the sector to contribute to improved food security and nutrition is all the more important as many of the jobs provided are located in poor and remote areas of the country. Across low and lower-middle income countries, food insecurity is predominantly rural, and smallholder farmers are particularly afflicted (OECD, 2013). This pattern is repeated in Indonesia: about two-thirds of Indonesia's poor live in coastal communities (BPS-Indonesia Statistik, 2008). A number of these communities are highly dependent on the sector. Fisheries also provide part-time and seasonal jobs that complement other activities.

Over the past decade, aquaculture has become the main driver for fish supply growth in Indonesia, expanding at an average rate of over 23% per year between 2005 and 2014 (FAO, 2016b). As a result, the share of aquaculture in total seafood production volume in Indonesia rose from just over 10% in 1960 to almost 50% in 2014 (rising to almost 70% if seaweed production is taken into account) (FAO, 2016c). The value of Indonesia's aquaculture production was USD 10.6 billion in 2014 (FAO, 2016c), contributing over half the total value of seafood produced in the country (OECD, 2016). According to the *OECD-FAO Outlook 2015-2024* projections, the strong growth of the aquaculture sector will be maintained in the coming decade, with production peaking at about 6 million tonnes by 2024 (excluding seaweed) while capture production, on the other hand, is expected to increase only marginally (OECD-FAO, 2016; Figure 7.6).

Box 7.1. Marine fisheries and aquaculture production in Indonesia

Indonesia has become the second-largest producer of fish, crustaceans and aquatic plants in the world by volume, after China, mainly thanks to a boom in aquaculture production: capture fisheries production reached 6.2 million tonnes (Figure 7.2) and aquaculture 15.6 million tonnes in 2015.

Figure 7.2. Fisheries and aquaculture production, 1950-2014

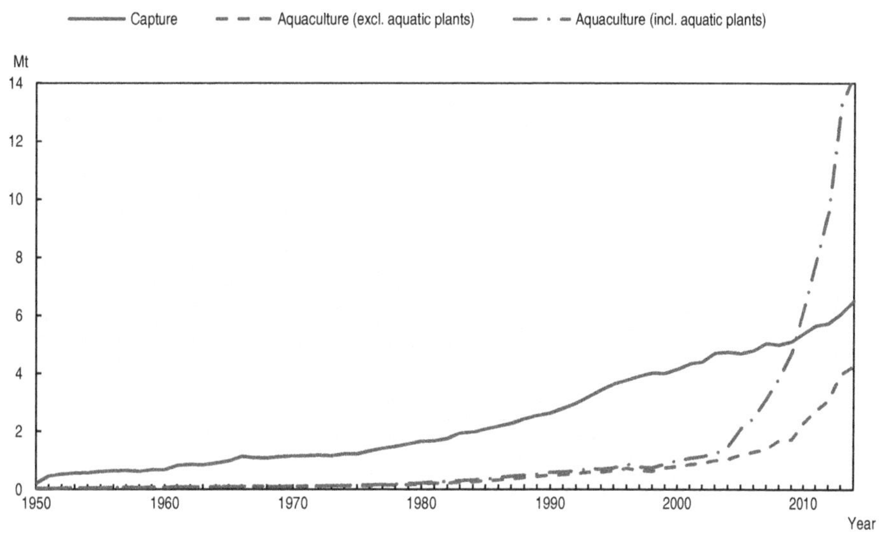

Source: FAO (2016a), *Food Balance Sheet Indonesia*, http://faostat.fao.org/default.aspx.

Artisanal fishing boats comprise the vast majority of the Indonesian marine fleet. They target mainly near-shore species such as reef fishes (grouper and snapper) and demersal fishes (scabbardfishes and hairtails), as well as small pelagic fishes (scads, jacks and mackerel). The artisanal fleet accounts for somewhere between a quarter and a third of the total catch, depending on data sources. Larger industrial vessels mainly target tuna and shrimp, as well as demersal fish and reef fish to a lesser extent. While they comprise only about 5% of the fleet, they account for the remaining vast majority of the total catch (Figures 7.3 and 7.4).

Figure 7.3. Composition of capture fisheries (marine and inland) by main species, 2014

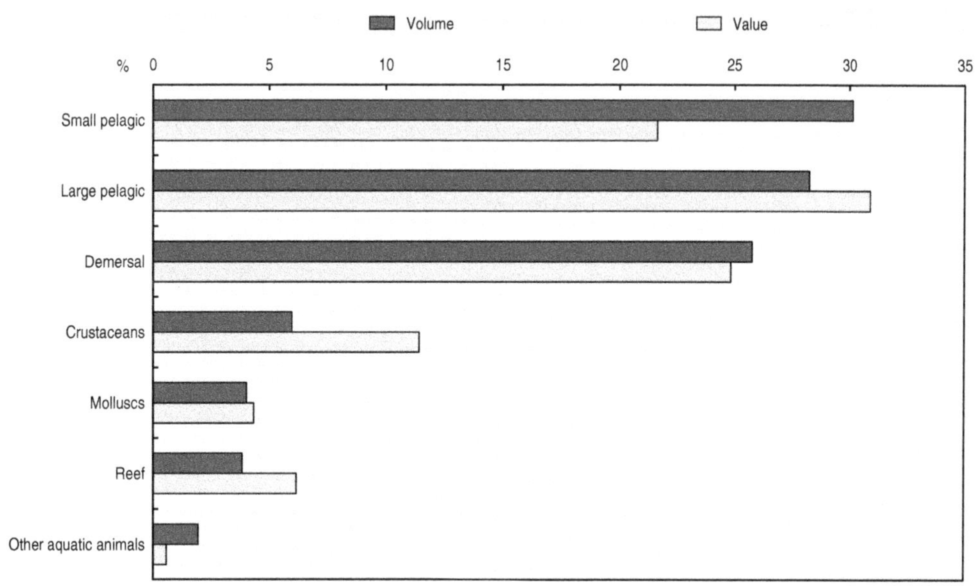

Source: MMAF (2015).

Figure 7.4. Capture fisheries by fishing segment, 1950-2010

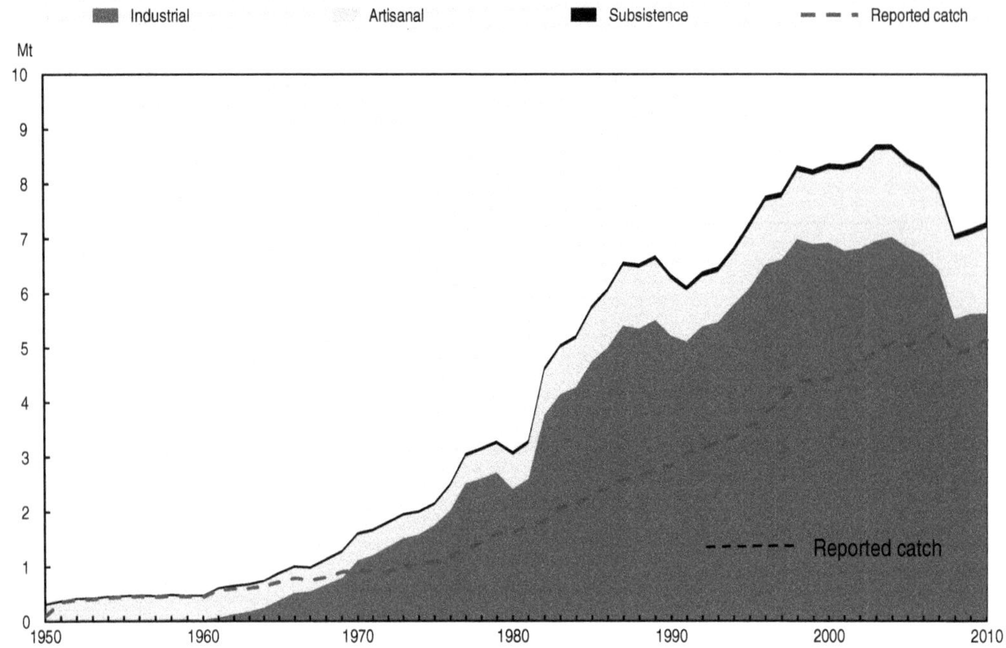

Source: Pauly and Budimartono (2015).

Aquaculture production takes place in a large number of ecosystems, including marine, brackish and freshwater, using nets, pens and ponds. By total volume, seaweed is the largest output, accounting for almost 70% of volume in 2014, although only 16% of value (Figure 7.5). Omnivorous and herbivorous species account for the majority of animal production, with carnivorous species accounting for only about 15% of production in 2014 (MMAF, 2016; Figure 7.5). Shrimp is the highest-value product, accounting for about a quarter of total value. Freshwater production in lakes and reservoirs is also significant, with tilapia and milkfish as its main products. While the bulk of aquaculture production originates in small-scale traditional farms, shrimp production is highly concentrated in large-scale production centres.

Figure 7.5. Composition of aquaculture production by main species, 2014

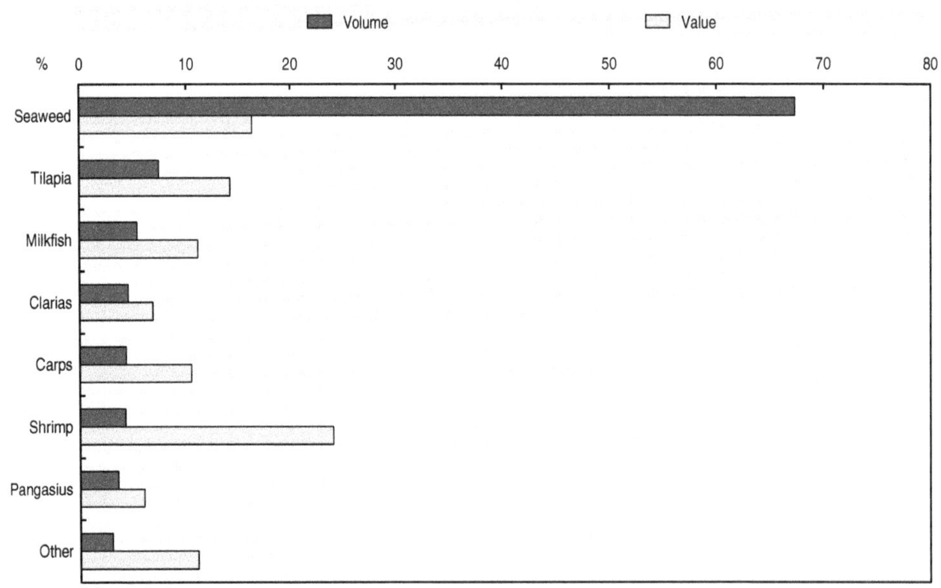

Source: Tran et al. (2017).

Figure 7.6. Projected fisheries and aquaculture production (excluding seaweed), 2015-24

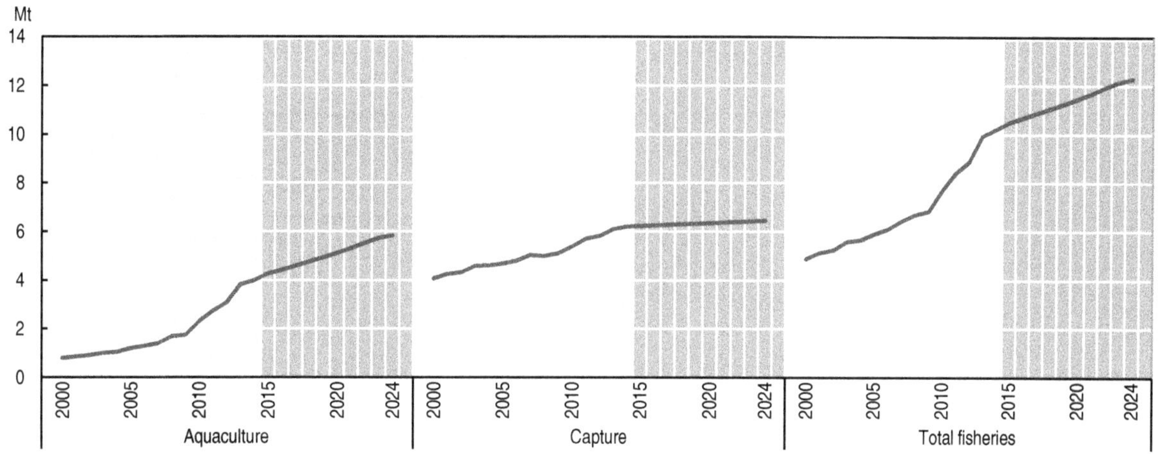

Source: OECD-FAO (2016).

Government focus on food sovereignty

Fisheries and aquaculture policies in Indonesia can be seen as the product of two main political orientations. First, like other governments in the ASEAN region, Indonesia has approached the problem of food security and nutrition primarily from the perspective of increasing the volumes of food available, coupled with the objective of relying on domestically-produced commodities as much as possible (OECD, 2016).[6] In this context, the main objective for the fisheries and aquaculture sector is to increase domestic production in order to increase the availability of seafood in the country, as well as to support the livelihoods of artisanal fishers and aquaculture producers.

Second, President Joko Widodo, known as President Jokowi, who was elected in July 2014, has announced his intention to position Indonesia as a regional maritime power, making maritime and fisheries policy a central priority for his administration.[7] In 2015, the budget of the Ministry of Marine Affairs and Fisheries (MMAF) was almost doubled, and unprecedented powers were given to its minister, Susi Pujiastuti, who is both in charge of maritime and fisheries policies and head of the newly created Presidential Task Force to Combat Illegal Fishing. Territorial issues feature prominently on the presidential political agenda, with implications for fisheries ranging from increased support for economic activities in the outer islands bordering neighbouring countries, to the promotion of a domestic ship-building industry.

The MMAF Strategic Plan 2015-2019, which provides strategic direction on the main issues facing the sector, embodies these political aims. While the plan describes the governmental strategy as being organised around three pillars: sovereignty, sustainability and prosperity, it focuses mainly on increasing domestic seafood production and contains annual production and revenue generation targets.

Under the Strategic Plan, government action focuses on the following:

- a strong and unprecedented stance against illegal fishing, which involved a doubling of the budget allocated to surveillance between 2014 and 2015 (Figure 7.7)
- the prioritisation of artisanal fishing, by closing Indonesian waters to all foreign-built industrial vessels and providing support to modernise the artisanal fleet
- the provision of support to artisanal aquaculture expansion, by creating artisanal co-operatives and developing the production of feed with locally-sourced ingredients, in line with objectives pursued by the previous administration's Blue Economy programme initiated in 2014
- the introduction of restrictive trade and investment policies, notably in the shipping industry, to protect domestic industries, producers and fishers from competition.

Figure 7.7. Budget allocation for the main entities of MMAF, 2014 and 2015

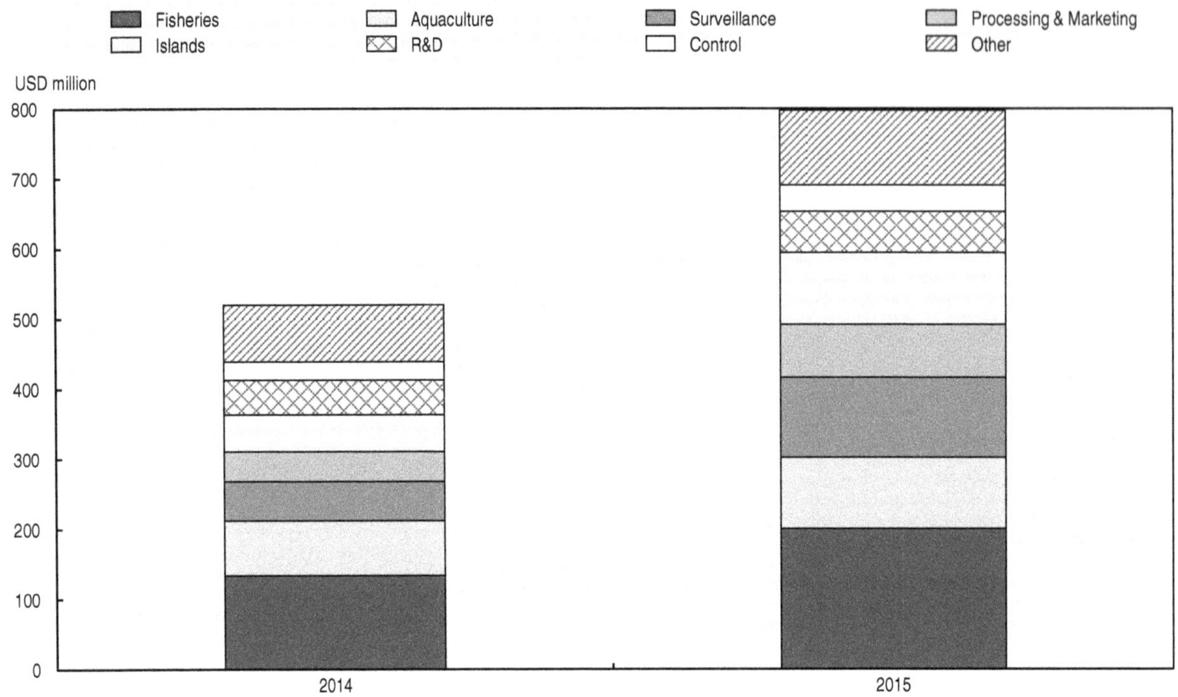

Note: MMAF: Indonesian Ministry of Marine Affairs and Fisheries.
Source: MMAF (2016).

7.2 Sustainable management of natural resources for future food security

Sections 7.2 and 7.3 consider the policies that are currently in place in Indonesia in light of the priorities identified by the OECD as key for improving food security, with a view to improving the potential of Indonesian fisheries and aquaculture policies to contribute sustainably to food security and nutrition.

OECD work on food security (OECD, 2013; Brooks and Matthews, 2015) and sustainable fisheries management (OECD, 2015) has highlighted three important policy pathways to improve food security and nutrition through better fisheries and aquaculture policies:

- First, ensuring the sustainability of resources on which the sector relies is a pre-requisite to any future contribution of the sector to food security and nutrition.
- Second, as the principal cause of food insecurity remains poverty, rather than insufficient food availability, income growth is central to lasting reductions in hunger. Governments should aim to unlock the potential of aquaculture and fisheries to sustainably generate the *incomes* needed to ensure food security, rather than support domestic food production.
- Finally, the role of trade in increasing the availability of seafood and lowering its price for consumers needs to be better recognised and supported, as the benefits of protection for fishers and aquaculture producers are too rarely compared with their cost for consumers and their effects on the food security of the poorest.

Fisheries and aquaculture policies in Indonesia can thus contribute to improved food security and nutrition if they focus on three key objectives: sustainable management of natural resources for the future, increased economic opportunities for fishers and aqua-farmers, and improved consumer access to seafood. This section focuses on the sustainability objective, while Section 7.3 addresses socio-economic objectives related to income generation and access to food.

Environmental impact of seafood production

Seafood production in Indonesia is achieved at high cost to the country's natural resources, creating risk that future production cannot be sustained. This is an area of concern, as ensuring the sustainability of the natural resources on which the sector relies is a prerequisite for any future contribution of the sector to food security and nutrition. Improving the health of fish stocks and coastal and inland water ecosystems has the potential to improve food availability, lower prices for consumers and increase incomes of fishers and aqua-farmers after transitional adjustments.

Fully-exploited and over-exploited fish stocks

Increased demand for seafood from a growing population, as well as the development of industrial fisheries, trawling in particular – which started in the early 1960s and grew rapidly – are responsible for a strong increase in pressure on Indonesian marine fishery resources in recent decades (Pauly and Budimartono, 2015). According to the findings of the Indonesia National Committee on Assessment of Fisheries Resources, the vast majority of Indonesian fisheries are defined as fully-exploited or over-exploited (Table 7.1; MMAF, 2016). The highest rate of over-exploitation is found for shrimp species, Indonesia's most valuable catch. Signs of declining catch per unit of effort (CPUE) have been reported for tunas, small pelagics, shrimp and coastal reef fish (DAFF, 2011), which indicates that the resource base is already affected. Exploitation rates vary across the waters of Indonesia, and some opportunity for expansion of catches in the eastern part of the country may exist, notably for small pelagics. Of all large tuna species, only skipjack remains somewhat underexploited (DAFF, 2011). According to calculations, only 15% of Indonesian landings come from under-exploited stocks. Another 32% of landings come from unassessed stocks, while 40% come from fully-exploited stocks and 11% from stocks in poor condition (CEA, 2014).

Table 7.1. Status of fisheries resources by major species category and Fishery Management Area (WPP), 2016

WPP	Shrimp	Demersal	Small pelagic	Large tuna species	Squid
Indian Ocean					
571	O	O	O	F	F
572	O	F	F	O	U
573	O	F	F	F	O
Pacific Ocean					
711	O	F	O	U	O
712	O	F	F	O	O
713	O	O	F	F	O
714	F	F	F	F	F
715	O	F	O	O	O
716	F	U	U	F	O
717	U	U	F	F	F
718	O	O	F	F	F

Notes: Status is coded as follows: U = under-exploited, F = fully exploited, O = over-exploited. WPPs (*Wilayah Pengelolaan Perikanan*, or fisheries management areas) are numbered from West to East.
Source*:* MMAF (2016).

Threatened ecosystems

Indonesia's coral reef and mangrove ecosystems are also threatened by seafood production. The Indonesian coast is home to around one-fifth of global coral reefs and more than half of global coral species, as well as 3 million hectares (ha) of mangrove forests, representing almost a quarter of mangrove forests in the world, with the greatest diversity of mangroves and sea grasses. Indonesian marine and brackish ecosystems offer a unique range of services, both to capture and aquaculture activities and many other sectors of activities. The vast majority of capture fisheries production comes from marine fisheries (Tran et al., 2017) and 80% of aquaculture production is from coastal areas (MMAF, 2016).[8] The coral reef ecosystem, apart from its function as marine biota, is also a carbon absorber, sea wave breaker and fish producer. The mangroves offer an important breeding ground for thousands of species as well as numerous products and ecosystem services, including firewood, recreation and tourism, erosion control, climate regulation, and protection against extreme weather events (Tran et al., 2017).

However, the resources of these ecosystems are being damaged by fishing techniques and overfishing, aquaculture environmental externalities, waste disposal, sea sand mining, and other destructive practices. In 2007, only a quarter of reefs were in healthy or excellent condition, with about a third in bad or very bad condition and the remaining in moderate condition (Burke et al., 2011). Between 1975 and 2005, Indonesia lost 40% of mangroves due to conversion into fish ponds, salt production, pollution or palm plantation for oil production (Adhuri, 2016). In particular, 95% of the mangrove forests and estuarine creeks that once lined the coast of Java have been developed into aquaculture ponds (Inside Indonesia, n.d.). Inland water bodies are also at risk, notably due to eutrophication around cage and pen farms. In some areas of Java, aquaculture in lakes is competing with freshwater for human use in urban areas (Rimmer et al., 2013).

Policies against illegal fishing

The Government of Indonesia has identified illegal, unregulated and unreported (IUU) fishing as the most serious challenge to fisheries sustainability, as well as a key economic and maritime security issue, as IUU fishing has created a major strain both on fish availability and on remuneration from fishing, and caused breaches of sea sovereignty.

IUU fishing places marine resources under significant pressure in Indonesia, where the practice is thought to account for up to a third of total catches. Reconstructing catch data for Central and Eastern Indonesia from 1950 to 2010 by combining official data with published and anecdotal knowledge on IUU fishing into coherent time series, Pauly and Budimartono (2015) find a total catch that is 39% larger than that reported by the FAO on behalf of Indonesia. They suggest that, in addition to illegal fishing, notably by foreign vessels, non-reported catches by domestic trawlers may also be significant, particularly in remote eastern regions, where numerous small landing sites mean that a large proportion of the catch is not included in official statistics. Eliminating IUU fishing could thus substantially reduce the pressure on resources, providing that IUU fishing is not entirely replaced by legal fishing.

In 2014, a Presidential Taskforce to Prevent and Combat IUU Fishing (hereafter "Taskforce") was established by President Jokowi to lead efforts to combat IUU. The Taskforce focuses on illegal fishing and associated crimes, as well as under-reported fishing, but is not concerned with unregulated fishing, which in fact encompasses much of the artisanal fishing activities over which MMAF has not yet implemented controls and restrictions. The Taskforce has built institutional co-operation by bringing together the navy, the police, tax authorities, the maritime authority and the Attorney General's office under the responsibility of the Taskforce's Minister Pudjiastuti. The budget allocated to surveillance of marine and fisheries resources doubled between 2014 and 2015, using funds previously allocated to fuel subsidies for the sector which had been reduced in late 2014 (CEA, 2014). This budget increase was the strongest across all budget spending categories of MMAF, which on average saw its allocation increase by over 60% between 2014 and 2015 (Figure 7.7).

Soon after the Taskforce was established, MMAF decided to close Indonesian waters to industrial fishing boats. The ministry took this decision after observing that the vast majority of industrial vessels fishing in the Indonesian exclusive economic zone (EEZ) operated in fraudulent circumstances, and that closing the EEZ would permit it to perform an in-depth audit of the situation at sea. Regulation 56/2014 introduced both a permanent moratorium on fishing by ex-foreign vessels operating within the EEZ – effectively shutting down access to the EEZ by the largest vessels, as most of these were foreign-built as well as operated by foreign companies and crews – and a six-month moratorium on fishing licences for all boats measuring more than 30 gross tonnage (GT). This second restriction was later extended until the end of 2015, with Regulation 10/2015. The compliance audit performed by the Taskforce in 2015 found that all of the 1 132 industrial vessels that were licenced in 2015 (all ex-foreign vessels) violated Indonesian laws and regulations in some way. Double flagging was a serious problem, as was marking down the gross tonnage of vessels. The government also estimated that between 5 000 and 10 000 vessels operating in Indonesian waters were using fraudulently copied licences; in some cases, one licence was being used by up to ten different vessels.

In addition, the Taskforce found that smuggling and other illegal activities, including human rights abuses and trafficking, occurred regularly (Santosa, 2016). A lack of compliance regarding logbook reporting was also noted – with less than 5% of the 245 fish ports integrated with logbook information systems – as well as the use of non-designated fish landing ports to land catch in violation of fishing licence requirements. Many falsified data entries for fish capture were also found. On this basis, by August 2016, 236 vessels had been seized and sunk, resulting in extensive media coverage.

Furthermore, the Taskforce noted that government officials were limited in their ability to detect, respond and punish violations, particularly due to a lack of vessel monitoring system (VMS) transmitters, on-board observers and patrol boats that covered the EEZ. Corruption meant that inspectors could also too easily be dissuaded from registering problems, meaning that sanctions were not acting as an effective deterrent. In addition, corporate criminal liability was rarely imposed, and serious violations were subject only to administrative sanctions (Santosa, 2016).

The Taskforce therefore concluded that post-landing regulations, such as those of the FAO Port State Measures Agreement (which Indonesia subsequently ratified through Presidential Decree 43/2016) were not sufficient, as most illegal activities took place at sea.

The Indonesian government has decided to re-authorise industrial fishing, but – in order to prevent IUU – under more restrictive conditions than had previously been in place. First, Regulation 57/2014 prohibits all transhipment at sea. Second, since December 2015, all boats above 30 GT have to use VMS (Ministerial Regulation 23/2014), and VMS data is shared with Global Fishing Watch, a data-processing system developed by a partnership between Google, SkyTruth and Oceana, which will help to monitor commercial fishing activity (and make information public). Third, MMAF is working to improve the vessel registration and fisheries licence system (Ministerial Regulation 30/2012 is currently being revised), while improving catch reporting documentation. Under the new regulation in preparation, licences will be delivered to domestically-built vessels of up to 200 GT, for one year, following a due-diligence procedure.

According to Minister Pujiastuti, the establishment of a "one-roof" enforcement system with the creation of the Taskforce, backed up by strong political will from the President as well as a strong communications strategy, have proved decisive in tackling the corruption that pervaded the sector and the political system, and allowing the adoption of policies that effectively closed Indonesian seas to a significant section of the fleet.

The government's efforts have been successful at making illegal fishing an inherently riskier activity in Indonesia, but the impact on food security remains uncertain. Neighbouring countries are considering implementing similar policies because illegal fishers have relocated their activities to their waters after leaving Indonesian waters. The actual extent of progress is, however, still difficult to estimate. Joint work between the sustainable fisheries group at UC Santa Barbara and MMAF estimates that the measures taken in 2015 led to a reduction of fishing effort of about 35%. However, this effort is believed to have since started to recover (MMAF, 2016). Minister Pujiastuti often refers to anecdotal indications that the availability and size of fish caught by artisanal fishers has increased, which suggest that stock health and fish availability is improving for these producers. Evidence is still scarce to support these assertions, but simulations indicate that there is potential. Catch of skipjack tuna could, for example, increase by 25% by 2035, if policies against IUU fishing were efficient and coupled with effective management of domestic fisheries, compared with an expected *decrease* by 81% in the absence of such measures (MMAF, 2016).

While the continued fight against IUU fishing is likely to help protect those Indonesian fisheries exploited by industrial boats, the direct link drawn between a decrease in IUU fishing activities and the increased availability of fish for artisanal fishers in coastal areas may only be true for certain species. Coastal and territorial waters are connected, and some species, like shrimp, spawn in the deep sea but are caught in coastal waters by both artisanal and industrial vessels. However, tuna, for example, spawns in Indonesian archipelagic waters before migrating to the Pacific, where they are caught by industrial ships. Reef fish such as groupers and snappers, which are mainly caught by small-scale boats, stay in archipelagic waters and do not swim beyond a few kilometres of the shore. Reef fish stock health and the availability of large tunas in coastal waters are thus unlikely to substantially increase with the measures taken to abolish industrial illegal fishing.

Strengthening domestic fisheries management

An overly open existing management system

To ensure the sustainability of a broader range of fisheries, domestic artisanal fisheries also need to be better monitored and regulated. The current system of controls indeed permits relatively open access to Indonesian waters for artisanal fishers because a) the large unlicensed share of the artisanal fleet is not subject to any controls or monitoring, and b) the types of restrictions used to manage the fleet under licence are not appropriate to ensure the sustainable management of fisheries. The Indonesian government recognises that the existing framework is insufficient for the sustainable management of domestic fisheries. Minister Pujiastuti has announced that the fight against illegal fishing was only a first step, and that the government now needs to tackle the politically and practically more difficult challenge of addressing unreported and unregulated fishing.

Natural resources are jointly managed by the central government, the provinces and districts. Since the adoption of Law No. 23/2014 on local governance, responsibilities for resource management have been decentralised according to geography and types of operations. MMAF is directly responsible for managing archipelagic and EEZ water resources, the 33 provincial authorities manage territorial water resources, and the 250 district authorities manage coastal resources within 4 miles of the coast. With respect to boat licences, vessels between 5 and 30 GT can thus obtain a licence from provincial authorities, while those of more than 30 GT and ex-foreign vessels of any size require a licence issued by the central government (the latter will no longer be granted licences under current rules). Boats under 5GT are not required to obtain a license.

This licencing system is insufficient to manage fishing, as the Indonesian fleet is largely dominated by small-scale boats of under 5 GT operating within four nautical miles of the coast. More than 500 000 such boats account for approximately 35-40% of landings (DAFF, 2011). These boats continue to expand in number, reaching about 30% more in 2012 than in the early 2000s (Stobutzki et al., 2014). Putting in place a simple and inexpensive system for registering such boats and monitoring their catch is a priority.

In addition, there are few obligations attached to the licences, particularly with regard to limiting the quantities fished. No output controls such as total allowable catch limits (TACs) or quotas are currently in use. MMAF has indicated that a roadmap is being developed by the government to introduce quotas, capacity rules, closed seasons and zoning laws. However, the MMAF Strategic Plan 2015-2019 lacks precision as to how this will be undertaken. Although the document was issued as a Ministerial Decree, which makes it legally binding, policy implications remain uncertain, as no clear definitions or binding indicators are attached to key objectives, notably in terms of sustainability (CEA, 2014).

Additional instruments that can be used to regulate fisheries resources are defined in Law No. 45/2009, which amends Law No. 31/2004 on fisheries. These include minimum size requirements – the latest of which, Ministerial Regulation 01/2015, concerns lobsters and crabs, which were harvested for cultivation in aquaculture facilities – as well as restrictions on allowable gear, the most recent being the moratorium on the use of seine nets and trawls imposed by Ministerial Regulation 02/2015. Particular areas are also closed to fishing, such as the Banda Sea, where fishing was banned by Ministerial Regulation 04/2015, and fishing seasons are defined for some species.[9] These fishing regulations, however, tend to be loosely implemented, with limited policy efforts in place to rectify this situation.

Measurable longer-term objectives needed

A varied set of controls based on effort and restrictions on the areas where fishing is permitted, such as those currently used in Indonesia, leads to opacity and high transaction costs. Alternatively, the management of resources should be shifted towards fisheries management and rebuilding plans – built around medium- to long-term horizon measurable objectives – that first target overfished species (Box 7.2).

Box 7.2. Designing rebuilding and management targets

A key issue for any system of regulation is whether fishers have an incentive to conserve fish stocks if they have no guarantee that other fishers will do the same. To be considered acceptable, regulations therefore need to be clear so that fishers can be confident that institutions in charge can, and will, enforce regulations.

Understanding the complexity of tasks and the time horizons required to measure the success of programmes is therefore key; this makes clarity regarding the objectives pursued crucial. There are many examples of rebuilding programmes that have failed to reach targets, and it is vital that programmes include options in the event that the stock does not respond in the way predicted. Sometimes, this is a result of data deficiencies, changes in understanding of the biological characteristics of the stock, refinements in the models used to predict rebuilding pathways, or just broader environmental changes such as climate change. The lesson learned in most cases is that plans are likely to take longer than originally anticipated. Communication on these difficulties is important for fisheries policy-making institutions in order not to lose credibility.

Traditionally, fisheries managers and scientists providing advice on stocks have focused on the maximum sustainable yield (MSY) as an appropriate management target. While stock management is a necessary objective of fisheries policy makers, maximising social welfare is the ultimate goal. Addressing risk and uncertainties should also be explicitly considered when designing management plans. Stock status targets other than MSY are thus increasingly being used, for example, to maximise profits instead of production under stock conservation constraints. Alternatively, policies may aim to rebuild stocks, reduce the risk of collapse or impose particular social or environmental norms.

In addition, rebuilding and management plans should not only be based on biological targets but incorporate social and economic principles throughout the design and implementation process in an integrated fashion, as opposed to sequentially or in isolation (OECD, 2014). Plans ideally need to address direct fisheries adjustment, local employment (including those in the processing and marketing sectors), regional impacts or the need for alternative employment and livelihood opportunities, as well as food security.

The OECD also calls for the adoption of the now globally-recognised Ecosystem Approach to Fisheries (EAF), as far as possible. The EAF is defined as "striv[ing] to balance diverse societal objectives, by taking into account the knowledge and uncertainties about biotic, abiotic and human components of ecosystems and their interactions, and applying an integrated approach within ecological meaningful boundaries" (FAO, 2003). This approach implies that decisions are made using a comprehensive, inclusive framework for all living resources, rather than reacting to the status of a single stock of fish. The main obstacle to implementing the EAF is that it substantially increases the level of information and analysis required, and this level of information is not yet available in Indonesia.

Law No. 31/2004 paved the way for the development of fisheries management plans on the basis of 11 fisheries management areas. Plan development is the responsibility of the central government, while their implementation falls in part on provinces and districts. To date, however, only three plans have been adopted, and 12 years later, the remaining eight are still at the drafting stage. According to Sutinen (2013), despite aiming to follow FAO recommendations to use an ecosystem-based approach to management, the plans are not tailored to protect or rebuild specific stocks, largely because no quotas or TACs have been adopted.

A major constraint on the use of output-based management instruments in Indonesia is the lack of accurate data on which biomass and optimal capacity calculations depend, particularly catch data. While stock status assessments should be undertaken annually in the 11 management areas by MMAF and associated research centres and universities, stocks were not evaluated between 2011 and 2016.

The task of producing up-to-date and accurate data is particularly difficult in the Indonesian context: catches are extremely diverse, with each family in the reconstructed catch, such as "tuna", representing dozens of species, and each species representing hundreds of stocks scattered over thousands of kilometres of complex geography. Therefore, a stock-by-stock or even species-by-species study of these catch data is complicated, with the exception of a few species of tuna, while multispecies stock assessments have proven notoriously difficult (Pauly and Budimartono, 2015). Alternatively, assessment of multispecies fish stocks should focus on changes in catch levels and composition, with rapid changes of these indicators providing strong evidence of a lack of sustainability (Pauly and Budimartono, 2015). Tracking catch levels and compositions in turn requires accurate catch data from all fisheries, including from small-scale boats, illegal fisheries, and discards, which are usually not included in catch statistics. The fact that most artisanal boats do not need a licence (if they are under 5 GT), and that MMAF does not receive information from provincial authorities on the number of boats between 5 and 30 GT that they licence, is a major constraint to adequate catch monitoring. Scientific data gathering could be complemented by the collection of fishers' knowledge on the health of stocks at district and provincial levels. In addition to providing a low-cost means of information gathering, the use of fishers' knowledge also has the potential to increase the acceptability of policy decisions.

Designing comprehensive management and rebuilding plans, with a view to adopting the Ecosystem Approach to Fisheries (EAF) in the longer term, also requires integrating biological information with information on the socio-economic characteristics of fisheries. Matching socio-economic information on fisheries with biological information on resources would be useful for identifying overfished stocks of greatest significance in terms of income, employment or food consumption in order to better target rebuilding efforts. In this perspective, the government is currently working jointly with the World Bank on the creation of a centralised data collection system which would replace the more than 200 datasets that are currently in use at MMAF.

Setting up transparent and inclusive decision-making processes to set TACs and quotas is also important to avoid conflicts of interest among users and policy makers while limiting transaction costs. Making clear how scientific information will be collected, how it will be used in the decision-making process, and making this information public, for example, can help build accountability and trust in the system. Working on the governance of the sector should also be a priority in order to build faith in the management system so that fishers – and stakeholders more generally – will understand that regulation is in their best interest, and that it will be enforced (Box 7.3). This would help to create confidence among investors that management policies will be effective in the long term and that the sector is a sustainable market to invest in.

Box 7.3. Legitimate and acceptable governance

Governance of fisheries and aquaculture, and conservation of the ecosystems that sustain these, can be understood as a process through which institutions, governments, and stakeholders in the sector elaborate, adopt and implement policies and management strategies.

The impact of governance on fisheries and aquaculture management, and ultimately on their performance, is increasingly recognised. Governance has an impact on both the nature of policy decisions and their acceptability to different stakeholders. An OECD report on rebuilding fisheries (OECD, 2012), for example, underscores how good governance is a key element to ensuring the success of rebuilding plans. While there is no optimal arrangement for fisheries governance frameworks, a few characteristics are important ingredients for success, notably legitimacy and acceptability (Menard, 2014):

Legitimacy relates to the characteristics of the institutional arrangements through which policies are elaborated and implemented: the transparency of procedures in the decision-making process, mechanisms guaranteeing the accountability of institutions with respect to decisions made, and room for scientific information and advice.

Acceptability concerns the perception by stakeholders of the capacity of institutional arrangements to deliver fair decisions. Acceptability relates to: the degree of delegation in decision-making processes; the space for stakeholders to share available information, express various positions including dissent, and be confident that these voices are explicitly part of the decision-making process; the capacity to integrate or change the norms and beliefs of stakeholders; and the existence of mechanisms of appeal. Building inclusive institutional arrangements oriented towards raising broad consensus is key to acceptability.

Towards more sustainable aquaculture production

Limiting environmental externalities of aquaculture production to preserve both the ecosystems on which the sector relies, and the services that these systems supply to many other sectors, requires both adequate regulation and public support to encourage adoption of best practices. In a report on green growth in fisheries and aquaculture, the OECD highlights that the best response to externalities depends on the direction of the externalities caused by aquaculture production (OECD, 2015). While self-interest should lead producers to respond to externalities that affect their own production (local water pollution harming productivity, for example), the more the externality impacts other aqua-farmers or other ecosystem users, the more that public regulation is necessary, as well as, potentially, support to facilitate the transition to sustainable practices. The industry thus has an interest in addressing externalities, as the negative effects reduce profitability and demand in the long run (Asche, 2011). In Indonesia, however, this will mostly hold true for intensive systems, where production must meet the standards expected by export markets and where sufficient capital exists to invest in improvements. Extensive production systems, on the other hand, supply local markets where quality is less an issue, and which are operated by agents with limited resources and, sometimes, limited capacity to switch to more sustainable production methods.

Aquaculture production externalities

Environmental externalities related to aquaculture production fall into two main categories:

- establishment impacts, which arise from the conversion of one type of land to another, such as from agricultural land or coastal habitats to ponds; and
- operational impacts, which arise from the day-to-day running of an aquaculture business.

The extensive nature of most aquaculture production in Indonesia means that, to date, Indonesia has mostly suffered from establishment impacts, which include the removal of pre-existing habitat as well as long-term impacts on water and land quality. Indeed, traditional farms, where extensive production is practiced, have relatively few operational impacts, as there is little water exchange and almost no use of fertilisers or feed inputs due to their prohibitive cost for small-scale aqua-farmers. However, clusters of small farms can have a significant impact within a watershed or enclosed waterbody, when concentration levels exceed the carrying capacity of the ecosystem.

On the other hand, most of Indonesia's shrimp production – its most valuable seafood export – comes from modernised farms. Semi-intensive or intensive farms have greater operational impacts, such as disease and nutrient release from aquaculture ponds and eutrophication near cages due to limited water circulation, possibly with implications for other sectors such as tourism in coastal areas or agriculture inland (Rimmer et al., 2013). Pellet feed and aeration are indeed necessary to support high stocking densities, as are fertiliser, chemical and probiotic use to limit disease. These interventions produce high levels of organic waste, and may have around twice the environmental impact per shrimp than those grown using less-intensive methods (Cao et al., 2011). Further operational impacts include the health and environmental effects of antibiotic use. Although antibiotics are usually given to prevent or combat disease, their use poses risks due to the high stocking density in intensive systems, which can result in leeching into local water supplies or residues found in the finished product. As intensity increases, Indonesia will become more vulnerable to both an abrupt shock – such as disease – and a slower degradation of the environment, leading to an ecosystem collapse. Getting aquaculture right at the policy level is therefore critical for the sustainability of future production.

Foreseen further growth of shrimp production also raises concerns for the sustainability of forage fish stocks. Carnivorous species such as shrimp are very demanding in fish meal and oil when raised in intensive conditions. Feed accounts for over 50% of production costs in some cases (Rana, Siriwardena and Hasan, 2009). Given the insufficient regulation of capture fisheries, this has the potential to have negative effects on stocks and on the production of edible supplies from marine capture fisheries. Substitutes for fish meal and oil are needed as the sub-sector expands, requiring research and innovation.

Strengthening aquaculture management

The regulation of aquaculture production is a shared responsibility of MMAF and the Ministry of Environment and Forestry, which is responsible for aquaculture environmental impact control. Three main policies are used: mandatory *ex ante* impact assessment, spatial limitations and restrictions on production practices.

The Environment and Forestry Law 23/2009 requires environmental impact assessments to be performed for the creation of any aquaculture development with an area of 50 ha or more for shrimp ponds, more than 500 floating net cages on lakes or reservoirs, or floating net cage installations in sea water covering more than 5 ha or consisting of over 1 000 cages. For smaller-scale farms, installation is regulated by the obligation, for all aqua-farmers, to request an aquaculture business licence directly from the ministry or from regional authorities, depending on the scale of the production unit. Nevertheless, a major weakness of this system is that, while the impact assessment of big aquaculture plants should be followed by regular monitoring, further evaluation is currently only completed in the case of a complaint from the local community.

In addition, a number of spatial limitations apply to all producers. The main regulations include Ministerial Decree KEP 28/MEN/2004, which regulates shrimp culture in brackish water ponds; Regulation 5/2014 on water quality standard of effluents for aquaculture ponds; and Law 26/2007 concerning spatial planning of land use. These regulations, for example, do not permit brackish water farms within a 100-metre green belt adjacent to coastal waterways, in order to protect mangroves (Rimmer et al., 2013).

However, spatial planning measures are undermined by lack of data. While careful site selection and spatial restrictions should be based on estimation of the carrying capacity of the environment in order to ensure that production does not exceed what the local ecosystem can support (Santosa, 2013), in practice, carrying capacities are administratively stipulated on the basis of poor information. This is because the reliable data necessary for trustworthy modelling output is lacking. Since spatial planning on both land and at sea has been decentralised to provincial and district levels (with Laws 54/2003 and 32/2004), no clear mandate for information gathering and monitoring has been given. As a result, neither the national government nor the local governments appear to be collecting information on the location of activities and monitoring of their impacts and performance, or on the use and performance of spatial planning at the local level. Such lack of information also means that best practices in terms of sustainability cannot be adequately identified and disseminated by extension services. Clear allocation of responsibilities between the central, provincial and district authorities, and improved co-operation and information sharing between these administrations would greatly improve the capacity of public authorities to promote the sustainable use of resources and help limit the cost of public action.

Finally, to address operational impact, the use of fish drugs is regulated by Ministerial Decrees 52/KEPMEN-KP/2014 (drug classification) and Ministerial Regulation PER.04/MEN/2012 (registration for use). The government has also recently established, with the support of the European Commission, a National Plan for Residue Monitoring to control the presence of antibiotics and other contaminants (on the basis of Ministerial Regulation 39/PERMEN-KP/2015). Presence of contaminants, notably chemical residues, has been a major issue for the export of Indonesian seafood, notably to the United States, where they have faced high rejection rates. Following a recommendation of the World Organisation for Animal Health (OIE), the government is currently developing a programme for monitoring water quality and antimicrobial resistance in fish and water ecosystems.

Fish disease monitoring and prevention has been relatively successful to date. Unlike other Asian countries such as Thailand or Vietnam, Indonesia has not experienced the wholesale decimation of the domestic aquaculture industry from disease. This is largely thanks to the amount of extensive production. The fish disease monitoring programme, based on Directorate Decrees 163/2014 and 11/2016, includes both zoning restrictions and controls, and communicates information on these online. Import restrictions on live and frozen shrimp from countries affected by AHPND (Acute Hepatopancreatic Necrosis Disease) also apply in accordance with the prescriptions of the OIE.

Certification schemes for greener production

The Indonesian government uses certification schemes to promote greener production processes and standardise inputs and practices for supply chain management by disseminating technical practices to producers in a structured and formalised fashion (Kusumawati and Bush, 2015). Sustainable seafood certification is indeed a useful means of communication about the sustainability of production practices of both capture and farmed seafood. Through the creation of market demand for sustainable practices, it creates additional incentives for investment in improved fishing and aquaculture practices (Potts et al., 2016).

The Indonesian government has introduced its own certification scheme, IndoGAP, based on biosafety, food safety and the environment. The standard certifies farmed carps, catfish and tilapia, brackish water shrimp and the marine farming of seabass, groupers and seaweed. MMAF spent over USD 580 000 in 2015 on providing assistance to certify hatcheries, and around USD 50 000 to assist freshwater and brackishwater farms certification. In the future, MMAF will need to outsource auditing to an independent third party in order to be fully compliant with international norms (FAO, 2011).

Private sustainability standards and certification schemes also operate in Indonesia, and have greater breadth and depth than national schemes, particularly for shrimp aquaculture (Gutierrez et al., 2016). Certified products include extensively-farmed organic shrimp, farmed tilapia and feed certified by private organisations such as the Aquaculture Stewardship Council (ASC) and Sustainable Fisheries Partnership. Tuna pole and line fishing is currently part of a Fishery Improvement Project (FIP) that will eventually lead to its certification by the Marine Stewardship Council (MSC). However, there is no market data on the percentage of total seafood products in Indonesia that are certified.

Private sustainability standards and certification schemes may be favourable for governments, because participating in these projects a good image of responsible governance. However, the main challenges associated with such schemes are that they may change power relationships between actors involved in seafood trade, and make market access more difficult for non-certified exporters, especially smallholders (Wijaya and Glasbergen, 2016). When possible, evidence suggests that the simplest approach may be for the government to contribute to the costs for a fishery to obtain private certification, rather than create additional certification schemes (OECD, 2011).

Vulnerability to climate change

Indonesia faces high risks due to multiple hazards directly linked to climate change, such as tsunamis, floods, landslides, and droughts, alongside the effects of sea-level rise on coastal areas (MER, 2015). Indirect effects of climate change occur through impacts on feed, seed, freshwater and other inputs (FAO, 2016b). The World Bank estimates that by 2100, climate change impacts will cost Indonesia 2.5-7% of its GDP and disproportionately affect poor, coastal communities that depend on agriculture, fisheries and forestry for their livelihoods (World Bank, 2017). Food security and water availability will be increasingly affected by unpredictable rainfall, temperature and salinity increases, which could have an impact on fish migratory routes, stock abundance, biodiversity and water quality. This could in turn reduce catches, cause physical destruction of aquaculture facilities, spread diseases and, ultimately, reduce the availability and affordability of seafood (De Silva and Soto, 2009).

The World Bank also estimates that Indonesia is relatively unready to cope with climate change impacts (World Bank, 2017). In fact, the country has a very developed national institutional policy framework to tackle climate change mitigation.[10] However, planning for adaptation is still a work in progress. The fisheries and aquaculture sectors will have to adapt to future changes in climate and ocean conditions. Adaptation strategies require marine science and technology, spatial planning, technical coastal adaptation, and adjustment of cultured and capture fisheries management. At the level of regulation, climate change forecasting should be taken into consideration when setting fishing pressure limits to sustainable levels, while weather-related risks should be considered when defining locations suitable for aquaculture (De Silva and Soto, 2009). Other adaptation measures include aquaculture relocating to less-exposed areas, fish health management, water recycling, feed efficiencies, developing better-adapted seed stock, improved monitoring and early warning systems and improving value-addition (FAO, 2016b).

Investment in research in these areas will prove crucial. Promising areas, which have shown success in other countries, include domesticating species resistant to higher salinity and temperatures (Shelton, 2014) and identifying preventative treatments for new diseases, animal physiology, better feeds and feeding practices, and technology transfers, especially to small-scale farmers (De Silva and Soto, 2009). Coastal management and capacity building can also help indirectly. Restoration of mangroves, for example, can buffer coastal communities from storm surges or erosion, while disaster risk management and early warning systems may help fishers and aqua-farmers to cope with disruptions in production.

Finally, aquaculture insurance or disaster-linked cash transfers can help limit bankruptcies resulting from losses caused by climatic events. As mandated by Law No. 19/2013, the government must develop agricultural insurance to cushion farmers from financial loss due to natural calamities, but there are as yet no provisions for fishers or aquaculture farmers.

7.3 Increasing the incomes of fishers and aqua-farmers while improving consumer access to seafood

Continued production growth under a business-as-usual scenario is not an option: as shown in Section 7.2, most fish stocks cannot absorb further increases in catch effort, and expansion of aquaculture production will only be sustainable if it consumes less land and water resources and if externalities are better controlled to avoid degradation of the ecosystems on which it depends. Increasing the economic returns to capture and aquaculture production is thus key to improving the incomes of fishers and aqua-farmers and, in so doing, their access to food and their resilience to food security risks in the long run.

Direct support: The need for decoupling

In an effort to improve the incomes of fishers and aqua-farmers, the Government of Indonesia subsidises fuel for the sector as well as the renewal of the artisanal fleet. The government also invests in artisanal aquaculture plants creation and supports the development of cheaper feed production based on locally-sourced ingredients. Direct support to fishers and aquaculture producers, which consists of fuel tax concessions and subsidies for vessels and equipment, or equipment and inputs for aquaculture, totalled more than two-thirds of government support to fisheries and more than three-quarters of government support to aquaculture in 2014 and 2015 (Figure 7.8).

Of these, fuel tax concessions (available to artisanal fishers and small-scale aqua-farmers only) cost more than 150 USD million in 2015, equating to more than half of total support, despite a significant reduction from previous years (Figure 7.8).

Most of the remaining direct support is allocated to aquaculture production equipment and the distribution of fishing boats. On the aquaculture side, floating cage construction received USD 450 000 dollars in 2015 and almost double this in 2016, while over USD 2.5 million was allocated to fish feed production and transport machinery and infrastructure in 2016. With respect to fisheries, a budget of over USD 200 million was allocated to the construction of 3 450 new fishing boats in 2016 (95% of which will be under 10 GT). These fishing boats will be distributed to fisher co-operatives, which primarily consist of artisanal fishers. At the same time, the government is investing in the modernisation of the artisanal fleet. A separate budget of almost USD 450 000 was provided in 2015 for the purchase of machines, gear, solar cells, submersible lamps and other equipment. Grant aid was also provided to assist fishermen and their livelihoods in the border areas of Indonesian territory by securing boats, purchasing gear and equipment (OECD, 2017).

Allocation of these subsidies is likely to be costly – given the transaction costs associated with selecting a small number of recipients among very large numbers of fishers and aqua-farmers – and to create incentives for corruption.

For fisheries, productive support is more problematic. While investments have the potential to increase the incomes of recipients, they also increase fishing effort in a context of insufficient management of resources (Section 7.2), leading to the common problem of declining productivity. This support is based on the assumption that policies on illegal fishing have improved fish stocks and will continue to do so, hence increasing the availability of fish that can be caught by artisanal fishermen. However, as mentioned earlier, the direct link between a decrease in IUU activities and the increased availability of fish for artisanal fishers in coastal areas may only be true for certain species. In addition, in the absence of effective controls on domestic artisanal fisheries, stocks remain prone to overfishing.

Figure 7.8. Support to fisheries and aquaculture, 2014 and 2015

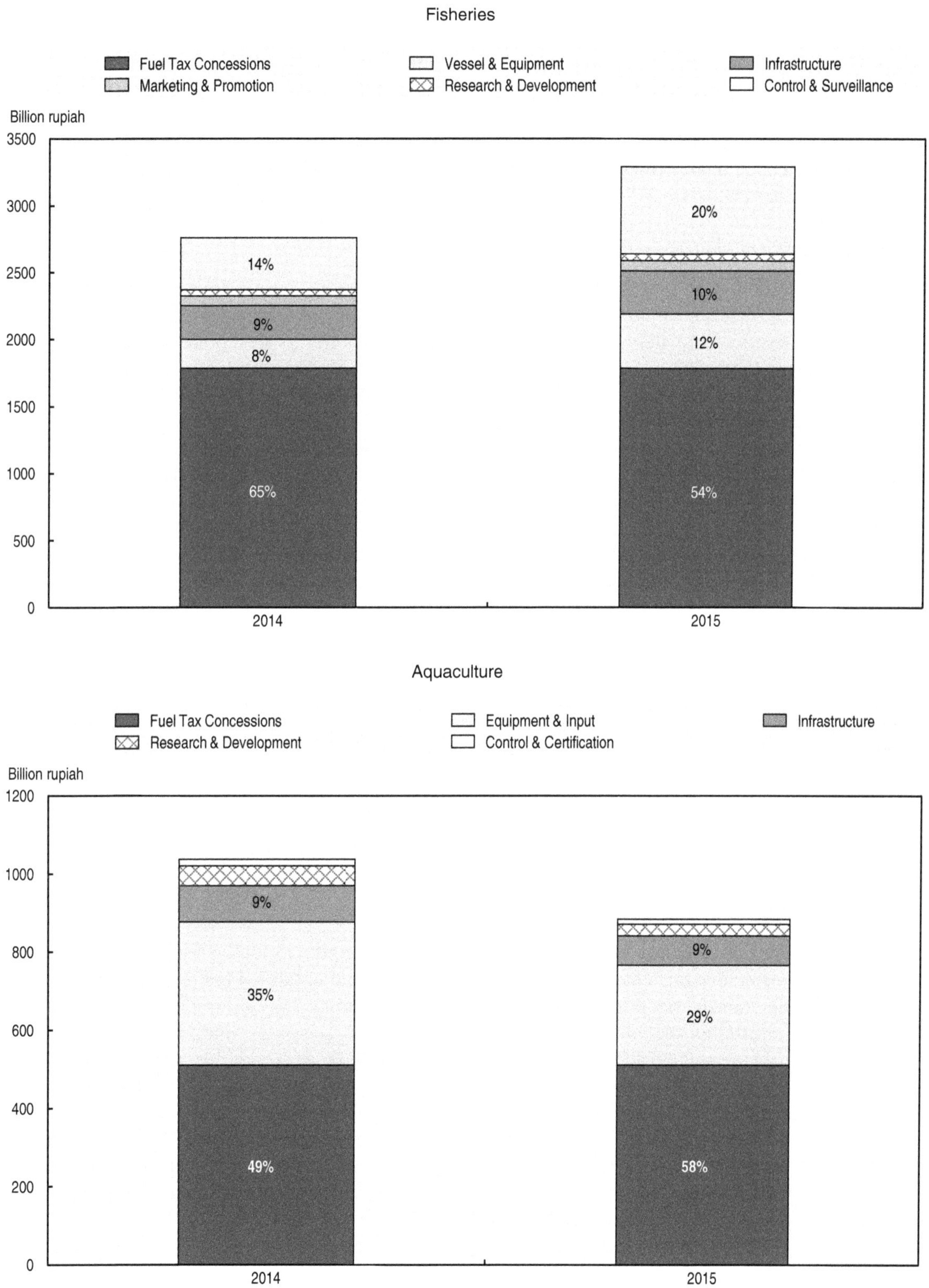

Source: OECD (2017), *Economic Indicators for Agriculture and Fisheries*, https://stats.oecd.org/Index.aspx?DataSetCode=FISH_FSE.

Offering direct support for livelihood diversification would be a better investment policy instead of support that risks encouraging new entrants to the sector. The development of alternative livelihood options for fishing and aqua-farming communities, notably through investment in the provision of new skills, can be a long process, however, and initial training may be needed to overcome traditional and conservative attitudes. Alongside primary services such as health care, clean water and sanitation, micro-enterprise programmes, vocational training such as teacher training, and support for trading in non-farm items such as handicrafts can be supported through the provision of capital assets, market information, credit facilities and education. Indonesia is experienced in implementing conditional cash transfers aimed at increasing participation in health and education services (USAID, 2013).

Sustainable resource management for stable income growth

While direct support is necessary in the short run to support the livelihoods of poor fishers and aqua-farmers, in the medium to long run, investing in the sustainable management of resources and the ecosystem would ensure more stable and increasing economic returns to stakeholders that depend on these resources. Improving the health of fish stocks is not only important for their survival – it also means that more fish can be caught and reduces the cost per catch. The composition of the fish caught can also change, to contain more high-value fish, thereby increasing the average price of the catch. Research conducted jointly by the World Bank and the FAO (Arnason, Kelleher and Willmann, 2009) as well as Costello et al. (2016) demonstrate the huge potential of good resource management to increase the productivity of capture fisheries, with tens of billions of USD to be gained annually worldwide. Costello et al. also show that the time horizons involved are not so long, with stock recovery targets predicted to be reached within less than ten years in half of the fisheries considered by their model. MMAF is currently working with Costello and his team to evaluate the gains associated with the elimination of IUU fishing and better management of domestic fisheries in Indonesia, with the objective of creating incentives for stakeholders and policy makers to fully engage in better management.

Setting total catch limits could have additional economic benefits, if associated with restrictions on access to resources or the allocation of fishing rights aimed at reducing overcapacity, which could further increase the catch per unit of effort and thus the returns of fishing. Such restrictions are currently incompatible with Indonesian legislation. Article 33 (paragraph 3) of the Constitution specifies that land, water, air and natural resources are controlled by the State, and responsibility for these resources cannot be transferred to individuals. On the basis of this article, the provision of the Coastal Management and Small Islands Act (No. 27/2007), that would have permitted user rights up to 12 miles from shore, was rejected by the High Court (16 June 2011 Decree). However, an important step towards the legal possibility to restrict access to resources has been taken by the government with the preparation of a law or regulation that will allow the use of rights or privileges in fisheries once approved by the Parliament.

Setting up good governance processes to allocate fishing rights or deliver access to fisheries is an important step to ensure that any changes are socially acceptable and durable. First, rights or access should be assigned as unambiguously as possible to avoid conflicts, bargaining and politicisation. Additionally, transparent rules regarding the allocation of fishing rights or access to individuals, co-operatives or communities ideally have to be discussed extensively ex-ante with these stakeholders, who also participate in the decision-making process on the basis of such rules. Community-based quota management should particularly be considered for the management of fisheries that are operated exclusively by small-scale fishers in areas that are both geographically diffuse and difficult to access, making it impossible to regulate these fisheries through individual quotas or rights as management and surveillance costs would be excessive. Because it is often difficult to implement user privileges when fishers have traditionally fished freely, MMAF will also need to plan a public education campaign and public hearings to explain the benefits of rights-based management to secure widespread support and compliance (Sutinen, 2013).

Better regulation of environmental externalities and the adoption of sustainable production practices also have the potential to increase the economic returns of aqua-farms in the medium to long term. Improved biosecurity measures in aquaculture reduce mortality while increasing the quality of products delivered for export as well as reducing needs for drugs and hence production costs. Well-managed certification schemes can also create economic premiums on the final product that enhance the economic benefits for participating fisheries and farms.

Transitioning to a rights-based fishing system – or to more sustainable aquaculture production modes – will not be easy, and requires complementary measures. Transitional costs are to be expected, and the careful anticipation of impacts and design of adaptive strategies will be necessary to avoid income disruptions. The profitability of the aquaculture sector is for example currently affected by some of the new fishing management measures. Harvest restrictions on baby crabs and lobster have interrupted their cultivation,[11] while the export of live mariculture products (notably grouper) has been complicated by the measures taken to counter IUU fishing, as their transport was largely undertaken by ex-foreign vessels now prohibited from operating in Indonesian waters. Similarly, restrictions to access fishing resources, and tighter regulations, inevitably create winners and losers and require stakeholders to trust that new policies will lead to a better future that is worth the transition costs – namely reduced catches in the short term, as well as potential exits from the sector. In countries such as Indonesia, where the fishing sector is an important source of livelihood for the poor, focusing on finding the highest-value use of the resource is particularly difficult and desirable only to the extent that, in the short term, the human costs of the transition can be addressed by complementary policies. Providing assistance with the costs and process of adjustment via complementary measures can, however, be effective, as can providing direct livelihood assistance to those who lose out from reform.

Facilitating public support for infrastructure and R&D

Public support can also help the transition to greater economic returns by establishing an enabling environment that is conducive to sustainable productivity growth. Two key aspects of such an enabling environment, directly under the responsibility of fisheries and aquaculture authorities, are infrastructure and research and development (R&D).

Efficient and well-developed transport, communication and energy provision infrastructure plays an important role in connecting fishers and aqua-farmers to market opportunities and knowledge, as well as specialised support services, thus reducing the cost of doing business and stimulating value creation (Waite et al., 2014). Transport infrastructure, such as suitable roads, waterways, railways, ports, wharf and landing docks; and energy infrastructure, which is important for light and cold-chain storage, are particularly critical given that fish is a perishable commodity and production centres may be located far from transport hubs for export as well as urban markets where the majority of seafood is consumed (75% of aquaculture production is still consumed domestically, according to Indonesia Investments [2014]). In addition, with a strategy focused on supporting artisanal fishing in national waters, leaving the industrial fleet to operate only in the high seas, a big challenge for the Indonesian government is to connect artisanal fishers with processors and distributors to substitute supply from industrial vessels with supply from artisanal fishers.

OECD work on agricultural productivity shows that investments in R&D, technology transfer and extension services can yield great returns (OECD, 2013). New techniques and innovations can spur growth through improvements in quality and quantity, by reducing costs of production or by creating increased private and social benefits. Ideally, innovation can enable more value to be derived from the same level of natural resources while reducing negative impacts on the environment. Technological innovations have played an important role for growth in every aspect of aquaculture operations, particularly in the development of efficient feed, bio-secure seed and disease reduction. Innovations have also led to the creation of new high-value products from capture fisheries waste, hence increasing the total value of fisheries production.

By making domestic seafood production more profitable, investment in infrastructure and R&D also increases the competitiveness of domestic products compared to imports, and promotes food self-sufficiency – a key objective of the Jokowi administration. However, unlike import restrictions, investment in infrastructure and R&D also benefit net consumers of seafood with lower prices, hence also improving food security.

Infrastructure to address high transport costs

Transport infrastructure is particularly expensive, as Indonesia is comprised of 6 000 inhabited islands spread across 5 000 km from east to west, making it difficult to efficiently connect products and people to markets. High inter-island transport costs contribute to large inter-regional price differences. Shipping costs within Indonesia are found to be more expensive than the costs of importing from Singapore or China (Tabor, 2015). Shipping Law 17/2008, which introduced cabotage principles limiting the movement of cargo between Indonesian ports and Indonesian-flagged vessels, added to inter-island transport costs.

According to data published by the Indonesian Chamber of Commerce and Industry (Kadin Indonesia), around 17% of a company's total expenditure in Indonesia is absorbed by logistics costs (Indonesia Investments, 2016a). In peer regional economies, this number is much lower: 5% in Japan, 8% in Malaysia, 10% in Thailand and 12.5% in China. According to the World Economic Forum Global Competitiveness Index 2015, since 2013, every transport category has receded in quality, indicating that Indonesia is not keeping up with the rest of the world in terms of transport development (World Economic Forum, 2015). Quality of port infrastructure was the lowest-rated category. Problems include a lack of equipment, low port efficiency and poor road access.

Energy infrastructure also requires critical improvement. Energy services and consumption are oriented disproportionately toward Indonesia's economic growth centres, and large expanses of remote areas still lack access to basic energy services (Tabor, 2015). In addition, demand is outstripping supply capacities in many parts of the national grid, due to electricity theft, underinvestment and infrastructure needs related to transmission and distribution.

The government is conscious of the need to prioritise infrastructure investments. In 2014, President Jokowi announced that infrastructure spending would increase throughout his first term, peaking at around 7.7% of GDP in 2017, with the biggest investment in maritime infrastructure. The government intends to build 24 new sea ports and 15 new airports by 2019. By 2014, over USD 19.2 million had been spent on the development of facilities at fishing ports as part of a three-year development programme (2011-14) (OECD, 2017). The investment targets improvements to clean water supply, the expansion of landing areas and facilities, better transportation to and from the port, renewable energy sources and solar panels. In 2015, a further USD 23.9 million was spent on developing fishing ports in Indonesia's outermost regions, while USD 666 000 was also spent on the installation of solar cells in fishing ports (OECD, 2017). Together, infrastructure projects accounted for 9-10% of government support in fisheries and aquaculture in 2014 and 2015 (Figure 7.8). Infrastructure needs can understandably place a huge burden on government budgets and, in future, Indonesia may wish to consider private sector participation through public-private partnerships (PPP).

Encouraging research and development

An important strategy for the Indonesian government would be to create the conditions for the R&D that enables innovation to take place. In the OECD context, an industry-wide approach to innovation activities – through industry organisations, producer organisations and specialised university laboratories – has been found to be conducive to better research outcomes (OECD, 2013). Facilitating PPPs can also enable the development of innovations (Subasinghe et al., 2000). However, private sector investment tends to be orientated towards larger enterprises (OECD, 2013). Finally, in Indonesia, small-scale aqua-farmers and fishers form the majority of producers. While small businesses can pioneer new systems, species, products and technologies, they often lack resources to innovate on their own. Public policy can thus be used to kick-start the required R&D.

Potential areas for innovation investment include research areas related to on-farm activity, such as water quality management, better feed efficiency and innovative production systems; longer-term research programmes on the use of genetics, for example; and reliable data collection and analysis to better inform management decision-making. Developing a strategy for the diffusion of innovation results can also bring broad benefits. In developing countries in particular, government investment is important to scale-up good aquaculture and management practices so that they become industry norms (Ross et al., 2013). Providing simple skills and technologies to small-scale aquaculture can strengthen their innovative and learning capacities.

In Indonesia, spending on R&D accounted for less than 2% of support to fisheries and less than 5% of support to aquaculture in 2014 and 2015, a relatively low figure compared with levels observed in many OECD countries (OECD, 2016). One key area identified by the government is feed. As mentioned in Section 7.2, fish feed accounts for the largest percentage of aquaculture operating costs, and reducing these costs would contribute to improved productivity. Research programmes therefore receive public funding to conduct experiments with local ingredients and test alternative protein sources such as seaweed and palm oil.

However, investment in research will only boost productivity to the extent that it does not lead to the substitution of imports with domestically-produced goods that are only competitive thanks to trade restrictions (USAID, 2013). Using research to develop products that can be more profitably sourced from other countries would effectively mean that public funds are used to create rents for input producers at the expense of both seafood producers and consumers.

Trade and foreign investment restrictions: Support at the cost of higher prices for consumers

As mentioned in Chapter 3, ASEAN countries often use trade restrictions to increase the demand for domestically-produced food products with the dual objective of increasing producer incomes and becoming more self-sufficient in food production. To this end, the Indonesian Food Law of October 2012 expanded the ways and means by which the government might influence domestic food markets, and extended coverage beyond rice. Applied tariff rates for seafood products range between 0 and 10%, but bound tariffs average 40%. Indonesia also restricts seafood imports on the basis of quality control and food safety reasons according to Regulation 46/2014. As a result, imports remain very limited, and have followed a much slower growth path than exports (Figure 7.9).

Figure 7.9. Indonesian imports and exports by value, 2014

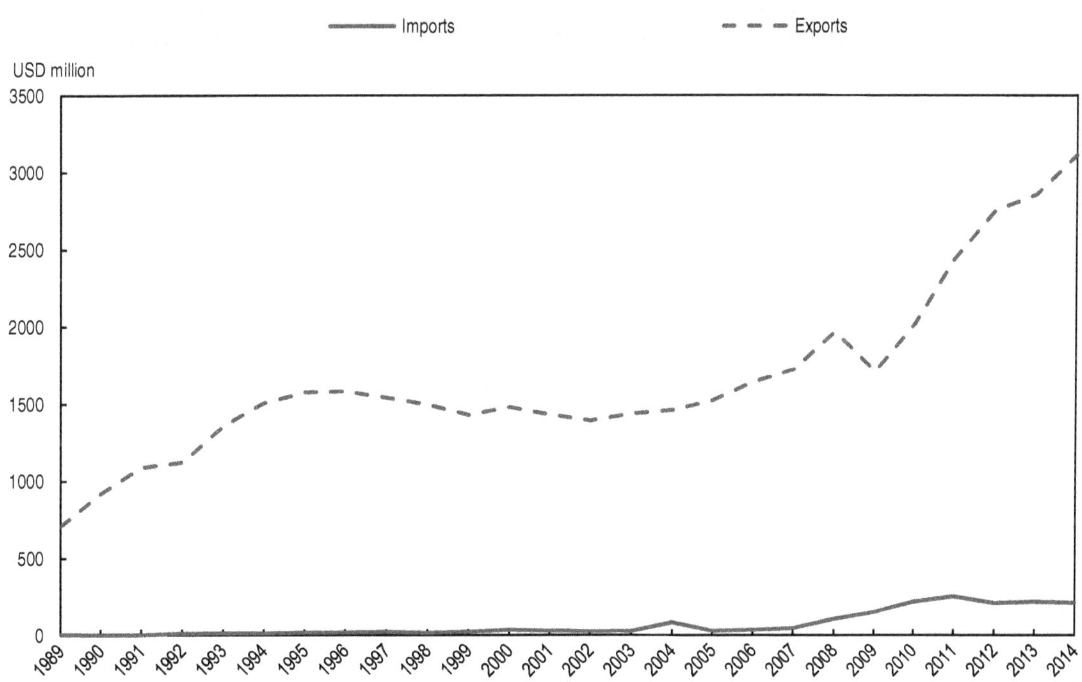

Source: UN DESA (2016), *UN Comtrade*, https://comtrade.un.org/.

Restricting imports raises producer prices and encourages more domestic production, which boosts the welfare of net sellers of the protected product: fishers and aquaculture producers. However, the benefits accrue mainly to the largest producers, who are the least likely to be food-insecure. At the same time, higher prices lower the welfare of net buyers – food consumers – as well as their access to food and ultimately their food security and nutrition. As the poorest and most food-insecure spend the highest proportion of their income on food, they are hurt the most (Chapter 4). Such restrictions thus contradict the government's stated objective to encourage fish consumption in the coming years.

Protection also reduces incentives to improve domestic productivity with a view to competing with foreign producers, and hence tend to lock domestic production systems into expensive forms of production. Moreover, as far as overfished capture species are concerned, restrictions on imports increase the demand for domestically-caught fish, hence worsening pressures on endangered stocks.

Restrictions on investment by foreign investors can have similar effects. The most recent Negative Investment List adopted by the government (Regulation 44/2015) limits the share of foreign investment in certain sectors of the economy. In capture fisheries, foreign capital is entirely prohibited. Because large vessels are not currently produced in Indonesia, this means that not many large fishing vessels will operate in Indonesian waters in the near to medium term. The Indonesian government aims to ensure that the artisanal sector catches the fish that would have otherwise been caught by the industrial fleet, hence increasing artisanal incomes. However, as mentioned in Section 7.1, the artisanal and industrial fleet partially target different species.

In addition, implications of these restrictions for national food security are ambiguous. Indeed, the processing sector may be unable to source from artisanal fishers the fish that they previously sourced from the industrial fleet, at least in the short run, resulting in potential job and income losses in the processing sector. In addition, to the extent that the industrial fleet was more cost-efficient than the artisanal fleet, the consumer price for fish caught by artisanal fishers could be higher than if it had been caught by industrial vessels. Restrictions on investment would consequently have distributional effects similar to those of import restrictions.

Investigating these issues in more detail, and trying to estimate the impacts of trade and investment restrictions for particular groups of producers or consumers, requires both detailed price data and household level data on production and consumption. The Government of Indonesia is currently improving its fisheries and aquaculture data collection system. Once the necessary data becomes available, investment in the abovementioned impact analysis would be a welcome priority.

7.4 The way forward: From food *sovereignty* to food *security*

Fisheries and aquaculture policies in Indonesia have the potential to improve food security and nutrition if they are focused on three key, closely-intertwined objectives: the sustainable management of natural resources for the future, sustainable increases to economic opportunities for fishers and aqua-farmers, and improved consumer access to seafood.

The sustainable exploitation of resources and ecosystems can both reduce the risks associated with the sector's contribution to food security and increase incomes for fishers and aqua-farmers in the long run. To transition to more sustainable seafood production, government efforts to combat illegal fishing should be complemented with better management of domestic fisheries, notably by adopting long-term measurable management targets for the main fisheries. Government efforts to improve catch data monitoring and centralise the data collection system of MMAF are an important prerequisite to better fisheries management. The challenge will be to also include small-scale fisheries in the data collection process. Establishing a legitimate and acceptable governance system that involves both stakeholders and scientists in target-setting and access allocation is the second key stepping-stone towards sustainable fisheries management. The analysis also identifies scope for better regulation of the aquaculture sub-sector, accompanied by more comprehensive impact assessment of large farm operations, more regular controls, and the clearer allocation of monitoring and information-gathering responsibilities between different government levels.

Further improvements to the incomes of fishers and aqua-farmers will predominantly have to come from increases in the value of seafood produced as well as reductions in production cost. The direct support currently provided by the government to individual fishers and aqua-farmers – mainly in the form of subsidies for fuel and vessel or productive equipment – with a view to increasing their incomes or reducing their costs biases production incentives and increases pressure on resources. To prevent this, it is recommended – particularly during periods of transition to more sustainable production practices – that the government instead directly support livelihoods via the provision of social safety nets that are targeted at vulnerable households, and help the conversion to more remunerative activities by means of education and training. Directing public support to R&D and infrastructure would also favour greater value creation, while at the same time reducing prices for consumers – thus improving access to food, the third key objective – and increasing food availability. Reliance on restrictive trade and investment barriers, on the other hand, supports incomes at the expense of reduced consumer access to seafood products, and therefore has the potential to be counterproductive to food security in the long run.

Notes

1. This chapter has been co-ordinated and drafted by Claire Delpeuch and Ingrid Kelling from the Natural Resources Policies Division of the Trade and Agriculture Directorate. Information and data on government programmes supporting the fisheries and aquaculture sectors was collected by Rohana Subasinghe, FutureFish, Sri Lanka. The authors would like to thank the Indonesian Ministry of Marine Affairs and Fisheries (MMAF) and the Presidential Task Force to Combat Illegal Fishing of Indonesia for their kind co-operation and, in particular, Maskur Maskur for their help with the support data collection process; Rita Octafiani for organising in-country missions and visits; and Sonny Koeshendrajan, Naftalia Siregar and Achmad Santosa for sharing information. Ria Fitriana, Charles Darwin University; Abdullah Habibi, WWF Indonesia; Daniel Pauly, University of British Columbia; and Nhuong Tran, WorldFish kindly commented on earlier drafts. Statistical support was provided by Fabiana Cerasa. The chapter was edited by Robert Akam and Clara Thompson-Lipponen.

2. The ten ASEAN member states are Brunei Darussalam, Cambodia, Indonesia, Lao People's Democratic Republic (PDR), Malaysia, Myanmar, Philippines, Singapore, Thailand and Viet Nam.

3. Indonesia's Law No. 18/2012 defines food security in a way that is very much in line with the FAO theoretical framework. It is described as the condition in which all people, in all households, at all times have sufficient food in both quantity and quality to enable them to live healthy, active, productive and sustainable lives, and that the food is safe, diverse, nutritious, equitably distributed and affordable, and does not conflict with religion, beliefs or culture (OECD, 2016).

4. Using a range of indicators, Chapter 1 of this publication provides a snapshot of the food security situation both in the ASEAN region and in comparison with other regions.

5. The FAO prevalence of food inadequacy index measures the percentage of the population that is at risk of not meeting the food intake requirements associated with normal physical activity, and therefore also includes those who, although they cannot be considered chronically undernourished, are likely to be conditioned in their economic activity by insufficient food.

6. The National Food Law, adopted in Indonesia in October 2012, expanded the ways and means by which the government might influence domestic food markets. The key objectives of the law are to increase food production and self-sufficiency; improve the welfare of farmers, fishers and food processors and their competitiveness; ease consumer access to food, especially for the food insecure; and improve nutrition outcomes by promoting more diversity in food consumption. Among the policy tools introduced are import and export restrictions; investment in research and development (R&D) and rural

infrastructure; better management of land and water resources; as well as promotion of consumer knowledge of the nutritional benefits of an adequate diet (USAID, 2013).

7. On 20 October 2014, President Jokowi declared "We have to work as hard as we can to restore Indonesia as a maritime power. The oceans, the seas, the straits and the bays are the future of our civilisation." (Noegroho, 2016).

8. It is estimated that inland fisheries account for about 6% of the total catch volume and 9% of catch value (Tran et al., 2017). About 200 000 boats operate in inland waters, mostly without engines (FAO, 2012b). Due to lack of available data, this chapter does not cover policies for inland water fishing.

9. Indonesia is also committed to establishing marine protected areas: over 20 million hectares by 2020 and 30 million hectares by 2030. In February 2016, Marine Protected Areas (MPAs) already covered over 17 million hectares. They are managed jointly by MMAF and the Ministry of Environment and Forestry. MPA policies are beyond the scope of this chapter.

10. Indonesia has signed the UN Convention on Biological Diversity (CBD), the Convention to Combat Desertification (CCD), the Framework Convention on Climate Change (UNFCCC) and the Kyoto Protocol. In addition, National Communications on Climate Change were adopted in 1999 and 2011, the National Commission on Clean Development Mechanisms was established in 2005, the National Action Plan on Climate Change Adaptation was adopted in 2007 and the Indonesian Climate Change Sectoral Roadmap was defined in 2010.

11. Artificial breeding for crabs is technically feasible, but in the absence of control mechanisms allowing differentiation between wild and cultured crabs, hatcheries have not yet been authorised. Lobster breeding, on the other hand, is not yet technically operational.

References

Adhuri, D. (2016), "Combining resource management and livelihood development: Some reflections from East Flores and Lombok", in *Proceedings of the Eighteenth Conference of the International Institute of Fisheries Economics and Trade, July 11-15 2016, Aberdeen, Scotland: Challenging New Frontiers in the Global Seafood Sector: A Northern Enlightenment*, A. L. Shriver (eds.), International Institute of Fisheries Economics and Trade (IIFET), Corvallis, Oregon, United States of America.

Arnason, R., K. Kelleher and R. Willmann (2009), *The Sunken Billions: The Economic Justification for Fisheries Reform,* The International Bank for Reconstruction and Development / The World Bank, Washington, DC.

Asche, F. (2011), *Green Growth in Fisheries and Aquaculture Production and Trade*, OECD internal document, Paris, www.oecd.org/tad/fisheries/green-growth-fisheries-aquaculture-production-trade-report.pdf (accessed on 17 November 2016).

ASEAN (2016), *Regional Report on Nutrition Security in ASEAN: Volume 1*, ASEAN Secretariat, Jakarta, and United Nations Children's Fund, Bangkok.

BPS-Indonesia Statistik (2008), *Badan Pusat Statistik* (database), Statistics Indonesia, Jakarta, www.bps.go.id/ (accessed on 12 September 2016).

Brooks, J. and A. Matthews (2015), "Trade dimensions of food security", *OECD Food, Agriculture and Fisheries Papers*, No. 77, OECD Publishing, Paris, http://dx.doi.org/10.1787/5js65xn790nv-en.

Burke, L., K. Reytar, M. Spalding and A. Perry (2011), *Reefs at Risk Revisited*, World Resources Institute, Washington, DC.

Cao, L., J.S. Diana, G.A. Keoleian and Q. Lai (2011), "Life cycle assessment of Chinese shrimp farming systems targeted for export and domestic sales", *Environmental Science and Technology*, Vol. 45/15, ASC Publications, Washington, DC, pp. 6531–6538.

CEA (2014), "Indonesia" in *Country Scoping Project*, California Environmental Associates, San Francisco, United States of America.

Costello, C., D. Ovando, T. Clavelle, C.K. Strauss, R. Hilborn, M.C. Melnychuk, T.A. Branch, S.D. Gaines, C.S. Szuwalski, R.B. Cabral, D.N. Rader and A. Leland (2016), "Global fishery prospects under contrasting management regimes", *PNAS*, Vol. 113/18, Proceedings of the National Academy of Sciences of the United States of America, Washington, DC, pp. 5125-5129.

DAFF (2011), *Net Returns: A Human Capacity Development Framework for Marine Capture Fisheries Management in South East Asia*, Department of Agriculture, Fisheries and Forestry, Canberra.

De Silva, S.S. and D. Soto (2009), "Climate change and aquaculture: Potential impacts, adaptation and mitigation", in K. Cochrane, C. De Young, D. Soto and T. Bahri (eds.), *Climate Change Implications for Fisheries and Aquaculture: Overview of Current Scientific Knowledge*, FAO Fisheries and Aquaculture Technical Paper No. 530, Food and Agriculture Organization, Rome.

FAO (2016a), *Food Balance Sheet Indonesia* (database), Food and Agriculture Organization, Rome, http://faostat.fao.org/default.aspx (accessed on 19 September 2016).

FAO (2016b), *The State of World Fisheries and Aquaculture 2016: Contributing to Food Security and Nutrition for All*, Food and Agriculture Organization, Rome.

FAO (2016c), *FAOSTAT* (database), http://faostat3.fao.org/home/E (accessed on 19 September 2016).

FAO (2015), *Regional Overview of Food Insecurity Asia and the Pacific: Towards a Food Secure Asia and the Pacific*, Food and Agriculture Organization, Bangkok.

FAO (2012a), *Fishery and Aquaculture Statistics 2010*, Food and Agriculture Organization, Rome.

FAO (2012b), *Fishery and Aquaculture Country Profiles: The Republic of Indonesia* (webpage), Food and Agriculture Organization, Rome, www.fao.org/fishery/facp/IDN/en.

FAO (2011), *Technical Guidelines on Aquaculture Certification*, Food and Agriculture Organization, Rome.

FAO (2003), *Fisheries Management: 2. The Ecosystem Approach to Fisheries*, FAO Technical Guidelines for Responsible Fisheries, Vol. 4/2, Food and Agriculture Organization, Rome.

Gutierrez, N.L., O. Defeo, S.R. Bush, D.S. Butterworth, C.A. Roheim and A.E. Punt (2016), "The current situation and prospects of fisheries certification and ecolabelling", *Fisheries Research*, Vol. 182, pp. 1-6.

HLPE (2014), *Food Losses and Waste in the Context of Sustainable Food Systems: A Report by the High Level Panel of Experts on Food Security and Nutrition*, High Level Panel of Experts on Food Security and Nutrition, Committee on World Food Security, Rome.

IFPRI (2014), *Global Hunger Index: The challenge of Hidden Hunger*, International Food Policy Research Institute, Welthungerhilfe, and Concern Worldwide, Bonn, Washington, DC and Dublin.

Indonesia Investments (2016a), *Infrastructure Development in Indonesia* (webpage), www.indonesia-investments.com/business/risks/infrastructure/item381 (accessed on 10 October 2016).

Indonesia Investments (2016b), *Poverty in Indonesia* (webpage), www.indonesia-investments.com/finance/macroeconomic-indicators/poverty/item301 (accessed on 8 November 2016).

Indonesia Investments (2014), *Export Target of Indonesia's Fishery Sector Revised on Weak Demand* (webpage), 28 March 2014, www.indonesia-investments.com/news/todays-headlines/export-target-of-indonesias-fishery-sector-revised-on-weak-global-demand/item1816K (accessed on 10 October 2016).

Inside Indonesia (no date), *Aquaculture in Adversity* (webpage) Jakarta, www.insideindonesia.org/aquaculture-in-adversity (accessed 12 September 2016).

Koeshendrajana, S. (2016), *Profiles of Indonesian Fisheries: Future Supply and Demand Scenarios of Fish in ASEAN Region*, Stakeholder Workshop, Penang, 7-8 June 2016, WorldFish, Penang, Malaysia.

Kusumawati, R. and S.R. Bush (2015), "Co-producing better management practice standards for shrimp aquaculture in Indonesia", *Maritime Studies*, Vol. 14/21, Springer, London.

Menard, C. (2014), *Institutional Aspects of Governance in Fisheries Management*, OECD internal document, Paris.

MER (2015), *Climate Change Profile: INDONESIA*, Netherlands Commission for Environmental Assessment, Dutch Sustainability Unit, The Netherlands.

MMAF (2016), Information received from Indonesian Ministry of Marine Affairs and Fisheries, Jakarta.

MMAF (2015), *Statistik Perikanan Tangkap Indonesia Menurut Provinsi 2015: Capture Fisheries Statistics of Indonesia by Province,* Directorate General of Capture Fisheries, Ministry of Marine Affairs and Fisheries, Jakarta.

Noegroho, A. (2016), *Improving Traceability and Chain of Custody to Meet Market and Export Requirement*, Second Bali Tuna Conference, Denpasar, Indonesia, 19 May 2016.

OECD (2017), "Fisheries Support Estimate", *Economic Indicators for Agriculture and Fisheries* (database), https://stats.oecd.org/Index.aspx?DataSetCode=FISH_FSE (data will be accessible in Q2 2017).

OECD (2016), *Indonesia Country Study, Review of Fisheries 2015*, OECD Publishing, Paris, http://dx.doi.org/10.1787/rev_fish_stat_en-2015-en.

OECD (2015), *Green Growth in Fisheries and Aquaculture*, OECD Green Growth Studies, OECD Publishing, Paris, http://dx.doi.org/10.1787/9789264232143-en.

OECD (2014), *Fishing for Development – Background Paper for Session 2: Rebuilding*. Joint meeting of the OECD Fisheries and Development Assistance Committees, the Food and Agriculture Organization of the United Nations and the World Bank, OECD, Paris, 10-11 April, 2014.

OECD (2013), *Global Food Security: Challenges for the Food and Agricultural System,* OECD Publishing, Paris, http://dx.doi.org/10.1787/9789264195363-en.

OECD (2012), *Rebuilding Fisheries: The Way Forward*, OECD Publishing, Paris, http://dx.doi.org/10.1787/9789264176935-en.

OECD (2011), *Fisheries and Aquaculture Certification*, OECD Publishing, Paris, http://dx.doi.org/10.1787/9789264119680-en.

OECD-FAO (2016), *OECD-FAO Agricultural Outlook 2016-2025*, OECD Publishing, Paris, http://dx.doi.org/10.1787/agr_outlook-2016-en.

Pauly, D. and V. Budimartono (eds.) (2015), *Marine Fisheries Catches of Western, Central and Eastern Indonesia, 1950-2010*, Working Paper No. 2015-61, University of British Columbia, Vancouver, Canada.

Phillips, M., P.J.G. Henriksson, N. Tran, C.Y. Chan, C.V. Mohan, U-P. Rodriquez, S. Suri, S. Hall, and S. Koeshendrajana (2015), *Program Report: 2015-39*, WorldFish, Penang, Malaysia.

Potts, J., A. Wilkings, M. Lynch and S. McFatridge (2016), *State of Sustainability Initiatives Review: Standards and the Blue Economy*, International Institute for Sustainable Development (IISD), Manitoba, Canada.

Rana, K.J., S. Siriwardena and M.R. Hasan (2009), *Impact of Rising Feed Ingredient Prices on Aquafeeds and Aquaculture Production*, Fisheries and Aquaculture Technical Paper No. 541, Food and Agriculture Organization, Rome.

Rimmer, M.A., K. Sugama, D. Rakhmawati, R. Rofiq and R.H. Habgood (2013), "A review and SWOT analysis of aquaculture development in Indonesia", *Aquaculture*, Vol. 5/4, pp. 255-279.

Ross, L.G., T.C. Telfer, L. Falconer, D. Soto and J. Aguilar-Manjarrez (eds.) (2013), "Site selection and carrying capacities for inland and coastal aquaculture: FAO/Institute of Aquaculture, University of Stirling, Expert Workshop, 6-8 December 2010", *FAO Fisheries and Aquaculture Proceedings No. 21*, Food and Agriculture Organization, Rome.

Santosa, M.A. (2016), *Indonesia's Approach to Tackling IUU Fishing: Strategy on the Prevention and Eradication of IUU Fishing and Post-Moratorium Policies*, presentation to WorldFish, Penang, Malaysia.

Shelton, C. (2014), "Climate change adaptation in fisheries and aquaculture: Compilation of initial examples", *FAO Fisheries and Aquaculture Circular No. 1088*, Food and Agriculture Organization, Rome.

Stobutzki, I., M. Stephan and K. Mazur (2014), *Overview of Indonesia's Capture Fisheries, 2013*, Australian Bureau of Agricultural and Resource Economics and Sciences, Canberra.

Subasinghe, R.P., P. Bueno, M.J. Phillips, C. Hough, S.E. McGladdery, and J.R. Arthur (eds.) (2000), *Aquaculture Development: Financing and Institutional Support*, Technical Proceedings of the Conference on Aquaculture in the Third Millennium, Bangkok, Thailand, 20-25 February 2000, pp. 259-263, Network of Aquaculture Centres in Asia-Pacific, Bangkok, and Food and Agriculture Organization, Rome.

Sutinen, J.G. (2013), *Final Report on Indonesian Fisheries Policy*, United States Agency for International Development, Washington, DC.

Tabor, S.R. (2015), "Constraints to Indonesia's economic growth", *ADB Papers on Indonesia No. 10*, Asian Development Bank, Manila.

Tran, N., U-P. Rodriguez, C.Y. Chan, M.J. Phillips, C.V. Mohan, P. J.G. Henriksson, S. Koeshendrajana, S. Suri and S. Hall (2017), "Indonesian Aquaculture Futures. Part 1: An analysis of fish supply and demand in Indonesia to 2010 and role of aquaculture using the AisaFish Model", *Marine Policy*, Vol. 79, pp.25-32.

UN DESA (2016), *UN Comtrade* (database), United Nations Department of Economic and Social Affairs, https://comtrade.un.org/ (accessed on 14 November 2016).

USAID (2013), *Indonesia's Food Law of 2012, Prospective Impact on Domestic Markets and Food Security*, United States Agency for International Development, Washington, DC.

Waite, R., M. Beveridge, R. Brummett, S. Castine, N. Chaiyawannakarn, S. Kaushik, R. Mungkung, S. Nawapakpilai and M. Phillips (2014), "Improving productivity and environmental performance of aquaculture performance", *World Resources Institute Working Paper*, World Resources Institute, Washington, DC.

WHO (2015), *Global Database on Child Growth and Malnutrition* (database), UNICEF-WHO-The World Bank Group, www.who.int/nutgrowthdb/estimates2014/en/ (accessed on 8 November 2016).

Wijaya, A. and P. Glasbergen (2016), "Toward a new scenario in agricultural sustainability certification? The response of the Indonesian national government to private certification", *The Journal of Environment & Development*, Vol. 25/ 2, pp. 219-246.

World Bank (2017), *Indonesia Dashboard*, The World Bank Group, http://sdwebx.worldbank.org/climateportalb/home.cfm?page=country_profile&CCode=IDN (accessed on 9 February 2017).

World Economic Forum (2015), *The Global Competitiveness Report 2015-2016*, World Economic Forum, Geneva.

ORGANISATION FOR ECONOMIC CO-OPERATION AND DEVELOPMENT

The OECD is a unique forum where governments work together to address the economic, social and environmental challenges of globalisation. The OECD is also at the forefront of efforts to understand and to help governments respond to new developments and concerns, such as corporate governance, the information economy and the challenges of an ageing population. The Organisation provides a setting where governments can compare policy experiences, seek answers to common problems, identify good practice and work to co-ordinate domestic and international policies.

The OECD member countries are: Australia, Austria, Belgium, Canada, Chile, the Czech Republic, Denmark, Estonia, Finland, France, Germany, Greece, Hungary, Iceland, Ireland, Israel, Italy, Japan, Korea, Latvia, Luxembourg, Mexico, the Netherlands, New Zealand, Norway, Poland, Portugal, the Slovak Republic, Slovenia, Spain, Sweden, Switzerland, Turkey, the United Kingdom and the United States. The European Union takes part in the work of the OECD.

OECD Publishing disseminates widely the results of the Organisation's statistics gathering and research on economic, social and environmental issues, as well as the conventions, guidelines and standards agreed by its members.

OECD PUBLISHING, 2, rue André-Pascal, 75775 PARIS CEDEX 16
(51 2017 03 1 P) ISBN 978-92-64-27238-5 – 2017

www.ingramcontent.com/pod-product-compliance
Ingram Content Group UK Ltd.
Pitfield, Milton Keynes, MK11 3LW, UK
UKHW051459290926
15471UKWH00008B/1298

9 789264 272385